Hyperspectral Remote Sensing of Agriculture and Vegetation

Hyperspectral Remote Sensing of Agriculture and Vegetation

Editors

Simone Pascucci
Stefano Pignatti
Raffaele Casa
Roshanak Darvishzadeh
Wenjiang Huang

MDPI • Basel • Beijing • Wuhan • Barcelona • Belgrade • Manchester • Tokyo • Cluj • Tianjin

Editors
Simone Pascucci
CNR IMAA, Tito Scalo (PZ)
Italy

Stefano Pignatti
CNR IMAA, Tito Scalo (PZ)
Italy

Raffaele Casa
University Via San Camillo de Lellis
Italy

Roshanak Darvishzadeh
University of Twente
The Netherlands

Wenjiang Huang
Chinese Academy of Sciences
China

Editorial Office
MDPI
St. Alban-Anlage 66
4052 Basel, Switzerland

This is a reprint of articles from the Special Issue published online in the open access journal *Remote Sensing* (ISSN 2072-4292) (available at: https://www.mdpi.com/journal/remotesensing/special_issues/hyperspectral_agri).

For citation purposes, cite each article independently as indicated on the article page online and as indicated below:

LastName, A.A.; LastName, B.B.; LastName, C.C. Article Title. *Journal Name* **Year**, *Volume Number*, Page Range.

ISBN 978-3-03943-907-2 (Hbk)
ISBN 978-3-03943-908-9 (PDF)

Contents

About the Editors

Simone Pascucci (Dr): Degree in Environmental Sciences in 2009 at the University La Tuscia of Viterbo. Remote sensing expert at the IMAA-CNR Institute of Methodologies for Environmental Analysis of Tito (Italy) since February 2010. Research has been focused on different aspects related to multi- and hyper-spectral remote sensing applications for environmental, agronomic, marine and urban environments. He participated in several field, airborne and oceanographic campaigns in charge of measurement and calibration/validation for satellite and airborne sensors conducted by the institute as a technical expert, both nationally and internationally. He has participated in several international funded (FP6 and FP7) projects as a leader of work packages. He has published several articles in international journals regarding applications of hyperspectral remote sensing. His work has also been published in peer-reviewed journals on precision agriculture applications using new-generation satellite and UAVs' multi- and hyper-spectral data. He is actually working on the development of the prototyping algorithms for retrieving soil and vegetation biophysical properties for the ASI PRISMA hyperspectral mission.

Stefano Pignatti (Dr): Senior researcher at the National Research Council of Italy at the Institute of Methodologies for Environmental Analysis (CNR IMAA). He has worked on hyperspectral remote sensing both from ground, airborne (VSWIR and TIR) and satellite platforms since 1990. His research activity includes hyperspectral data calibration, the exploitation of hyperspectral remote sensing in environmental applications, such as the retrieval of the biophysical and biochemical properties of agricultural vegetation, the identification of vegetation stressing conditions, the monitoring of natural vegetation traits and the retrieval of the structural topsoil characteristics including the organic content. He is coordinating national and international research projects involving the processing of hyperspectral data for environmental issues. He is also involved in the development of the prototyping algorithms for retrieving biochemical and biophysical properties of vegetation suited for the ASI PRISMA hyperspectral mission.

Raffaele Casa (Prof): Full Professor at the University of Tuscia (Viterbo, Italy) of Agronomy (BSc level course) and Experimental Methods in Agriculture (MSc level course). He received his Ph.D. at the University of Dundee (Scotland, UK) for a research on the application of multiangular hyperspectral remote sensing to agricultural crops stress monitoring. He carried out several research periods at international institutions (University of Wageningen, Netherlands; University of Edinburgh, Dundee and Aberdeen, Scotland; INRA Avignon France; Evora, Portugal; Beijing, China) in the framework of research projects, or as a visiting professor. He has been awarded twice Marie Curie IEF fellowships from the European Commission, respectively for a 1 year stay in Scotland (1999–2000) and for a 1 year stay at INRA Avignon (2008–2009). He has been a member of the Mission Advisory Board of the Sentinel 2 Mission of the European Space Agency (ESA) and of the Earth Observation Scientific Advisory Committee of the Italian Space Agency (ASI). He has been a member of the Group of Experts on Precision Agriculture at the Italian Ministry of Agriculture (2015–2016) and co-authored the policy document on the National Guidelines for Precision Agriculture in Italy. He was a member of the Directive Committee of the Italian Society of Agronomy (SIA) in 2016–2019. He is a member of the Editorial Board of the Remote Sensing (MDPI) journal, of the Italian Journal of Agronomy and of the Italian Journal of Agrometeorology.

His research has focused on the application of remote sensing and precision agriculture to the improvement of the agronomic management of crops for more than 20 years. Prof Casa has an overall H-index of 24 with 1650 citations of 71 papers indexed in the Scopus database (Elsevier). He is a member of the scientific technical committee of IBF Servizi SpA (participated by ISMEA, Bonifica Ferraresi, E-geos and A2A), the main company in Italy that offers services in precision agriculture.

Roshanak Darvishzadeh (Dr., Prof): Associate Professor at the Department of Natural Resources (NRS), Faculty of Geo-information Science and Earth Observation. She has a Ph.D. in hyperspectral remote sensing of vegetation from Wageningen University and ITC. Her research interests lie in quantitative remote sensing for mapping and modelling plants' biochemical and biophysical properties at leaf and canopy levels. Her research activities include analysis and exploration of various remote sensing data types at the ground, airborne and satellite platforms for assessing vegetation growth, health, disease, and stress with applications in natural as well as agricultural ecosystems. She has practical experience in the use of lab and field instruments for measuring vegetation properties as well as quantitative methods for monitoring plant traits. Her career includes several years of working in remote sensing and GIS modelling in private and public sectors before moving to academic work. She is a member of several national and international professional associations. Besides research and coordination of international projects, she is involved in educational activities, including lecturing, course development, mentoring and supervision of MSc and Ph.D. students.

Wenjiang Huang (Dr., Prof): Director of Key Laboratory of Digital Earth Science, Director of SINO-UK Crop Pest and Disease Forecasting & Management Joint Laboratory, working in Aerospace Information Research Institute, Chinese Academy of Science (AIR-CAS). He has a Ph.D. in Physical Geography and acted as PI for more than 40 major scientific fundings from Global Earth Observation (GEO), Ministry of Science and Technology (MOST), National Science Foundation of China (NSFC), Ministry of Agriculture and Rural Affairs (MARA) and Chinese Academy of Sciences (CAS). His research interests cover quantitative and remote sensing for agriculture, especially on crops, data fusion (multi-scale, multi-sensor and multi-temporal) for precision agricultural applications, and monitoring crop diseases using remote sensing technology. His work has been published in more than 200 SCI journal papers focused on remote sensing for crop biophysical and biochemical variable inversion and remote sensing for crop pest and disease monitoring. He has awarded more than 10 science and technology awards, including the National Science and Technology Progress Award. For more details, please visit the website (http://www.rscrop.com).

Editorial

Special Issue "Hyperspectral Remote Sensing of Agriculture and Vegetation"

Simone Pascucci [1,*], Stefano Pignatti [1], Raffaele Casa [2], Roshanak Darvishzadeh [3] and Wenjiang Huang [4]

[1] Institute of Methodologies for Environmental Analysis (CNR IMAA), National Research Council, 85050 Tito Scalo, PZ, Italy; stefano.pignatti@imaa.cnr.it

[2] Department of Agricultural and Forestry scieNcEs (DAFNE), Tuscia University Via San Camillo de Lellis, 01100 Viterbo, Italy; rcasa@unitus.it

[3] ITC—Faculty of Geo-Information Science and Earth Observation, Department of Natural Resources, University of Twente, PO Box 217, 7500 AE Enschede, The Netherlands; r.darvish@utwente.nl

[4] Aerospace Information Research Institute, Chinese Academy of Sciences, Beijing 100094, China; huangwj@aircas.ac.cn

* Correspondence: simone.pascucci@cnr.it

Received: 2 November 2020; Accepted: 3 November 2020; Published: 9 November 2020

Abstract: The advent of up-to-date hyperspectral technologies, and their increasing performance both spectrally and spatially, allows for new and exciting studies and practical applications in agriculture (soils and crops) and vegetation mapping and monitoring atregional (satellite platforms) andwithin-field (airplanes, drones and ground-based platforms) scales. Within this context, the special issue has included eleven international research studies using different hyperspectral datasets (from the Visible to the Shortwave Infrared spectral region) for agricultural soil, crop and vegetation modelling, mapping, and monitoring. Different classification methods (Support Vector Machine, Random Forest, Artificial Neural Network, Decision Tree) and crop canopy/leaf biophysical parameters (e.g., chlorophyll content) estimation methods (partial least squares and multiple linear regressions) have been evaluated. Further, drone-based hyperspectral mapping by combining bidirectional reflectance distribution function (BRDF) model for multi-angle remote sensing and object-oriented classification methods are also examined. A review article on the recent advances of hyperspectral imaging technology and applications in agriculture is also included in this issue. The special issue is intended to help researchers and farmers involved in precision agriculture technology and practices to a better comprehension of strengths and limitations of the application of hyperspectral measurements for agriculture and vegetation monitoring. The studies published herein can be used by the agriculture and vegetation research and management communities to improve the characterization and evaluation of biophysical variables and processes, as well as for a more accurate prediction of plant nutrient using existing and forthcoming hyperspectral remote sensing technologies.

Keywords: hyperspectral remote sensing for soil and crops in agriculture; hyperspectral imaging for vegetation; plant traits; high-resolution spectroscopy for agricultural soils and vegetation; hyperspectral databases for agricultural soils and vegetation; hyperspectral data as input for modelling soil, crop, and vegetation; product validation; new hyperspectral technologies; future hyperspectral missions

---

1. Introduction

The use of hyperspectral technology for an optimal quantification of crop and soil biophysical variables at various spatial scales is an important aspect in agricultural management practices and monitoring [1,2]. Moreover, there is a great interest to update (i.e., research of new variables) and optimize the retrieval of crop biophysical variables using drone and available satellite data [2–17], as well as future high spatial resolution hyperspectral satellites. To this aim, the exploitation of different approaches for assimilation of the retrieved biophysical parameters into agricultural models is also of primary interest. As it would allow deriving agronomical proxy variables addressing the issues of the multi-scale and multivariate nature of the retrieved variables [6,7,11–19]. For example, a complete and updated knowledge of the spatial distribution of leaf area index (LAI), pigments like chlorophyll content and nitrogen can support sustainable agricultural practices and optimize related costs, through optimal use of fertilizer, pesticides and water that are strictly subdued to an improvement of crop yields and quality. Hyperspectral imaging has great potential for applications in agriculture, particularly precision agriculture, owing to their ample spectral information sensitive to different plant and soil biophysical and biochemical properties [11–25]. Multiple platforms (satellites, airplanes, unmanned aerial vehicle (UAVs), and close-range platforms) have become more widely available in recent years for collecting hyperspectral data with different spatial (from centimeter to decameter), temporal, and spectral resolutions. These platforms have different strengths and limitations in terms of spatial coverage, flight endurance, flexibility, operational complexity, and costs. These factors need to be evaluated when choosing the hyperspectral platform(s) for specific research purposes, e.g., increasing productivity, expanded coverage, and reduced use of fertilizers, pesticides, and water. Further technological developments are also needed to overcome some of the limitations, such as the short battery endurance in UAV operations and high cost of hyperspectral sensors [4].

All in all, hyperspectral remote sensing (RS) represents an attractive and efficient technology capable of estimating soil and crop biophysical variables of interest from regional to intra-field scales.

Research advances are still required to validate methods and applications for the estimation of additional crop biophysical variables and proxy agronomical products [14–25] and for their assimilation into spatially distributed agricultural models (e.g., grains quality, pest and disease dynamic, water-driven, and crop growing models), also by comparing different assimilation approaches [10–24].

This special issue was set up to highlight and diffuse the recent advances in hyperspectral RS studies and their practical applications for agriculture (soils and crops) mapping and monitoring from regional to within-field scales. Our objectives as guest Editors were to encourage studies and applications on this topic and to assemble high-quality, peer-reviewed research and review articles in a special issue of *Remote Sensing* dedicated to this theme. We accepted manuscripts concerned with all aspects of hyperspectral RS (optical domain) for crop and natural vegetation. This included hyperspectral studies of agricultural soils, crops, as well as other vegetation types using the ground, drone, air-, and space-borne platforms (VIS-NIR, SWIR, and TIR). With various focus on: field, and laboratory hyperspectral measurements for monitoring agriculture and vegetation; retrieval of plant traits at leaf and canopy level from hyperspectral measurements; new methods for hyperspectral data processing and atmospheric compensation techniques; hyperspectral sensors calibration and products validation for agriculture and vegetation monitoring; statistical and computational methods for hyperspectral data analysis in agriculture and vegetation applications; integration or combined use of hyperspectral data from the optical domain with other Earth Observation (EO) technologies; modelling of soils, crops, and vegetation using hyperspectral data; next-generation hyperspectral technologies and missions, platforms, and sensors for agriculture and vegetation.

A total of 18 manuscripts were submitted and peer-reviewed by fifty anonymous, scrupulous reviewers. Of these, 11 manuscripts achieved the level of quality and innovation expected by *Remote Sensing* and were at the end published in this special issue. A total of 77 authors contributed to these 11 articles and hailed from six different nations: Brazil (26 authors), Canada (8), Australia (5), Finland (3), China (23), UK (1), Iran (3), Belgium (1), Spain (1), Poland (3), Ethiopia (1), and USA (2).

2. Overview of Contributions

The works composing this Special Issue cover a wide range of topics, from the use of high spectral resolution hyperspectral LiDAR (light detection and ranging) for vegetation parameters extraction, to the estimation of chlorophyll content in peanut leaf, to the estimation of heavy metal contents in grapevine foliage, to the application of UAV-based multiangle hyperspectral data in fine vegetation classification, to the use of artificial neural networks for modeling hyperspectral response of water-stress induced lettuce plants, to different classification methods and algorithms for agricultural biophysical variables retrieval, plants and invasive species retrieval, and to predict nutrient content. They are presented below in chronological order of acceptance.

First, Jiang et al. [25] employed and evaluated the use of high spectral resolution hyperspectral LiDAR (Acousto-optical Tunable Filter HSL-AOTF-HSL, active and non-contact instrument), with 10 nm spectral resolution, for leaf vegetation red edge parameters extraction. The results were compared with the referenced value from a standard SVC© HR-1024 spectrometer (Spectra Vista Corporation,

Poughkeepsie, NY 12603 USA) for validation. Green leaf parameter differences between HSL and SVC results were minor, which supported the notion that HSL was practical for extracting the employed parameter as an active method. This paper is just the beginning of using the high spectral resolution HSL for vegetation index detection, which might inspire the estimation of other vegetation parameters or biochemical content using this advanced LiDAR technique.

Second, the estimation of peanut leaf chlorophyll content with dorsiventral leaf adjusted ratio index (DLARI), performed by Xie et al. [26]. The study is one of the first attempts to assess the impact of spectral differences among dorsiventral leaves caused by leaf structure on leaf chlorophyll content (LCC) retrieval. The authors' objectives were to (1) analyze spectral differences in the adaxial and abaxial surfaces of peanut leaves; (2) identify the optimal wavelengths of the modified Datt (MDATT) index for estimating peanut LCC; (3) develop a novel index based on a four-band combination to reduce spectral differences in dorsiventral leaves for improving LCC retrieval; (4) compare the performance of the indices developed in this study with those widely used in the literature. The reliability of narrow-band indices can be influenced by a range of phenotypic characteristics. Further work is required to assess the application of DLARI to estimate LCC for other crop species. The robust wavelength regions proposed (715–820 nm) should provide a good starting point for optimizing the index for other crop species.

Third, [27], in their work applied five treatments of heavy metal stress (Cu, Zn, Pb, Cr, and Cd) to grapevine seedlings and hyperspectral data (350–2500 nm) and heavy metal contents were collected based on in-field and laboratory experiments. The partial least squares (PLS) method was used as a feature selection technique, and multiple linear regressions (MLR) and support vector machine (SVM) regression methods were applied for modelling purposes. Based on the PLS results, visible and red-edge regions were found most suitable for estimating heavy metal contents in the present study. The authors pointed out that each heavy metal has a special effect, leading to distinct responses depending on the plant species (including leaf color changes, chlorosis, necrosis, dwarfism, giant, leaf, and root spreading, etc.).

Fourth, Yan et al. [28] applied UAV-based multi-angle remote sensing for fine vegetation classification by combining a bidirectional reflectance distribution function (BRDF) model for multi-angle remote sensing and object-oriented classification methods. Bands of high importance for the fine classification of vegetation included the blue band (466– nm), green band (494–570 nm), red band (642–690 nm), red-edge band (694–774 nm), and near-infrared band (810–882 nm). The importance of the BRDF characteristic parameters are discussed in detail and the research results promote the application of multi-angle remote sensing technology in vegetation information extraction and provide important theoretical significance and application value for regional and global vegetation and ecological monitoring.

Fifth, [29] evaluated the hyperspectral response of water-stress induced lettuce (*Lactuca sativa* L.) using artificial neural networks (ANN). Hyperspectral response was measured four times, during 14 days of stress induction, with an ASD Fieldspec HandHeld spectroradiometer (325–1075 nm). Both reflectance and absorbance measurements were calculated. Different biophysical parameters were also evaluated. The performance of the ANN approach was compared against other machine learning algorithms. Authors' results showed that the ANN approach could separate the water-stressed lettuce from the non-stressed group with up to 80% accuracy at the beginning of the experiment. Absorbance data offered better accuracy than reflectance data to model it. This study demonstrated that it is possible to detect early stages of water stress in lettuce plants with high accuracy based on an ANN approach applied to hyperspectral data. The methodology has the potential to be applied to other species and cultivars in agricultural fields.

Sixth, a review of waveband selection in hyperspectral classification of plants was performed by [30]. The authors reviewed the last 22 years of hyperspectral vegetation classification literature that evaluate the overall waveband selection frequency, waveband selection frequency variation by taxonomic, structural, or functional groups. The influence of feature selection choice by comparing methods as stepwise discriminant analysis (SDA), support vector machines (SVM), and random forests (RF) is studied. They concluded that characteristics of plant studies influence the wavebands selected for classification and advised caution when relying upon waveband recommendations from the literature to guide waveband selections or classifications for new plant discrimination applications. In this regard, recommendations appear to be weakly generalizable between studies.

Seventh, Sabat-Tomala et al. [31] study proposed a comparison of SVM and RF algorithms for invasive and expansive species classification using airborne hyperspectral data (HySpex Visible and Near Infrared-VNIR-1800 scanners and a Shortwave Infrared-SWIR-384 scanner; Hyspex NEO, Oslo, Norway). These invasive species are considered a threat to natural biodiversity because of their high adaptability and low habitat requirements. Maps of the spatial distribution of analyzed species were obtained; high accuracies were observed for all data sets and classifiers. In particular, the authors verified whether the expansive/invasive *Rubus* spp., *Calamagrostis epigejos*, and *Solidago* spp. were characterized by a specific set of spectral characteristics that allowed them to be distinguished from the surrounding species, which altogether created a mix of fuzzy, covered patterns. Moreover, an analysis of the impact of the number of pixels in training data set on the classification accuracy was performed. The accuracy assessment method presented in the paper confirmed that all analyzed species can be identified in heterogeneous habitats through hyperspectral airborne remote sensing.

Eight, machine learning (ML) algorithms were applied by [32] to predict macro- and micronutrient nutrient content (N, P, K, Mg, S, Cu, Fe, Mn, and Zn) in Valencia-Orange from leaf hyperspectral measurements. A Fieldspec ASD FieldSpec® HandHeld 2 (Malvern PANalytical Ltd, Malvern, WR14 1XZ, United Kingdom) spectroradiometer was used and the surface reflectance and first-derivative spectra from the spectral range of 380 to 1020 nm (640 spectral bands) was evaluated. K-Nearest Neighbor (kNN), Lasso Regression, Ridge Regression, Support Vector Machine (SVM), Artificial Neural Network (ANN), Decision Tree (DT), and Random Forest (RF) ML algorithms were tested. The methods were assessed based on Cross-Validation and Leave-One-Out. The Relief-F metric of the algorithms' prediction was used to determine the most contributive wavelength or spectral region associated with each nutrient. RF model was the most suitable to model most of them. The results indicate that, for the Valencia-orange leaves, surface reflectance data is more suitable to predict macronutrients, while first-derivative spectra are better linked to micronutrients.

Ninth, a review article provided by [33] analyzed the recent advances of hyperspectral imaging technology and applications in agriculture. Due to limited accessibility outside of the scientific community, hyperspectral images have not been widely used in precision agriculture. In recent years, different mini-sized and low-cost airborne hyperspectral sensors (e.g., Headwall Micro-Hyperspec, Cubert UHD 185-Firefly) have been developed, and advanced space-borne hyperspectral sensors have also been or will be launched (e.g., PRISMA, DESIS, EnMAP, HyspIRI). Hyperspectral imaging is becoming more widely available to agricultural applications. Meanwhile, the acquisition, processing, and analysis of hyperspectral imagery remain a challenging research topic (e.g., large data volume, high data dimensionality, and complex information analysis). The imaging platforms and sensors (airplane, UAV, satellite, close-range ground- or lab-based) together with analytic methods used in the literature, were discussed. Performances of hyperspectral imaging for different applications (e.g., crop biophysical and biochemical properties' mapping, soil characteristics, and crop classification) were also evaluated. This review intended to assist agricultural researchers and practitioners to better understand the strengths and limitations of hyperspectral imaging to agricultural applications and promote the adoption of this valuable technology. Recommendations for future hyperspectral imaging research for precision agriculture were also presented.

Tenth, Zhang et al. [34] presented a study on the detection of canopy chlorophyll content for three growth stages of corn using continuous wavelet transform (CWT) analysis. The reflectance spectrum increased in the 325–400 and 761–970 nm regions as the growth stage advanced and the growth period shifted. The reflectance decreased in the 401–700 and 971–1075 nm regions as the growth stage advanced. The characteristic bands related to chlorophyll content in the spectral data and the wavelet energy coefficients were screened using the maximum correlation coefficient and the local correlation coefficient extrema, respectively. A partial least square regression (PLSR) model was established. Results showed that bands selected via local correlation coefficient extrema in a wavelet energy coefficient created a detection model with optimal accuracy.

Last, a different study in terms of application is proposed by [35], who studied the nutrient content of tef (*Eragrostis* tef), an understudied plant that has importance due to both food and forage benefits, and investigated the replicability of methods across two study sites situated in different international and environmental contexts [35]. The research aims were to (1) determine whether calcium, magnesium, and protein of both the tef plant and grain can be predicted using hyperspectral data and PLSR model through waveband selection, and (2) compare the replicability of models across varying environments. Results suggest the method can produce high nutrient prediction accuracy for both the plant and grain in individual environments, but the selection of wavebands for nutrient prediction was not comparable across study areas. Results using PLSR model with hyperspectral data from non-milled grains were generally positive, and wavebands for protein prediction generally agreed with other studies. While more research is needed to determine whether these consistencies are true positives or are affected by other factors. This study contributes to the gap in the literature related to non-milled grains. Therefore, there is a need for greater attention to methods and results replication in remote sensing, specifically hyperspectral analyses, in order for scientific findings to be repeatable beyond the plot level.

3. Concluding Remarks

Hyperspectral remote sensing for studying agriculture and natural vegetation is a challenging research topic that will remain of great interest for different sciences communities for the next decades. As a matter of fact, Space agencies, on a worldwide basis, have ongoing programs to develop hyperspectral satellite missions to assure global coverage at high spatial resolution that will have a noteworthy impact on agricultural and natural vegetation monitoring studies. The eleven manuscripts collected in this special issue and, therefore, represent some meaningful progress in the application of hyperspectral EO data for agricultural and vegetation research themes. Further work in this area is required in view of the recent advances and funding opportunities in this field. We expect that the studies published herein will help the agriculture and vegetation research and management communities to better characterize and assess biophysical variables and processes, as well as more effectively predict plant nutrient using upcoming hyperspectral remote sensing technologies.

Author Contributions: All authors have read and agreed to the published version of the manuscript.

Funding: This research received no external funding.

Acknowledgments: The guest editors would like to thank the authors who contributed to this special issue and the reviewers who helped to improve the quality of the special issue by providing constructive recommendations to the authors. We would like to especially thank all 77 contributing authors who dedicated their research and time to this special issue. Likewise, we want to explicitly thank the fifty reviewers for their great work and effort put in this often-underestimated task. The careful, suitable and original comments provided by the reviewers improved each of the papers published in this special issue. A special thanks goes to Quenby Qu.

Conflicts of Interest: The guest editors declare no conflict of interest.

References

1. Gara, T.W.; Darvishzadeh, R.; Skidmore, A.K.; Wang, T.; Heurich, M. Evaluating the performance of PROSPECT in the retrieval of leaf traits across canopy throughout the growing season. *Int. J. Appl. Earth Obs. Geoinf.* **2019**, *83*, 101919. [CrossRef]
2. Weiss, M.; Jacob, F.; Duveiller, G. Remote sensing for agricultural applications: A meta-review. *Remote Sens. Environ.* **2020**, *236*, 111402. [CrossRef]
3. Sahoo, R.N.; Ray, S.S.; Manjunath, K.R. Hyperspectral remote sensing of agriculture. *Curr. Sci.* **2015**, *108*, 848–859.
4. Adão, T.; Hruška, J.; Pádua, L.; Bessa, J.; Peres, E.; Morais, R.; Sousa, J. Hyperspectral Imaging: A Review on UAV-Based Sensors, Data Processing and Applications for Agriculture and Forestry. *Remote Sens.* **2017**, *9*, 1110. [CrossRef]
5. Mariotto, I.; Thenkabail, P.S.; Huete, A.; Slonecker, E.T.; Platonov, A. Hyperspectral versus multispectral crop-productivity modeling and type discrimination for the HyspIRI mission. *Remote Sens. Environ.* **2013**, *139*, 291–305. [CrossRef]
6. Marshall, M.; Thenkabail, P. Advantage of hyperspectral EO-1 Hyperion over multispectral IKONOS, GeoEye-1, WorldView-2, Landsat ETM+, and MODIS vegetation indices in crop biomass estimation. *ISPRS J. Photogramm.* **2015**, *108*, 205–218. [CrossRef]
7. Transon, J.; d'Andrimont, R.; Maugnard, A.; Defourny, P. Survey of Hyperspectral Earth Observation Applications from Space in the Sentinel-2 Context. *Remote Sens.* **2018**, *10*, 157. [CrossRef]
8. Zarco-Tejada, P.J.; Suarez, L.; Gonzalez-Dugo, V. Spatial resolution effects on chlorophyll fluorescence retrieval in a heterogeneous canopy using hyperspectral imagery and radiative transfer simulation. *IEEE Geosci. Remote Soc.* **2013**, *10*, 937–941. [CrossRef]
9. Lu, B.; He, Y.; Dao, P.D. Comparing the Performance of Multispectral and Hyperspectral Images for Estimating Vegetation Properties. *IEEE J. Stars* **2019**, *12*, 1784–1797. [CrossRef]
10. Darvishzadeh, R.; Wang, T.; Skidmore, A.K.; Vrieling, A.; O'Connor, B.; Gara, T.W.; Ens, B.J.; Marc, P. Analysis of Sentinel-2 and RapidEye for Retrieval of Leaf Area Index in a Saltmarsh Using a Radiative Transfer Model. *Remote Sens.* **2019**, *11*, 671. [CrossRef]
11. Darvishzadeh, R.; Skidmore, A.K.; Abdullah, H.; Cherenet, E.; Ali, A.M.; Wang, T.; Nieuwenhuis, W.; Heurich, M.; Vrieling, A.; O'Connor, B.; et al. Mapping leaf chlorophyll content from Sentinel-2 and RapidEye data in spruce stands using the invertible forest reflectance model. *Int. J. Appl. Earth Obs. Geoinf.* **2019**, *79*, 58–70. [CrossRef]
12. Xie, R.; Darvishzadeh, R.; Skidmore, A.K.; Heurich, M.; Holzwarth, S.; Gara, T.W.; Reusen, I. Mapping leaf area index in a mixed temperate forest using Fenix airborne hyperspectral data and Gaussian processes regression. *Int. J. Appl. Earth Obs. Geoinf.* **2020**, *95*, 1–13. [CrossRef]
13. Ali, A.M.; Skidmore, A.K.; Darvishzadeh, R.; van Duren, I.C.; Holzwarth, S.; Mueller, J. Retrieval of forest leaf functional traits from HySpex imagery using radiative transfer models and continuous wavelet analysis. *Isprs J. Photogramm. Remote Sens.* **2016**, *122*, 68–80. [CrossRef]
14. Casa, R.; Castaldi, F.; Pascucci, S.; Palombo, A.; Pignatti, S. A comparison of sensor resolution and calibration strategies for soil texture estimation from hyperspectral remote sensing. *Geoderma* **2013**, *197*, 17–26. [CrossRef]
15. Castaldi, F.; Palombo, A.; Santini, F.; Pascucci, S.; Pignatti, S.; Casa, R. Evaluation of the potential of the current and forthcoming multispectral and hyperspectral imagers to estimate soil texture and organic carbon. *Remote Sens. Environ.* **2016**, *179*, 54–65. [CrossRef]
16. Castaldi, F.; Palombo, A.; Pascucci, S.; Pignatti, S.; Santini, F.; Casa, R. Reducing the Influence of Soil Moisture on the Estimation of Clay from Hyperspectral Data: A Case Study Using Simulated PRISMA Data. *Remote Sens.* **2015**, *7*, 15561–15582. [CrossRef]
17. Castaldi, F.; Chabrillat, S.; Jones, A.; Vreys, K.; Bomans, B.; van Wesemael, B. Soil Organic Carbon Estimation in Croplands by Hyperspectral Remote APEX Data Using the LUCAS Topsoil Database. *Remote Sens.* **2018**, *10*, 153. [CrossRef]
18. Casa, R.; Pascucci, S.; Pignatti, S.; Palombo, A.; Nanni, U.; Harfouche, A.; Laura, L.; Di Rocco, M.; Fantozzi, P. UAV-based hyperspectral imaging for weed discrimination in maize. In *Precision Agriculture '19*; Stafford, J.V., Ed.; Wageningen Academic Publishers: Wageningen, The Netherlands, 2019; pp. 24–35.

19. Yue, J.; Feng, H.; Yang, G.; Li, Z. A comparison of regression techniques for estimation of above-ground winter wheat biomass using near-surface spectroscopy. *Remote Sens.* **2018**, *10*, 66. [CrossRef]
20. Casa, R.; Castaldi, F.; Pascucci, S.; Basso, B.; Pignatti, S. Geophysical and Hyperspectral Data Fusion Techniques for In-Field Estimation of Soil Properties. *Vadose Zone J.* **2013**, *12*, vzj2012.0201. [CrossRef]
21. Casa, R.; Castaldi, F.; Pascucci, S.; Pignatti, S. Potential of hyperspectral remote sensing for field scale soil mapping and precision agriculture applications. *Ital. J. Agron.* **2012**, *7*, 43. [CrossRef]
22. Upreti, D.; Pignatti, S.; Pascucci, S.; Tolomio, M.; Huang, W.; Casa, R. Bayesian Calibration of the Aquacrop-OS Model for Durum Wheat by Assimilation of Canopy Cover Retrieved from VENµS Satellite Data. *Remote Sens.* **2020**, *12*, 2666. [CrossRef]
23. Wang, Z.; Skidmore, A.K.; Darvishzadeh, R.; Wang, T. Mapping forest canopy nitrogen content by inversion of coupled leaf-canopy radiative transfer models from airborne hyperspectral imagery. *Agric. For. Meteorol.* **2018**, *253*, 247–260. [CrossRef]
24. Zheng, Q.; Huang, W.; Ye, H.; Dong, Y.; Shi, Y.; Chen, S. Using continous wavelet analysis for monitoring wheat yellow rust in different infestation stages based on unmanned aerial vehicle hyperspectral images. *Appl. Opt.* **2020**, *59*, 8003–8013. [CrossRef]
25. Jiang, C.; Chen, Y.; Wu, H.; Li, W.; Zhou, H.; Bo, Y.; Hyyppä, J. Study of a high spectral resolution hyperspectral LiDAR in vegetation red edge parameters extraction. *Remote Sens.* **2019**, *11*, 2007. [CrossRef]
26. Xie, M.; Wang, Z.; Huete, A.; Brown, L.A.; Wang, H.; Xie, Q.; Ding, Y. Estimating Peanut Leaf Chlorophyll Content with Dorsiventral Leaf Adjusted Indices: Minimizing the Impact of Spectral Differences between Adaxial and Abaxial Leaf Surfaces. *Remote Sens.* **2019**, *11*, 2148. [CrossRef]
27. Mirzaei, M.; Verrelst, J.; Marofi, S.; Abbasi, M.; Azadi, H. Eco-Friendly Estimation of Heavy Metal Contents in Grapevine Foliage Using In-Field Hyperspectral Data and Multivariate Analysis. *Remote Sens.* **2019**, *11*, 2731. [CrossRef]
28. Yan, Y.; Deng, L.; Liu, X.; Zhu, L. Application of UAV-Based Multi-Angle Hyperspectral Remote Sensing in Fine Vegetation Classification. *Remote Sens.* **2019**, *11*, 2753. [CrossRef]
29. Osco, L.P.; Ramos, A.P.M.; Moriya, É.A.S.; Bavaresco, L.G.; Lima, B.C.D.; Estrabis, N.; Imai, N.N. Modeling hyperspectral response of water-stress induced lettuce plants using artificial neural networks. *Remote Sens.* **2019**, *11*, 2797. [CrossRef]
30. Hennessy, A.; Clarke, K.; Lewis, M. Hyperspectral Classification of Plants: A Review of Waveband Selection Generalisability. *Remote Sens.* **2020**, *12*, 113. [CrossRef]
31. Sabat-Tomala, A.; Raczko, E.; Zagajewski, B. Comparison of Support Vector Machine and Random Forest Algorithms for Invasive and Expansive Species Classification Using Airborne Hyperspectral Data. *Remote Sens.* **2020**, *12*, 516. [CrossRef]
32. Osco, L.P.; Ramos, A.P.M.; Faita Pinheiro, M.M.; Moriya, É.A.S.; Imai, N.N.; Estrabis, N.; Li, J. A Machine Learning Framework to Predict Nutrient Content in Valencia-Orange Leaf Hyperspectral Measurements. *Remote Sens.* **2020**, *12*, 906. [CrossRef]
33. Lu, B.; Dao, P.D.; Liu, J.; He, Y.; Shang, J. Recent Advances of Hyperspectral Imaging Technology and Applications in Agriculture. *Remote Sens.* **2020**, *12*, 2659. [CrossRef]
34. Zhang, J.; Sun, H.; Gao, D.; Qiao, L.; Liu, N.; Li, M.; Zhang, Y. Detection of Canopy Chlorophyll Content of Corn Based on Continuous Wavelet Transform Analysis. *Remote Sens.* **2020**, *12*, 2741. [CrossRef]
35. Flynn, K.C.; Frazier, A.E.; Admas, S. Nutrient Prediction for Tef (Eragrostis tef) Plant and Grain with Hyperspectral Data and Partial Least Squares Regression: Replicating Methods and Results across Environments. *Remote Sens.* **2020**, *12*, 2867. [CrossRef]

Publisher's Note: MDPI stays neutral with regard to jurisdictional claims in published maps and institutional affiliations.

 *remote sensing*

Article

Study of a High Spectral Resolution Hyperspectral LiDAR in Vegetation Red Edge Parameters Extraction

Changhui Jiang [1], Yuwei Chen [1,2,*], Haohao Wu [2], Wei Li [2], Hui Zhou [3], Yuming Bo [4], Hui Shao [1,5], Shaojing Song [6], Eetu Puttonen [1] and Juha Hyyppä [1]

[1] Center of Excellence of Laser Scanning Research, Finnish Geospatial Research Institute, Masala FI-02430, Finland
[2] Key Laboratory of Quantitative Remote Sensing Information Technology, Chinese Academy of Sciences, Beijing 100094, China
[3] Electronic Information School, Wuhan University, Wuhan 430079, China
[4] School of Automation, Nanjing University of Science and Technology, Nanjing 210094, China
[5] Department of Electronics Engineering, Anhui Jianzhu University, Hefei 230601, China
[6] Department of Communication and Information Engineering, Shanghai Polytechnic University, Shanghai 200216, China
* Correspondence: yuwei.chen@nls.fi

Received: 3 July 2019; Accepted: 22 August 2019; Published: 26 August 2019

Abstract: Non-contact and active vegetation or plant parameters extraction using hyperspectral information is a prospective research direction among the remote sensing community. Hyperspectral LiDAR (HSL) is an instrument capable of acquiring spectral and spatial information actively, which could mitigate the environmental illumination influence on the spectral information collection. However, HSL usually has limited spectral resolution and coverage, which is vital for vegetation parameter extraction. In this paper, to broaden the HSL spectral range and increase the spectral resolution, an Acousto-optical Tunable Filter based Hyperspectral LiDAR (AOTF-HSL) with 10 nm spectral resolution, consecutively covering from 500–1000 nm, was designed. The AOTF-HSL was employed and evaluated for vegetation parameters extraction. "Red Edge" parameters of four different plants with green and yellow leaves were extracted in the lab experiments for evaluating the HSL vegetation parameter extraction capacity. The experiments were composed of two parts. Firstly, the first-order derivative of the spectral reflectance was employed to extract the "Red Edge" position (REP), "Red Edge" slope (RES) and "Red Edge" area (REA) of these green and yellow leaves. The results were compared with the referenced value from a standard SVC© HR-1024 spectrometer for validation. Green leaf parameter differences between HSL and SVC results were minor, which supported that notion the HSL was practical for extracting the employed parameter as an active method. Secondly, another two different REP extraction methods, Linear Four-point Interpolation technology (LFPIT) and Linear Extrapolation technology (LET), were utilized for further evaluation of using the AOTF-HSL spectral profile to determine the REP value. The differences between the plant green leaves' REP results extracted using the three methods were all below 10%, and the some of them were below 1%, which further demonstrated that the spectral data collected from HSL with this spectral range and resolution settings was applicable for "Red Edge" parameters extraction.

Keywords: hyperspectral LiDAR; Red Edge; AOTF; vegetation parameters

---

1. Introduction

The remote sensing community has demonstrated the effectiveness of hyperspectral imagers and LiDAR to obtain spectral and spatial information [1–5]. The hyperspectral imager is capable of obtaining consecutive and abundant spectral profiles of targets, which has been employed in vegetation parameter extraction, food production prediction, target classification, etc. [1–5]. However, the hyperspectral imager relies on environmental illumination conditions, so poor lighting will affect the hyperspectral information acquisition. LiDAR is an active sensor invented to acquire spatial information. In LiDAR a laser source emits monochromatic laser beams to a target, and thus, the ranging information is obtained through measuring the travel time of the laser beam [6,7]. With a scanning operation, LiDAR is able to obtain spatial information from the environment. Besides this, the power of the reflected signal from the target in LiDAR can be obtained with ranging operation. With careful calibration, the power of the reflected signal is measured and termed as intensity. Researchers have carried out some investigations using the intensity of a single wavelength to obtain some textures of the targets, for instance, rock analysis in outcrop models, landcover classification, etc. [1–8].

Restricted by the monochromatic laser source, intensity information of the back-scattered laser pulse or the spectral information from a traditional single wavelength LiDAR is much less efficient than a passive spectrometer [9–11]. Recently, two approaches were investigated for the fusion of spatial and spectral data. The first approach ias to combine spectral and spatial data from two standalone instruments into the same framework, and this method was employed in forest area classification, urban species classification, automatic building extraction, and outcrop analysis [12–18]. The disadvantage is that the data registration is complicated and time-consuming, and the coordinate transformation between the two instruments will probably introduce additional errors [12–18].

The second approach refers to the integration of ranging with spectral measuring functions into a single sensor or instrument. Hyperspectral LiDAR (HSL) or Multispectral LiDAR (MSL) were developed as active sensors to obtain spectral and spatial information simultaneously. Basically, there are two solutions to develop an HSL or MSL. The first solution is to combine several monochromatic laser sources of different wavelengths together. Since more channels mean more laser sources at different spectral wavelengths, it was hard to combine tens or hundreds of monochromatic laser sources together in this framework [19–21]. The second solution is to develop the HSL through employing a super-continuum (SC) laser source replacing the above monochromatic laser sources of different wavelengths, and the SC laser source is able to emit ultra-wideband coherent laser transmissions with spectral ranging from approximately 400 nm to 2500 nm [19–21]. Scientists from the Finnish Geospatial Research Institute (FGI) proposed the SC laser source-based spectral measurement concept in 2007 [22]. The first results with the prototype instrument were presented with a discussion of improvements and applications in laser-based hyperspectral remote sensing [22]. Further, in 2010, a two-channel multispectral LiDAR with 600 nm and 800 nm spectral wavelengths was developed and demonstrated, which was capable of distinguishing between a vegetation target (Norway spruce) and inorganic material using the Normalized Difference Vegetation Index (NDVI) parameter [23]. In 2012, the first full-waveform HSL with eight spectral channels was constructed by FGI. The novel instrument produced 3D point clouds with spectral back-scattered reflectance data [24]. Then, HSL was investigated in vegetation content estimation, leaf level chlorophyll estimation, leaf biochemical content estimation, landcover classification, and artificial object classification [25–30]. However, compared with the hyperspectral imager, these HSLs had restricted and discrete spectral bands and channels. For broadening the applications of HSL, attention should be paid to develop a HSL enabling continuous spectral band collection with higher spectral resolution [31–33].

As an active instrument to acquire abundant spectral profiles, HSL usually has limited spectral bands and coverage, and a more universal and practical HSL with fine spectral resolution and coverage is of great significance for non-contact and active vegetation parameter extraction. Motivated by this, in this paper, an Acousto-optical Tunable Filter HSL (AOTF-HSL) with 10 nm spectral resolution covering 500–1000 nm was developed, and the HSL was evaluated by comparing the selected "Red Edge" (RE) vegetation parameter-related results from AOTF-HSL with those obtained using an SVC HR-1024 spectrometer.

In this research, leaves from four different plants were measured to evaluate the capacity of using the spectral from the AOTF-HSL for vegetation RE-related parameter extraction. In the vegetation research community, the important "Red Edge" position (REP) parameter is closely related to various physical and chemical parameters of vegetation, and it is commonly employed to indicate the growing states of the vegetation and monitor the plant activity [34,35]. Thus, the RE related parameters were selected as the representative for evaluating the HSL in vegetation applications. As shown in Figure 1, REP refers to the position of an inflection point of the first derivative of reflectance values, and it usually locates in the red spectrum band [33–36]. REP result comparison between the HSL and SVC spectrometer could provide a preliminary evaluation of the utility of HSL in vegetation parameter extraction.

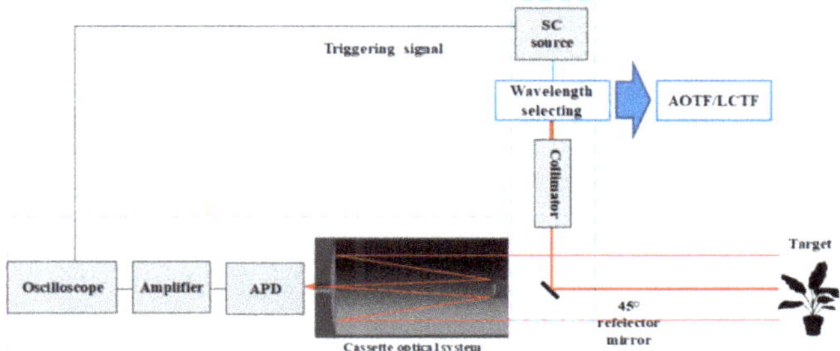

Figure 1. Schematic diagram of a tunable Hyperspectral LiDAR system based on AOTF.

In addition, three most common used methods (First-order Reflectance Slope (FRS), Linear Four-point Interpolation technology (LFPIT) and Linear Extrapolation technology (LET)) were investigated in this research for fully and furtherly evaluating the HSL capacity in spectral profiles acquirement and vegetation parameters extraction.

The contribution of this paper was summarized as follows:

1. This paper presents a more universal and applicable HSL with high spectral resolution to obtain vegetation spectral profiles, and three different RE position extraction methods were firstly employed for addressing the acquired HSL spectral profiles;
2. This paper is just the beginning of using the high spectral resolution HSL for vegetation index detection, which might inspire estimation of other vegetation parameters or biochemical content using this advanced HSL.

The remainder of this paper is organized as follows: Section 2 presents the system design of the AOTF-HSL and the REP determination methods in detail; Section 3 presents the results and analysis of the laboratory experiments concerning the RE-related parameters measurements, result comparisons between different methods and the analysis; and then the conclusions are drawn in Section 4.

2. Materials and Methods

This section is divided into two subsections, the first subsection is the AOTF-HSL design, components and description; and the second subsection presents the REP determination methods, including the calculation equations.

2.1. AOTF-HSL Design and Components

Figure 1 presents the design and diagram of the AOTF-HSL, which employs a super-continuum (SC) laser source covering 450–2350 nm. Figure 2 presents the relationship between the wavelength and power density of the employed SC laser source. An AOTF is installed in front of the SC source. AOTF is capable of consciously selecting and filtering laser beam with 10 nm spectral resolutions from 430–1450 nm. After the laser beam passing through AOTF, the emitted broadband laser beam is collimated and then reflected towards the target by a reflecting mirror. A Cassegrain telescope optical system is employed to collect the energy of the reflected laser pulses from the targets. An APD sensor module with an integrated amplifier is placed on the focal point of the Cassegrain telescope to collect the back-scattered laser echoes and transform them to electronic signals, which are sampled and recorded by a linked high-speed oscilloscope (20 GHz sampling rate, which equals 7.5 mm range resolution). Spectral information can be extracted from the recorded raw waveform. Meanwhile, the triggering signal indicating the emission of SC source is collected by the linked high-speed oscilloscope. Distance information is obtained by measuring the time difference between the triggering signal and reflected signal from the target. More details on the spectral and ranging information acquiring capacity could be found in our recent paper [32]. The remainder of this section will present the specifications of different parts of the AOTF-HSL.

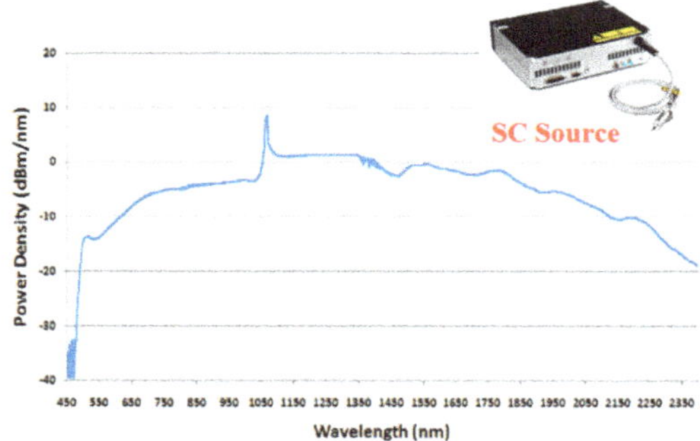

Figure 2. The supercontinuum laser source and the power density against the wavelength.

2.1.1. SC Source

In order to increase the spectral range of the AOTF-HSL, SC source (Figure 2) employed in this paper is with a large single pulse energy (>19 uJ) and the pulse width is two nanoseconds with power density distribution shown in Figure 3. The total power of the selected SC model is more than 8W, and the M^2 factor of SC source is better than 1.1 with a 200 MHz maximum repeating rates. With such a powerful laser source, the effective range of AOTF-HSL is improved to several tens meters based on the previous experiments [32].

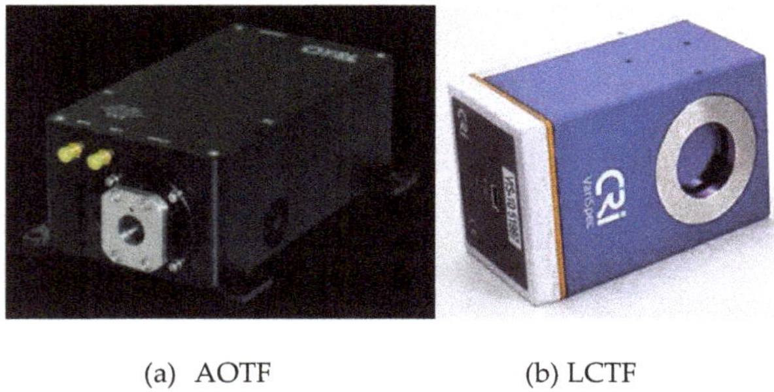

(a) AOTF   (b) LCTF

Figure 3. Employed filter device. (**a**) AOTF, (**b**) LCTF.

2.1.2. AOTF vs. LCTF

An AOTF (Figure 3) is an electro-optical device that functions as an electronically tunable excitation filter to simultaneously modulate the intensity and wavelength of the laser beam. The AOTF relies on a specialized birefringent crystal whose optical properties vary upon interaction with an acoustic wave. Changes in the acoustic frequency alter the diffraction properties of the crystal, enabling very rapid (in microsecond level) wavelength tuning, limited only by the acoustic transit time across the crystal. The selected wavelength is determined by Equation (1):

$$\lambda = \frac{\Delta n V}{f}\left[\sin^2 2\theta_i + \sin^4 \theta_i\right]^{1/2},$$

(1)

where the selected wavelength (λ) is a function of the difference of the refractive indexes due to birefringent Δn, the frequency of the applied RF signal f, the incident angle θ_i and the variable speed of acoustic waves in the crystal material V.

Previously, a feasibility study for detecting REP values was also carried out with a Liquid Crystal Tunable Filter (LCTF)-based HSL (LCTF-HSL) [31]. The major parameter specifications of LCTF and AOTF are shown in Table 1. The AOTF has a wider spectrum, covering from 430 nm to 1450 nm, while the LCTF just covers 400 nm to 720 nm. The AOTF is capable of continuously selecting or tuning the spectral resolution from 2 nm to 10 nm, and the LCTF just has three independent selections of 7, 10 and 20 nm. Moreover, the response time of AOTF is 10 μs, which is at least three orders of magnitudes quicker than the LCTF (the typical response time is 50 ms). Obviously, AOTF is much more preferable while selecting a filter in HSL. Figure 3 presents the AOTF and LCTF devices employed in this research.

Table 1. The AOTF vs LCTF Major Parameters Specifications.

Parameter	AOTF	LCTF (VariSpec VIS)
spectral range	430–1450 nm	400–720 nm
Response time	10 μs	50 ms
spectral resolution	2–10 nm	7/10 or 20 nm

2.1.3. Collimator

After the transmission of the laser beam from the laser source, a collimator is necessary to collimate the beam. In a traditional monochromatic LiDAR, the collimator design is comparatively simple, since only a single wavelength is considered in the laser beam collimation. However, in HSL, the wider spectral range of the laser beams should be taken into consideration in the laser beam collimation operation. An achromatic Galileo-type collimator is utilized for HSL system development with a beam expansion ratio of 1:5. Figure 4 shows the changing reflectance rate over the spectral wavelengths of the employed collimator. It can be observed that the collimator has a stable and low reflectance rate with the spectral wavelength ranging from approximately 650nm to 1050nm with broadband anti-reflection coating technique.

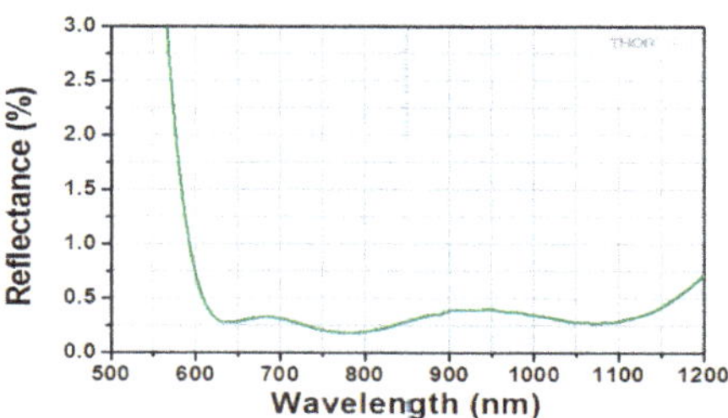

Figure 4. Transmittance of laser beam expander.

2.1.4. Reflector

The main function of the reflector installed on the optical axis of the receiving Cassette telescope is employed to steer the spectrally tuned laser beam toward the target. In order to mitigate the loss of laser energy, this paper takes full consideration of the devices with high reflection efficiency in the range from visible band (VIS) to near infrared spectrum band (NIS) during the selection of reflector.

As shown in Figure 5, the reflectivity of the selected reflector is over 94% in the spectrum range of 500 nm to 1000 nm.

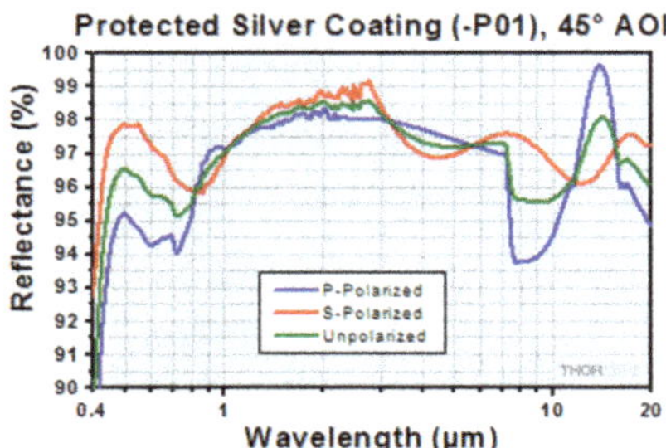

Figure 5. Reflectivity of the reflector.

2.2. REP Extraction Methods

The first-order differential of the spectral reflectance is the most commonly used method to extract the REP values. Since the spectral profiles are sampled with $\Delta\lambda$, Equation (2) is employed for calculating the first differential of the spectral reflectance. The equation is given in detail as:

$$\rho'(\lambda_i) = [\rho(\lambda_{i+1}) - \rho(\lambda_{i-1})]/2\Delta\lambda, \tag{2}$$

where, λ_i is the corresponding spectral wavelength, $\rho'(\lambda_i)$ is the first-order differential of the spectral reflectance, $\rho(\lambda_{i+1})$ and $\rho(\lambda_{i-1})$ are reflectance values at spectral wavelength of λ_{i+1} and λ_{i-1} respectively, $2 \times \Delta\lambda$ is the spectral increment between λ_{i+1} and λ_{i-1}. In this AOTF-HSL system, 10 nm spectral resolution is selected, thereby λ_i is sampled as 670, 680, 690, 700, 710, 720, 730, 740 and 750 nm in the spectral band.

Apart from this, another two RE parameters, RE slope and RE area, are also included. REP refers to the position of the RE in spectral wavelength, and REP slope is the spectral reflectance slope of the REP. REA refers to the area surrounded by the first derivative of the spectral reflectance, and it is calculated by the accumulating the reflectance slope with the spectral range from 680 nm to 750 nm [32–37]. Illustrative figures of these RE related parameters can be found in [32–37].

In addition, another two methods, Linear Four-point Interpolation Technology (LFPIT) and Linear Extrapolation Technology (LET), are employed for REP determination. Firstly, LFPIT is based on Equations (3) and (4). In this experiment, the method employs four wavelength data for calculating the REP. As illustrated, the 670 nm and 780 nm spectral information is used to calculate the reflectance at the REP, and the 700 nm and 740 nm spectral data are for determining the REP. In Equations (3) and (4), R_{670} and R_{780} are the corresponding reflectance values at 670 nm and 780 nm respectively. R_{700} and R_{740} are the corresponding reflectance values at 700 nm and 740 nm respectively:

$$R_{REP} = \frac{(R_{670} + R_{780})}{2}, \tag{3}$$

$$\lambda_{REP} = 700 + 40\frac{(R_{REP} - R_{700})}{R_{740} - R_{700}}, \tag{4}$$

Secondly, the LET method can be represented by Equations (5)–(7), and the REP is determined using the two extrapolation equations. Equation (5) is the extrapolation of the reflectance for spectral wavelengths ranging from 680 nm to 700 nm. Equation (6) is the extrapolation of the reflectance for spectral wavelengths ranging from 725 nm to 760 nm. Then, the REP is determined using the parameters (m_1, m_2, c_1 and c_2) from Equations (5) and (6). In the following equations, the FDR_1 and FDR_2 are the spectral reflectance slope of the spectral ranging 680 nm and 700 nm and 725 nm and 760 nm. m_1, m_2, c_1 and c_2 are the parameters for describing the spectral reflectance slope and determining the REP. The REP determination using this method is as Equation (7):

$$FDR_1 = m_1\lambda + c_1, \tag{5}$$

$$FDR_2 = m_2\lambda + c_2, \tag{6}$$

$$\lambda_{REP} = \frac{-(c_1 - c_2)}{m_1 - m_2}, \tag{7}$$

3. Results

As aforementioned, four different plants with green or yellow leaves were used in the laboratory experiments for the AOTF-HSL testing. Figure 6 shows these measured plants, namely dracaena (Figure 6a), aloe (Figure 6b), rubber plant (Figure 6c) and radermachera (Figure 6d). The AOTF-HSL and the SVC spectrometer are employed to measure the REP, the corresponding reflectance REP slope and REA. As aforementioned in Section 2, the LCTF-HSL is not sufficient for determining the REA parameters according to the LCTF parameters.

(**a**) Dracaena   (**b**) Aloe

(**c**) Rubber plant   (**d**) Radermachera

Figure 6. Four different plants employed in lab experiment. (**a**) Dracaena, (**b**) Aloe, (**c**) Rubber plant, (**d**) Radermachera.

3.1. REP, RE Slope and REA Measurement Results and Analysis

The Dracaena leaves spectral measurement results are shown in Figures 7–10. In Figure 7, the first-order derivative spectral profiles of green leaf and yellow leaf from AOTF-HSL and SVC are presented. Figure 8 shows the results of aloe green and yellow leaf, and the remaining Figures 9 and 10 are the spectral slope curves for rubber plant and radermachera green leaf, respectively.

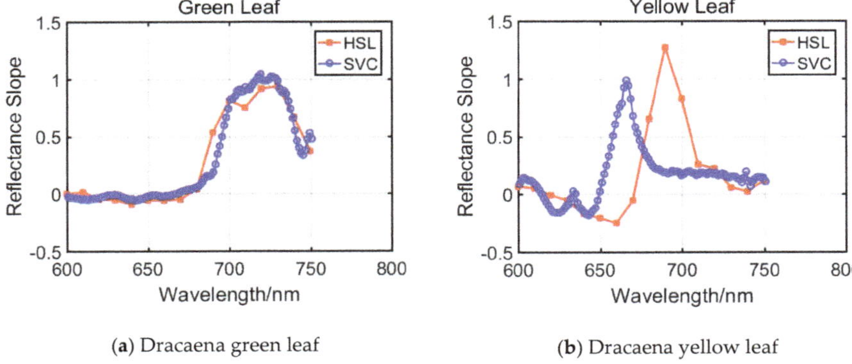

(**a**) Dracaena green leaf   (**b**) Dracaena yellow leaf

Figure 7. First derivative of the spectral reflectance versus spectral values of dracaena green and yellow leaf measured by the AOTF-HSL and the SVC spectrometer. (**a**) Dracaena green leaf, (**b**) Dracaena yellow leaf.

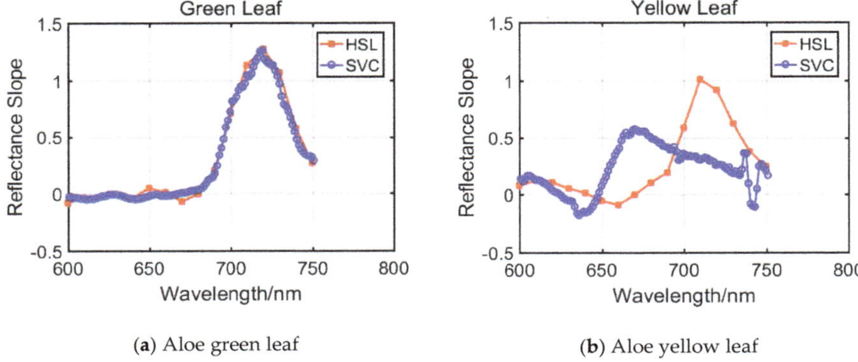

(**a**) Aloe green leaf   (**b**) Aloe yellow leaf

Figure 8. First derivative of the spectral reflectance versus spectral values of Aloe green and yellow leaf measured by AOTF-HSL and SVC. (**a**) Aloe green leaf, (**b**) Aloe yellow leaf.

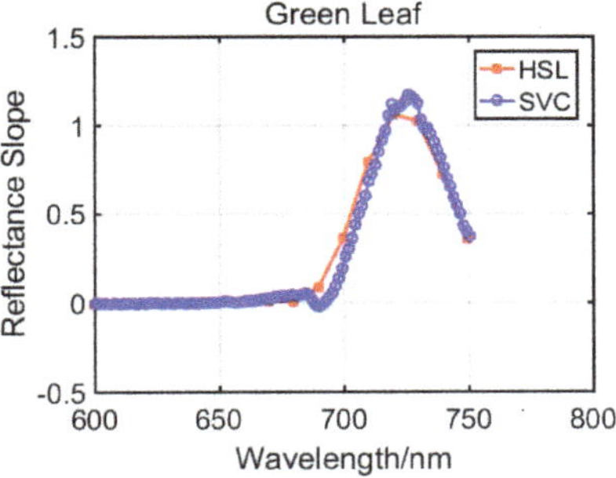

Figure 9. Rubber plant green leaf.

Among these figures, it can be seen from Figures 7a, 8a, 9 and 10 that spectral slope curves of green leaves of the selected plants extracted from AOTF-HSL coincide with its corresponding referenced measurements from the SVC spectrometer. Tables 2–4 list the quantitative results of the corresponding vegetation parameter results and the differences between the active and passive methods; the percentages are calculated using difference values dividing the corresponding referenced results from SVC. The differences of REP of all green leaf test cases are below 1%, however, for the REP slope, only the aloe green leaf REP slope difference percentage is below 1%. Specifically, the REP slope of the dracaena green leaf is large than 10%. For green leaf REA results, dracaena green leaf difference is also the largest, and it is more than 5%.

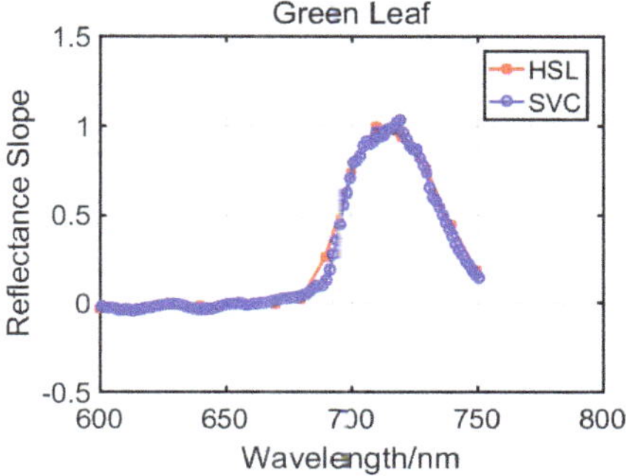

Figure 10. Radermachera green leaf.

Figures 7b and 8b present the spectral slope curves representing Dracaena yellow leaf and aloe yellow leaf, respectively. Compared with green leaf curves, both the yellow leaf curves from AOTF-HSL measurements are distinctive from that of SVC. Further analysis is also given in Tables 2–4, where the REP differences are 9.45 nm and 24.1 nm, respectively, and the corresponding difference is considerably higher than in the green leaf cases.

In addition, from comparing the REP between green and yellow leaves from the same plant, it can be seen by comparing the curves of Figures 7a, 7b, 8a and 8b that REP has an obvious shift towards a shorter wavelength, named "blue shift", since the yellow leaves have lower chlorophyll content and this tendency or REP behavior is consistent with previous research results.

Table 2. "Red edge" position measuring results based on FRS.

	REP (nm)		
	AOTF-HSL	**SVC**	**Difference**
Dracaena Green Leaf	725	718.85	6.15 (0.85%)
Dracaena Yellow Leaf	695	685.55	9.45 (1.3%)
Aloe Green Leaf	725	718.85	6.15 (0.85%)
Aloe Yellow Leaf	715	690.9	24.1 (3.4%)
Rubber Green Leaf	725	725.45	0.45 (0.06%)
Radermachera Green Leaf	715	718.85	3.85 (0.53%)

Table 3. "Red edge" slope measuring results.

	REP slope		
	AOTF-HSL	**SVC**	**Difference**
Dracaena Green Leaf	0.95	1.07	0.12 (11.2%)
Dracaena Yellow Leaf	1.44	1.03	0.41 (39.8%)
Aloe Green Leaf	1.29	1.3	0.01 (0.08%)
Aloe Yellow Leaf	1.1	0.59	0.51 (86%)
Rubber Green Leaf	1.09	1.17	0.08 (6.8%)
Radermachera Green Leaf	1	1.07	0.07 (6.5%)

Table 4. "Red edge" area measuring results.

	REA		
	AOTF-HSL	**SVC**	**Difference**
Dracaena Green Leaf	47.94	51.36	3.42 (6.7%)
Dracaena Yellow Leaf	36.67	26.34	10.33 (39.2%)
Aloe Green Leaf	52.54	51.1	1.43 (2.8%)
Aloe Yellow Leaf	41.63	31.53	10.1 (32.0%)
Rubber Green Leaf	44	43.68	0.32 (0.7%)
Radermachera Green Leaf	43.39	42.01	1.38 (3.2%)

3.2. Comparison of REP Results from Differnent Calculating Methods

In Section 3.1, the REP is calculated using the FRS technique. Actually, there are other methodologies for interpolating the reflectance REP, for instance, Linear Four-point Interpolation Technology (LFPIT), and Linear Extrapolation Technology (LET) [37,38]. For further validating and evaluating the proposed active remote sensing method, the AOTF-HSL spectral information is employed to calculate the REP based on LFPIT and LET.

With the parameters listed in Table 1, the selected wavelength spectral reflectance in LFPIT and LET beyond the wavelength scope of the LCTF-HSL, AOTF-HSL is more applicable than LCTF-HSL. In this section, only the spectral profiles collected by AOTF-HSL are used for extraction of these parameters. Tables 5–10 give the results of the four REP determination methods. Among them, four tables (Tables 5, 7, 9 and 10) are the REP results from the green leaves, and the other two (Tables 6 and 8) are the REP results from yellow leaves of dracaena and dloe.

Firstly, in the aspect of the green leaves results, the AOTF-HSL and SVC give a similar REP using the four different results, with differences all below 1% except for the rubber plant REP using the LFPIT method. The lower spectral resolution of HSL may account for the minor differences between AOTF-HSL and SVC. In this paper, the AOTF-HSL spectral resolution is 10 nm, and the SVC spectral resolution is better than 2 nm.

Table 5. The three different methods' "Red edge" position results for dracaena green leaf.

	REP (nm)		
	AOTF-HSL	**SVC**	**Difference**
LFPIT	717.80	722.32	4.52 (0.63%)
LET	709.42	712.24	2.82 (0.40%)
FRS	725	718.85	6.15 (0.85%)

Table 6. The three different methods' "Red edge" position results for dracaena yellow leaf.

	REP		
	AOTF-HSL	**SVC**	**Difference**
LFPIT	636.79	692.43	55.64 (8.7%)
LET	628.65	679.91	51.26 (8.2%)
FRS	695	685.55	9.45 (1.3%)

Table 7. The three different methods' "Red edge" position results for aloe green leaf.

	REP		
	AOTF-HSL	**SVC**	**Difference**
LFPIT	718.68	715.13	−3.55 (0.49%)
LET	720.32	719.95	−0.37 (0.05%)
FRS	725	718.85	6.15 (0.85%)

Table 8. The three different methods' "Red edge" position results for aloe yellow leaf.

	REP		
	AOTF-HSL	**SVC**	**Difference**
LFPIT	715.61	674.27	−41.34 (5.8%)
LET	721.12	669.01	−52.11 (7.2%)
FRS	715	690.9	24.1 (3.4%)

Table 9. The three different methods' "Red edge" position results for rubber plant.

	REP		
	AOTF-HSL	**SVC**	**Difference**
LFPIT	712.63	748.79	36.16 (5.1%)
LET	722.60	726.10	3.5 (0.5%)
FRS	725	725.45	0.45 (0.06%)

Table 10. The three different methods' "Red edge" position results for radermachera green leaf.

	REP		
	AOTF-HSL	**SVC**	**Difference**
LFPIT	712.14	709.25	−2.89 (0.4%)
LET	717.91	718.67	0.76 (0.1%)
FRS	715	718.85	3.85 (0.53%)

Secondly, in aspect of the yellow leaves (Tables 6 and 8), the differences of the results between AOTF-HSL and SVC calculated by the LFPIT and LET methods are all more than 5%, and the figures are slightly higher than the FPS-derived REP results. Compared with the green leaves, the differences between AOTF-HSL and SVC are larger in the yellow leaf cases. In addition, the three REP-derived methods give similar results in the green leaf cases, including AOTF-HSL and SVC, however, there is an obvious difference in the yellow leaves results. Especially, in the AOTF-HSL REP of the dracaena yellow leaf case, the LEPIT and LET methods afford similar results, but they are different from the FRS results. As aforementioned, the spectral resolution has an influence on the operation of LET and LEPIT, and in this experiment, the spectral resolution of AOTF-HSL is selected as 10 nm, which might affect the calculation of the results.

4. Discussion

The development of active measuring methods to obtain spectral information is of great significance. The AOTF-HSL technique presented in this paper has the unique characteristic of collecting a continuous spectrum in the visible and near infrared (VNIR) regions with 10 nm spectral resolution, which is the best spectral resolution in any published paper. However, limited by the HSL hardware configuration and the data processing capacity, compared with SVC spectrometer, the spectral resolution is still slightly restricted. We presume this is the major reason contributing to the slight differences between AOTF-HSL and SVC in green leaves measuring results. In addition, for REP slope and REA, these results are quite distinctive and the percentages are all larger than 30%. We think that these following reasons might account for this phenomenon:

1. Green leaves have more uniform spectral reflectivity over their surface, since the contents affecting "Red Edge" related parameters are distributed evenly on them; in contrast to this, yellow leaves have uneven distributions of these contents as Figure 7a,b present and the reflectivity varies for different parts of the yellow leaf;
2. As aforementioned in Section 2, the hardware design, optics system, and the measurement distance determine the diameter of the laser pulse footprint for sampling, which is approximately

1 cm in this experiment with a field of view (FOV) of 0.2 mill radian. The sampled area of the spectrometer is larger (resulting in a 5.5 cm radius footprint with a 25∘ field of view). Area coverage by the laser pulse has different reflectivity due to the non-uniformity of the yellow leaves.

For the FRS method, the spectral resolution affects the calculation results. Due to the higher spectral resolution of the SVC, the spectral profile collected by SVC could contribute to more detailed descriptions of the reflectance slope changes. In the LET and LPFIT methods, the parameters are calculated using the selected reflectance at several specific spectral wavelengths, and some of them are not consistent with the selected wavelength of the AOTF-HSL. Some spectral reflectance values are calculated using the average values of the intensities from the nearby spectral wavelength. For instance, in the LET method, the 725 nm reflectance is calculated through averaging that of 720 nm and 730 nm, which might slightly affect the results. Moreover, in the LET method, only three spectral reflectance values from AOTF-HSL are employed to fit the FDR1 (680, 690 and 700 nm), which might bias the final results. Thus, we think following future works are necessary for extending the HSL enabling vegetation index estimation and determination:

1. HSL with finer spectral resolution is anticipated to improve the performance in vegetation index or parameter extraction, and the ultimate HSL will have similar spectral resolution with the referenced SVC spectrometer. With better spectral resolution, it is of great significance to estimate the vegetation content and produce more comparable measurements. HSL spectral profiles covering the 500–1000 nm wavelength band with 2 nm resolution is anticipated for future work, whose resolution is more feasible to produce reliable results for vegetation-related applications.
2. In this paper, the influence of the spectral resolution on the REP or further vegetation resolution is not investigated, limited by the hardware design. A 10 nm spectral resolution is employed in this paper, which is determined by the LiDAR raw measurement processing capacity; actually, the spectral resolution of the AOTF-HSL can be adjusted from 2 nm to 10 nm, it is of great significance for exploring the influence of the spectral resolution on the performance of the above REP extraction method.
3. REP is one of the most important indicators for vegetation health monitoring, but there are still some other vegetation indexes for presenting vegetation growth or content; more work would be carried out on using HSL to extract these vegetation indexes.

5. Conclusions

This paper investigated an AOTF-based 10 nm spectral resolution HSL, and for which the consecutive spectral bands were tunable covering 500–1000 nm. An AOTF was installed after the super-continuum laser source and consecutively selecting the wavelength of the passing laser beam. 51-channel spectral information was acquired. Then, four different plants with four green leaves and two yellow leaves were measured by the AOTF-HSL and with corresponding referenced spectral profiles collected by an SVC spectrometer. For green leaves, the "Red Edge"-related parameters extracted using the first-order reflectance slope (FRS) from the AOTF-HSL and SVC spectrometer were trivial, and this demonstrated the AOTF-HSL was capable of measuring vegetation "Red Edge"-related parameters. Moreover, the observed shift behaviours were consistent with previous research results, which also supported the fact that measurement results from the AOTF-HSL are reliable. In addition, another two methods, LFPIT and LET, were employed for calculating the REP for further comparing the results between HSL and SVC. The results were similar to those from the FRS method, which further demonstrated the effectiveness of the HSL in this vegetation index detection and measurement.

Author Contributions: C.J. wrote this paper and processed the data; Y.C. proposed the idea, revised the paper designed this HSL and offered the funding; W.L. set up the HSL, the idea, collected the data, revised the paper and offered the funding; H.Z. discussed the idea and revised the paper; H.J. offer the funding, revised the paper and the guided the HSL design. Y.B., H.S., S.S. and E.P. reviewed the paper and provided valuable advice for paper writing and revision.

Funding: This research was financially supported by the Academy of Finland projects "Centre of Excellence in Laser Scanning Research (CoE-LaSR) (307362)" and "New laser and spectral field methods for in situ mining and raw material investigations (project 292648). Additionally, the Chinese Academy of Science (181811KYSB20160113), by the Chinese Ministry of Science and Technology (2015DFA70930) and Shanghai Science and Technology Foundations (18590712600) are acknowledged.

Conflicts of Interest: The authors declare no conflict of interest.

References

1. Plaza, A. Recent advances in techniques for hyperspectral image processing. *Remote Sens. Environ.* **2009**, *113*, S110–S122. [CrossRef]
2. Dalponte, M.; Bruzzone, L.; Gianelle, D. Fusion of hyperspectral and lidar remote sensing data for classification of complex forest areas. *IEEE Trans. Geosci. Remote Sens.* **2008**, *46*, 1416–1427. [CrossRef]
3. Ravikanth, L.; Jayas, D.S.; White, N.D.; Fields, P.G.; Sun, D.W. Extraction of spectral information from hyperspectral data and application of hyperspectral imaging for food and agricultural products. *Food Bioprocess Technol.* **2017**, *10*, 1–33. [CrossRef]
4. Camps-Valls, G.; Bruzzone, L. Kernel-based methods for hyperspectral image classification. *IEEE Trans. Geosci. Remote Sens.* **2005**, *43*, 1351–1362. [CrossRef]
5. Fernandez-Diaz, J.; Carter, W.; Glennie, C.; Shrestha, R.; Pan, Z.; Ekhtari, N.; Singhania, A.; Hauser, D.; Sartori, M. Capability assessment and performance metrics for the Titan multispectral mapping LiDAR. *Remote Sens.* **2016**, *8*, 936. [CrossRef]
6. Burton, D.; Dunlap, D.B.; Wood, L.J.; Flaig, P.P. Lidar intensity as a remote sensor of rock properties. *J. Sediment. Res.* **2011**, *81*, 339–347. [CrossRef]
7. Chen, Y.; Jiang, C.; Hyyppä, J.; Qiu, S.; Wang, Z.; Tian, M.; Bo, Y. Feasibility Study of Ore Classification Using Active Hyperspectral LiDAR. *IEEE Geosci. Remote Sens. Lett.* **2018**, *99*, 1–5. [CrossRef]
8. Song, J.-H. Assessing the possibility of land-cover classification using lidar intensity data. *Int. Arch. Photogramm. Remote Sens. Spat. Inf. Sci.* **2002**, *34*, 259–262.
9. Chust, G.; Galparsoro, I.; Borja, A.; Franco, J.; Uriarte, A. Coastal and estuarine habitat mapping, using LIDAR height and intensity and multi-spectral imagery. *Estuar. Coast. Shelf Sci.* **2008**, *78*, 633–643. [CrossRef]
10. Kaasalainen, S.; Hyyppa, H.; Kukko, A.; Litkey, P.; Ahokas, E.; Hyyppa, J.; Kaasalainen, M. Radiometric calibration of LiDAR intensity with commercially available reference targets. *IEEE Trans. Geosci. Remote Sens.* **2009**, *47*, 588–598. [CrossRef]
11. Hata, A.; Wolf, D. Road marking detection using LIDAR reflective intensity data and its application to vehicle localization. In Proceedings of the IEEE 17th International Conference on Intelligent Transportation Systems (ITSC), Qingdao, China, 8–11 October 2014; pp. 584–589.
12. Eitel, J.U.; Magney, T.S.; Vierling, L.A.; Dittmar, G. Assessment of crop foliar nitrogen using a novel dual-wavelength laser system and implications for conducting laser-based plant physiology. *ISPRS J. Photogramm. Remote Sens.* **2014**, *97*, 229–240. [CrossRef]
13. Gaulton, R.; Danson, F.M.; Ramirez, F.A.; Gunawar, O. The potential of dual-wavelength laser scanning for estimating vegetation moisture content. *Remote Sens. Environ.* **2013**, *132*, 32–39. [CrossRef]
14. Zimble, D.A. Characterizing vertical forest structure using small-footprint airborne LiDAR. *Remote Sens. Environ.* **2003**, *87*, 171–182. [CrossRef]
15. Douglas, E.S.; Strahler, A.; Martel, J.; Cook, T.; Mendillo, C.; Marshall, R.; Chakrabarti, S.; Schaaf, C.; Woodcock, C.; Li, Z.; et al. DWEL: A dual-wavelength echidna lidar for ground-based forest scanning. In Proceedings of the 2012 IEEE International Geoscience and Remote Sensing Symposium, Munich, Germany, 22–27 July 2012; pp. 4998–5001.
16. Asner, G.P.; Knapp, D.E.; Kennedy-Bowdoin, T.; Jones, M.O.; Martin, R.E.; Boardman, J.W.; Field, C.B. Carnegie airborne observatory: In-flight fusion of hyperspectral imaging and waveform light detection and ranging for three-dimensional studies of ecosystems. *J. Appl. Remote Sens.* **2007**, *1*, 013536. [CrossRef]
17. Alonzo, M.; Bookhagen, B.; Roberts, D.A. Urban tree species mapping using hyperspectral and lidar data fusion. *Remote Sens. Environ.* **2014**, *148*, 70–83. [CrossRef]
18. Sohn, G.; Dowman, I. Data fusion of high-resolution satellite imagery and LiDAR data for automatic building extraction. *ISPRS J. Photogramm. Remote Sens.* **2007**, *62*, 43–63. [CrossRef]

19. Du, L.; Gong, W.; Shi, S.; Yang, J.; Sun, J.; Zhu, B.; Song, S. Estimation of rice leaf nitrogen contents based on hyperspectral LIDAR. *Int. J. Appl. Earth Obs. Geoinf.* **2016**, *44*, 136–143. [CrossRef]
20. Wang, Z.; Chen, Y.; Li, C. A Hyperspectral LiDAR with Eight Channels Covering from VIS to SWIR[C]. *IEEE Int. Geosci. Remote Sens. Symp.* **2018**, 4293–4296.
21. Shi, S.; Song, S.; Gong, W.; Du, L.; Zhu, B.; Huang, X. Improving backscatter intensity calibration for multispectral LiDAR. *IEEE Geosci. Remote Sens. Lett.* **2015**, *12*, 1421–1425. [CrossRef]
22. Kaasalainen, S.; Lindroos, T.; Hyyppa, J. Toward hyperspectral lidar: Measurement of spectral backscatter intensity with a supercontinuum laser source. *IEEE Geosci. Remote Sens. Lett.* **2007**, *4*, 211–215. [CrossRef]
23. Chen, Y.; Räikkönen, E.; Kaasalainen, S.; Suomalainen, J.; Hakala, T.; Hyyppä, J.; Chen, R. Two-channel hyperspectral LiDAR with a supercontinuum laser source. *Sensors* **2010**, *10*, 7057–7066. [CrossRef] [PubMed]
24. Hakala, T.; Suomalainen, J.; Kaasalainen, S.; Chen, Y. Full waveform hyperspectral LiDAR for terrestrial laser scanning. *Opt. Express* **2012**, *20*, 7119–7127. [CrossRef] [PubMed]
25. Nevalainen, O.; Hakala, T.; Suomalainen, J.; Mäkipää, R.; Peltoniemi, M.; Krooks, A.; Kaasalainen, S. Fast and nondestructive method for leaf level chlorophyll estimation using hyperspectral LiDAR. *Agric. For. Meteorol.* **2014**, *198*, 250–258. [CrossRef]
26. Nevalainen, O.; Hakala, T.; Suomalainen, J.; Kaasalainen, S. Nitrogen concentration estimation with hyperspectral LiDAR[J]. *ISPRS Ann. Photogramm. Remote Sens. Spat. Inf. Sci.* **2013**, *2*, 205–210. [CrossRef]
27. Matikainen, L.; Karila, K.; Hyyppä, J.; Litkey, P.; Puttonen, E.; Ahokas, E. Object-based analysis of multispectral airborne laser scanner data for land cover classification and map updating. *ISPRS J. Photogramm. Remote Sens.* **2017**, *128*, 298–313. [CrossRef]
28. Puttonen, E.; Hakala, T.; Nevalainen, O.; Kaasalainen, S.; Krooks, A.; Karjalainen, M.; Anttila, K. Artificial target detection with a hyperspectral LiDAR over 26-h measurement. *Opt. Eng.* **2015**, *54*, 013105. [CrossRef]
29. Niu, Z.; Xu, Z.; Sun, G.; Huang, W.; Wang, L.; Feng, M.; Gao, S. Design of a new multispectral waveform LiDAR instrument to monitor vegetation. *IEEE Geosci. Remote Sens. Lett.* **2015**, *12*, 1506–1510.
30. Wang, L.; Sun, G.; Niu, Z.; Gao, S.; Qiao, H. Estimation of leaf biochemical content using a novel hyperspectral full-waveform LiDAR system. *Remote Sens. Lett.* **2014**, *5*, 693–702.
31. Li, W.; Jiang, C.; Chen, Y.; Hyyppä, J.; Tang, L.; Li, C.; Wang, S.W. A Liquid Crystal Tunable Filter-Based Hyperspectral LiDAR System and Its Application on Vegetation Red Edge Detection. *IEEE Geosci. Remote Sens. Lett.* **2018**, *16*, 291–295. [CrossRef]
32. Chen, Y.; Li, W.; Hyyppä, J.; Wang, N.; Jiang, C.; Meng, F.; Li, C. A 10-nm Spectral Resolution Hyperspectral LiDAR System Based on an Acousto-Optic Tunable Filter. *Sensors* **2019**, *19*, 1620. [CrossRef]
33. Malkamäki, T.; Kaasalainen, S.; Ilinca, J. Portable hyperspectral lidar utilizing 5 GHz multichannel full waveform digitization. *Opt. Express* **2019**, *27*, A468–A480. [CrossRef] [PubMed]
34. Pu, R. Extraction of red edge optical parameters from Hyperion data for estimation of forest leaf area index. *IEEE Trans. Geosci. Remote. Sens.* **2003**, *41*, 916–921.
35. Yao, F. Hyperspectral models for estimating vegetation chlorophyll content based on red edge parameter. *Trans. Chin. Soc. Agric. Eng.* **2009**, *25*, 123–129.
36. Zarco-Tajeda, P.J. Detection of chlorophyll fluorescence in vegetation from airborne hyperspectral CASI imagery in the red edge spectral region. *IEEE Int. Geosci. Remote Sens. Symp.* **2004**, *1*, 598–600.
37. Miller, J.R.; Hare, E.W.; Wu, J. Quantitative characterization of the vegetation red edge reflectance. An inverted-Gaussian reflectance model. *Remote Sens.* **1990**, *11*, 1755–1773. [CrossRef]
38. Shafri, H.Z.; Hamdan, N. Hyperspectral imagery for mapping disease infection in oil palm plantationusing vegetation indices and red edge techniques. *Am. J. Appl. Sci.* **2009**, *6*, 10.

Article

Estimating Peanut Leaf Chlorophyll Content with Dorsiventral Leaf Adjusted Indices: Minimizing the Impact of Spectral Differences between Adaxial and Abaxial Leaf Surfaces

Mengmeng Xie [1,†], Zhongqiang Wang [1,†], Alfredo Huete [2], Luke A. Brown [3], Heyu Wang [4], Qiaoyun Xie [2], Xinpeng Xu [5] and Yanling Ding [1,*]

[1] Key Laboratory of Geographical Processes and Ecological Security in Changbai Mountains, Ministry of Education, School of Geographical Sciences, Northeast Normal University, Changchun 130024, China; jiemm393@nenu.edu.cn (M.X.); wangzq027@nenu.edu.cn (Z.W.)

[2] Faculty of Science, University of Technology Sydney, Sydney, NSW 2007, Australia; alfredo.huete@uts.edu.au (A.H.); qiaoyun.xie@uts.edu.au (Q.X.)

[3] School of Geography and Environmental Science, University of Southampton, Highfield, Southampton, SO17 1BJ, UK; l.a.brown@soton.ac.uk

[4] Agronomy College, Shenyang Agricultural University, Shenyang 110866, China; wangheyu666@stu.syau.edu.cn

[5] Ministry of Agriculture Key Laboratory of Plant Nutrition and Fertilizer, Institute of Agricultural Resources and Regional Planning, Chinese Academy of Agricultural Sciences (CAAS), Beijing 100081, China; xuxinpeng@caas.cn

* Correspondence: dingyl720@nenu.edu.cn

† These authors contributed equally to this work.

Received: 16 August 2019; Accepted: 11 September 2019; Published: 15 September 2019

Abstract: Relatively little research has assessed the impact of spectral differences among dorsiventral leaves caused by leaf structure on leaf chlorophyll content (LCC) retrieval. Based on reflectance measured from peanut adaxial and abaxial leaves and LCC measurements, this study proposed a dorsiventral leaf adjusted ratio index (DLARI) to adjust dorsiventral leaf structure and improve LCC retrieval accuracy. Moreover, the modified Datt (MDATT) index, which was insensitive to leaves structure, was optimized for peanut plants. All possible wavelength combinations for the DLARI and MDATT formulae were evaluated. When reflectance from both sides were considered, the optimal combination for the MDATT formula was $(R_{723} - R_{738})/(R_{723} - R_{722})$ with a cross-validation R^2_{cv} of 0.91 and $RMSE_{cv}$ of 3.53 $\mu g/cm^2$. The DLARI formula provided the best performing indices, which were $(R_{735} - R_{753})/(R_{715} - R_{819})$ for estimating LCC from the adaxial surface ($R^2_{cv} = 0.96$, $RMSE_{cv} = 2.37$ $\mu g/cm^2$) and $(R_{732} - R_{754})/(R_{724} - R_{773})$ for estimating LCC from reflectance of both sides ($R^2_{cv} = 0.94$, $RMSE_{cv} = 2.81$ $\mu g/cm^2$). A comparison with published vegetation indices demonstrated that the published indices yielded reliable estimates of LCC from the adaxial surface but performed worse than DLARIs when both leaf sides were considered. This paper concludes that the DLARI is the most promising approach to estimate peanut LCC.

Keywords: leaf chlorophyll content; DLARI; MDATT; adaxial; abaxial; spectral reflectance; peanut

---

1. Introduction

Peanut (*Arachishypogaea* L.) is one of the major food legumes as well as oilseed crops being grown in 118 countries (or regions) around the world on about 28 million ha of land [1], and offers multiple benefits to meet human nutritional needs as well as being an important resource in the context of food security and hunger issues [2]. Leaf chlorophyll content (LCC) is an important indicator of plant photosynthesis [3], nutritional state [4], and stress [5]. Determination of LCC is crucial for crop management and precision agriculture practices [6].

Spectral vegetation indices, which are defined with the objective of enhancing spectral sensitivity to vegetation properties, have long been popular for estimating vegetation's biophysical and biochemical variables [7,8]. Decades of research have gone into determining wavelength regions sensitive to LCC in order to develop indices to maximize the accuracy of retrieval for different types of plants [9–11]. Datt [12] developed a three-band index for retrieval of LCC in higher plants based on the different response of reflectance at 710 nm and 850 nm to LCC. Sims and Gamon [13] analyzed nearly 400 leaf samples from 53 species and found that the mSR705 (simple ratio) and mND705 (normalized difference) were relatively insensitive to species and leaf structural variation. Gitelson et al. [14] proposed an index ($R_{nir}R_{red\text{-}edge} - 1$), which is an effective LCC predictor for maple, chestnut, wild vine, and beech leaves.

The pinnate leaves of peanut are highly sensitive to excess solar radiation and drought stress [15]. Field observations show that under strong solar irradiance, peanut easily turns the abaxial surface of leaves upwards. As a result, the spectral reflectance recorded by satellites or spectroradiometers may represent a mixture of the adaxial and abaxial surfaces in different proportions. The differences in optical properties of dorsiventral leaves due to the structural difference among the two sides have been well documented [16,17]. Baránková et al. [18] found that light incident from the adaxial side is more effectively absorbed than light incident from the abaxial side of green tobacco leaves. Lu and Lu [19] reported the lower reflectance of the adaxial white poplar surfaces compared to the abaxial faces in the 400–700 nm spectrum but reported an inversion of this effect in the near infrared wavelengths (700–1000 nm).

Leaf optical properties are a vital factor in determining the sensitivity of vegetation indices to LCC [13]. However, to the best of our knowledge, few studies have considered the influence of abaxial leaf reflectance on the retrieval of biochemistry and structure parameters. In one of the few studies carried out, Lu et al. [20] extended the wavelengths in the Datt's index to incorporate spectral reflectance from 400 nm to 1000 nm. They found that the modified Datt's index (MDATT) efficiently reduced the effects of bifacial leaf structure and improved the retrieval of white poplar and Chinese elm LCC. However, several characteristics of peanut leaves, such as leaf hair, wax, palisade, and spongy tissues, substantially differ from woody plants. Thus, the applicability of the MDATT to peanut LCC retrieval requires further investigation. In addition, the structural effects were mostly removed by MDATT but partially remained [21].

Theoretically, multiple-band indices can incorporate a larger amount of information and have the potential to improve retrieval accuracy [22–26]. For example, the mSR705 and mND705, which were developed by adding a band (R445) to the exiting two-band indices SR705 and ND705, effectively improved sensitivity to LCC [13]. Similarly, three-band indices such as the MERIS terrestrial chlorophyll index (MTCI) [22] and OLCI terrestrial chlorophyll index (OTCI) has been successfully used to retrieve chlorophyll content at the canopy scale (i.e., chlorophyll content) [27–29]. To date, very few studies have been conducted to assess the potential of vegetation indices based on four or more bands for improving LCC retrieval accuracy.

To address these gaps, this paper focuses on the development and optimization of new and existing indices that are insensitive to spectral differences among two sides of peanut leaves. The objectives of this work were to (1) analyze spectral differences in the adaxial and abaxial surfaces of peanut leaves; (2) identify the optimal wavelengths of the MDATT for estimating peanut LCC; (3) develop a novel index based on a four-band combination to reduce spectral differences in dorsiventral leaves for improving LCC retrieval; (4) compare the performance of the indices developed in this study with those widely used in the literature.

2. Materials and Methods

2.1. Data Collection

Ground data collection was carried out over a farmland area in Changchun, Jilin Province, China (44°42′27″N, 124°53′08″E), which is located in the temperate and monsoon climate zone with a typical continental climate. The field size was approximately 2 ha. During the peanut growing season in 2018, three field campaigns were conducted on 27 July (acicula forming stage), 19 August (bearing pod stage), and 19 September (maturity stage) to collect peanut leaves, respectively. In each campaign, 20 plots were randomly selected. One or two plants were selected at each plot, and leaves that were fully expanded, homogenous in color, and showing no visible signs of damage were detached from the top to the bottom of the canopy [30]. They were, immediately packed and sealed into plastic bags and placed inside a cooler (the interior temperature of the cooler was 0 °C) to avoid desiccation and decomposition of the chlorophyll by light. In each campaign, we collected 28 leaves. Thus, a total number of 84 leaves were used for spectral measurement and chlorophyll extraction. All the measurements, including the spectral measurements and chlorophyll extraction, were carried out within 4 h of leaf harvesting to minimize changes in chlorophyll content. Figure 1 illustrates the phenomenon of peanut leaves changing orientation under strong solar irradiance, making the view of the canopy a mixture of the adaxial and abaxial sides (Figure S1).

Figure 1. Photographs of the peanut canopy in the field. Only the adaxial surface is visible under low solar irradiance (**a**), while both the adaxial and abaxial surfaces are visible under high solar irradiance (**b**).

The reflectance of the adaxial and abaxial surfaces was measured using an ASD FieldSpec® 3 portable spectroradiometer (Analytical Spectral Devices, Boulder, CO, USA) and a contact probe, equipped with an internal halogen source and directly attached to the leaf surface using a leaf clip accessory. The spectrometer can collect data in the 350–2500 nm spectral region, with a sampling interval of 1.4 nm in the 350–1000 nm wavelength range and 2 nm in the 1000–2500 nm wavelength range. The average of 10 separate measurements from each sample was recorded. To reduce errors associated with variations in illumination geometry, the contact probe was pressed to the leaf surface, which was illuminated by the internal light source, ensuring a consistent illumination geometry.

The LCC was determined from the same leaf samples used for reflectance measurements. Circular discs with a diameter of 6 mm were cut from each leaf. Leaf discs were extracted in the dark at room temperature for 24 h with 95% ethanol and shaken repeatedly to ensure chlorophyll was completely extracted, as indicated by the completely white appearance of the disc [31]. The absorbance of each extract was measured at 663 nm and 645 nm using a UV757CRT ultraviolet-visible spectrophotometer (Shanghai Precision Scientific Instruments Corporation, Shanghai, China). The LCC ($\mu g/cm^2$) was then calculated according to the equations provided by Arnon [32]. In total, measurements of LCC and reflectance were collected for 84 leaves. The LCC values ranged from 21.50 to 70.55 $\mu g/cm^2$ with a mean value of 40.78 $\mu g/cm^2$ and a standard deviation of 11.68 $\mu g/cm^2$.

2.2. Data Analysis

2.2.1. Construction of a Dorsiventral Leaf Adjusted Ratio Index

A semi-empirical leaf reflectance model was proposed by Baret et al. [33], and can be expressed as:

$$R = R_s + S \exp(-\Sigma k_i C_i) \tag{1}$$

where R_s is the reflectance of the leaf surface, S represents scattering effects of the leaf mesophyll structure, and k_i and C_i are the extinction coefficient and chlorophyll concentration, respectively. R_s and S are thought to be the main factors introducing variability between adaxial and abaxial leaf reflectance, as they depend on the differences in the leaf surface and internal mesophyll structure Datt [12]. Based on this model and the principle that there is no absorption at 850 nm by any leaf pigment (i.e., $\Sigma k_i C_i = 0$), Datt [12] proposed an index of $(R_{850} - R_{710})/(R_{850} - R_{680})$. The index removed R_s and S. Lu et al. [21] extended the wavelengths in the Datt's index to 400–1000 nm, which can also remove R_s and S. The formula of MDATT is:

$$\text{MDATT} = \frac{\left(R_{\lambda_1} - R_{\lambda_2}\right)}{\left(R_{\lambda_1} - R_{\lambda_3}\right)} = \frac{\exp\left(-K_{\text{chl}(\lambda_1)}C_{\text{chl}}\right) - \exp\left(-K_{\text{chl}(\lambda_2)}C_{\text{chl}}\right)}{\exp\left(-K_{\text{chl}(\lambda_1)}C_{\text{chl}}\right) - \exp\left(-K_{\text{chl}(\lambda_3)}C_{\text{chl}}\right)} \tag{2}$$

where C_{chl} is the chlorophyll content and $k_{\text{chl}(\lambda1)}$, $k_{\text{chl}(\lambda2)}$, and $k_{\text{chl}(\lambda3)}$ are the specific absorption coefficients for Chl at wavelengths λ_1, λ_2, and λ_3, respectively.

Using the MDATT as a basis, we constructed a dorsiventral leaf adjusted ratio index (DLARI) by substituting the R_{λ_1} term with an additional wavelength (R_{λ_4}) in the denominator:

$$\text{DLARI} = \frac{\left(R_{\lambda_1} - R_{\lambda_2}\right)}{\left(R_{\lambda_3} - R_{\lambda_4}\right)} = \frac{\exp\left(-K_{\text{chl}(\lambda_1)}C_{\text{chl}}\right) - \exp\left(-K_{\text{chl}(\lambda_2)}C_{\text{chl}}\right)}{\exp\left(-K_{\text{chl}(\lambda_3)}C_{\text{chl}}\right) - \exp\left(-K_{\text{chl}(\lambda_4)}C_{\text{chl}}\right)} \tag{3}$$

Like the MDATT, the DLARI is also expected to suppress the effects of R_s and S. The efficiency of incorporating an additional band is evaluated in the following sections.

2.2.2. Published Vegetation Indices

The accuracy of LCC retrieval using the MDATT and DLARI indices was compared with that of 12 published vegetation indices that were originally proposed for chlorophyll content retrieval and have proven to be highly correlated with LCC [34]. The types of indices included single-band index (e.g., 1/R700), two-band indices (e.g., VOG1 = R740/R720), three-band indices (e.g., DATT = (R850 − R710)/(R850 − R680)), as well as four-band indices (e.g., VOG2 = (R734 − R747)/(R715 + R720)). The formulae of the indices are provided in Table 1. In addition, two MDATTs proposed by Lu et al. [21] were used to evaluate the optimal wavelengths in the MDATT for peanut LCC estimation. The first MDATT in Table 1 with wavelengths at 691 nm, 745 nm, and 736 nm was developed from the adaxial surface reflectance of woody plants, whilst the second MDATT with wavelengths at 721 nm, 744 nm, and 714 nm was based on a mixture of reflectance for both adaxial and abaxial surfaces.

Table 1. The list of the published vegetation indices compared in the study.

Index	Abbreviation	Formula	Scale	Reference
Gitelson's index	Gitelson	1/R700	Leaf	[35]
Vogelmann's first Index	VOG1	R740/R720	Leaf	[36]
Carter' Index	Carter	R710/R760	Leaf	[37]
MERIS Terrestrial Chlorophyll Index	MTCI	(R740 − R705)/(R705 − R665)	Canopy	[22]
Modified Simple Ratio	mSR705	(R750 − R445)/(R705 − R445)	Leaf	[13]
Modified Normalized Difference Vegetation Index	mND705	(R750 − R705)/(R750 + R705 − 2 × R445)	Leaf	[13]
Datt' Index	DATT	(R850 − R710)/(R850 − R680)	Leaf	[12]
Maccioni' Index	Maccioni	(R780 − R710)/(R780 − R680)	Leaf	[38]
Vogelmann's second Index	VOG2	(R734 − R747)/(R715 + R720)	Leaf	[36]
Red-Edge Position Index	REP	700 + 40 × (((R670 + R780)/2 − R700)/(R740 − R700))	Leaf	[39]
Modified Datt Index	Lu's MDATT	(R691 − R7745)/(R7691 − R745)	Leaf	[21]
Modified Datt Index	Lu's MDATT	(R721 − R744)/(R721 − R714)	Leaf	[21]

2.2.3. Model Calibration and Validation

All possible band combinations based on the MDATT and DLARI derived from adaxial, abaxial, and the bifacial (i.e., including both adaxial and abaxial reflectance measurements) datasets were correlated with LCC, respectively. The waveband combinations with the highest coefficient of determination (R^2) for each dataset were selected as the optimal indices and used to model LCC. Relationships between the measured LCC and indices were established using empirical regression analysis. The form of fitting functions (e.g., linear, exponential, logarithmic) relating the indices to LCC appeared to have a marginal impact when compared to the impact of band selection [24]. Therefore, we restricted the fitting method to ordinary least-squares linear regression.

The performance of the index-based models was evaluated using the R^2 and root mean square error (RMSE) with respect to the biochemically measured LCC. In order to avoid dependence on a single random partitioning of the datasets and guarantee that all samples were used for both training and validation, a repeated 10 fold cross-validation was used to evaluate the performance of each index [40]. The dataset was split into 10 consecutive folds, and each fold was then used once for validation while the remaining 9 folds formed the training dataset. This process was repeated 50 times, and combined R^2_{cv} and $RMSE_{cv}$ values were calculated as the mean of those from each repetition.

3. Results

3.1. Spectral Differences Between Adaxial and Abaxial Surfaces

The mean spectral reflectance of the adaxial and abaxial surfaces is shown in Figure 2a. The reflectance of the adaxial surface was much lower than that of the abaxial surface in the visible region (400 to 690 nm). This is because light incident from the adaxial side was more effectively absorbed than light incident from the abaxial leaf side [18]. In contrast, the adaxial reflectance was higher than that of the abaxial surface between 750 and 1400 nm. This was partly because the palisade structure at the adaxial side of the leaf contributed higher reflected radiation than the spongy structure at the abaxial side [17]. Spectral differences among the two leaf surfaces were small in the red-edge region (690 to 750 nm), especially between 718 and 732 nm, where differences in reflectance were less than 5% (Figure 2b). Differences were also less substantial at wavelengths longer than 1400 nm. Variations in the internal structure of the adaxial and the abaxial surface also contributed to these differences (Figure 3). The adaxial surface (Figure 3a) was characterized by increased waxes than the abaxial surface (Figure 3b). Rayleigh scattering by waxes is known to contribute to the higher reflectance of the adaxial surface at NIR wavelengths [41].

The correlation between LCC and the reflectance of both surfaces are plotted in Figure 4. The reflectance of both surfaces in the blue region (400 to 500 nm) and the main chlorophyll absorption region near 680 nm demonstrated the least sensitivity to LCC. High sensitivity to LCC was observed at 570 nm and near 710 nm. The strongest correlation was at 710 nm (the correlation coefficient is −0.91) and 704 nm (the correlation coefficient is −0.85) for the adaxial and abaxial surfaces, respectively. The abaxial reflectance at wavelengths over 750 nm also demonstrated strong correlations. Overall, the correlations between LCC and abaxial reflectance were stronger than those between LCC and adaxial reflectance.

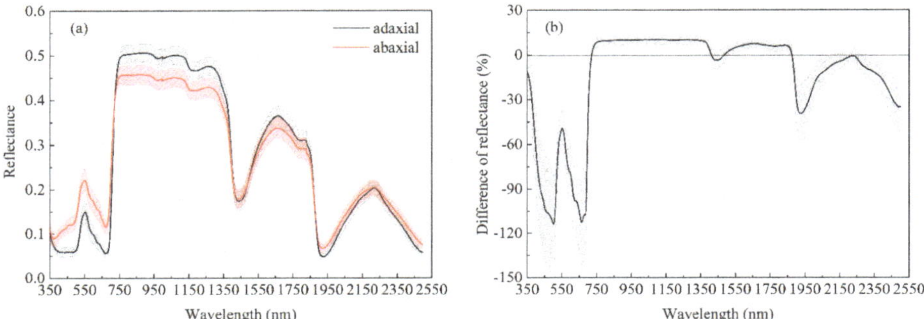

Figure 2. The spectral reflectance of the adaxial and abaxial surfaces (**a**) and the associated difference in reflectance among the two sides (**b**). (The solid line represents the mean of sampled reflectance and the shaded zone represents standard deviation).

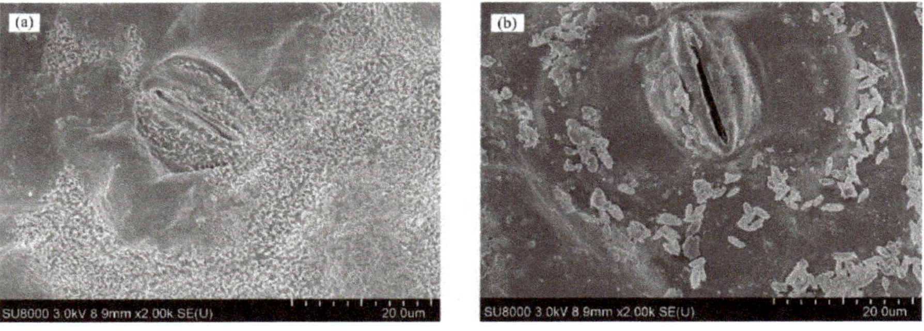

Figure 3. Optical micrographs of the adaxial (**a**) and abaxial (**b**) surface of peanut leaves.

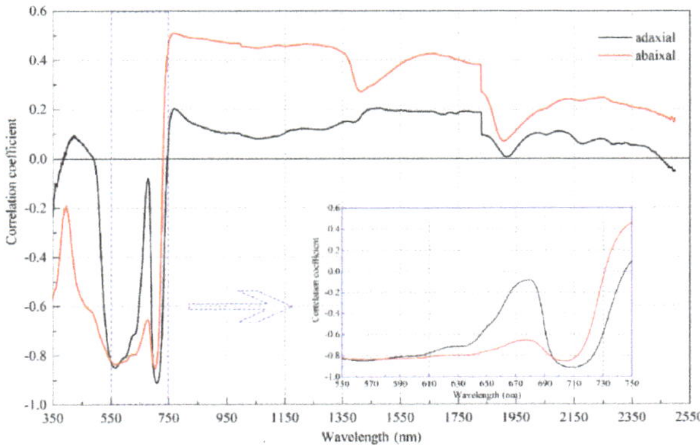

Figure 4. The correlation between LCC and the reflectance of the adaxial and abaxial surfaces from 350 to 2500 nm.

3.2. Relationships Between Optimal MDATT Indices and Peanut LCC

The relationship between LCC and the MDATT using band combinations ranging from 400 nm to 1000 nm was assessed for each dataset. The maximum R^2 was determined by fixing λ_2 and λ_3 as single values and changing λ_1 from 400 nm to 1000 nm. For the sake of concise display, only R^2 values greater than 0.8 were considered and they are shown in Figure 5, where the x-axis represents λ_3 and the y-axis λ_2. From this figure, robust wavelength regions for each band of the MDATT can be identified. For the adaxial dataset, the most sensitive region (red color in Figure 5a) ranged from 700 nm to 800 nm for λ_2, 400 nm to 800 nm for λ_3, and 650 nm to 750 nm for λ_1 (Figure 5d). The robust regions for the abaxial dataset were similar to those for the adaxial dataset (Figure 5b,e), but the most sensitive area (red color in Figure 5b) was reduced, and covered approximately 750 nm for λ_2, 730 nm for λ_3, and 690–750 nm for λ_1. For the bifacial dataset (Figure 5c,e), the sensitive wavelength regions were further reduced. The R^2 values greater than 0.88 were demonstrated when λ_2 and λ_3 were between 700 nm to 750 nm and λ_1 was between 660 nm and 710 nm.

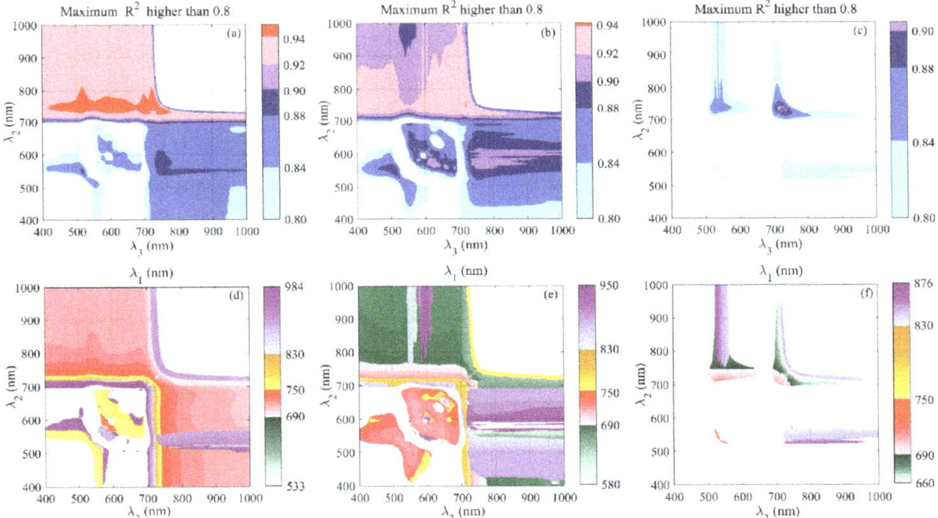

Figure 5. The maximum R^2 values associated with the MDATT band combinations ranging from 400 nm to 1000 nm (**a–c**) and its corresponding λ_1 (**d–f**). The left column is for the adaxial dataset, the middle column is for the abaxial dataset, and the right column is for the bifacial dataset. For the sake of concise display, only R^2 values greater than 0.8 were considered.

According to the principle of selecting indices demonstrating the highest R^2 values, three optimal MDATT indices for each dataset were determined and are presented in Table 2. The wavelengths λ_1, λ_2, and λ_3 of the best performing MDATT index for all three datasets were concentrated in the region of 701 to 747 nm. For the adaxial surface, the index incorporating reflectance values at 701 nm, 742 nm, and 740 nm demonstrated an R^2 of 0.95, whilst for the abaxial dataset, the best index incorporated reflectance values at 718 nm, 747 nm, and 720 nm ($R^2 = 0.94$). In the case of the bifacial dataset, the best index incorporated reflectance values at 723 nm, 738 nm, and 722 nm, reaching an R^2 of 0.91. The optimal indices for the adaxial and abaxial surfaces employed at least one wavelength that was highly correlated to LCC ($r < -0.60$), i.e., 701 nm, 718 nm, and 720 nm (Figure 4). The reflectance values at 722 nm and 723 nm, which were used by the optimal index for the bifacial dataset, demonstrated the minimum differences between adaxial and abaxial surfaces (Figure 2b).

Table 2. Cross-validation results for the MDATT indices in the case of the adaxial, abaxial, and bifacial datasets.

Index	Dataset	Wavelength Region Considered (nm)	Optimal Wavelengths (nm)	R^2	R^2_{cv}	RMSE$_{cv}$ (μg/cm^2)
	Adaxial reflectance	400–1000	λ_1: 701; λ_2: 742; λ_3: 740	0.95	0.95	2.52
MDATT	Abaxial reflectance	400–1000	λ_1: 718; λ_2: 747; λ_3: 720	0.94	0.94	2.69
	Bifacial reflectance	400–1000	λ_1: 723; λ_2: 738; λ_3: 722	0.91	0.91	3.53

The linear models established using the MDATT indices are shown in Figure 6. They were randomly selected from one of the 50 training datasets used in the repeated 10 fold cross-validation. The results indicated that for the adaxial surface, the index $(R_{701} - R_{742})/(R_{701} - R_{740})$ achieved the highest retrieval accuracy (R^2_{cv} = 0.95, RMSE$_{cv}$ = 2.52) (Figure 6a,d), followed by the index $(R_{718} - R_{747})/(R_{718} - R_{720})$ for the abaxial surface (R^2_{cv} = 0.94, RMSE$_{cv}$ = 2.69) (Figure 6b,e). Both performed better than the index $(R_{723} - R_{738})/(R_{723} - R_{722})$ for the bifacial dataset (R^2_{cv} = 0.91, RMSE$_{cv}$ = 3.53) (Figure 5c,f). The observed and predicted values fell close to the 1:1 line, indicating that the optimized MDATT indices were stable predictors of LCC. The optimal index for the bifacial dataset demonstrated a lower correlation with LCC and was not as accurate an LCC predicator as the other two indices.

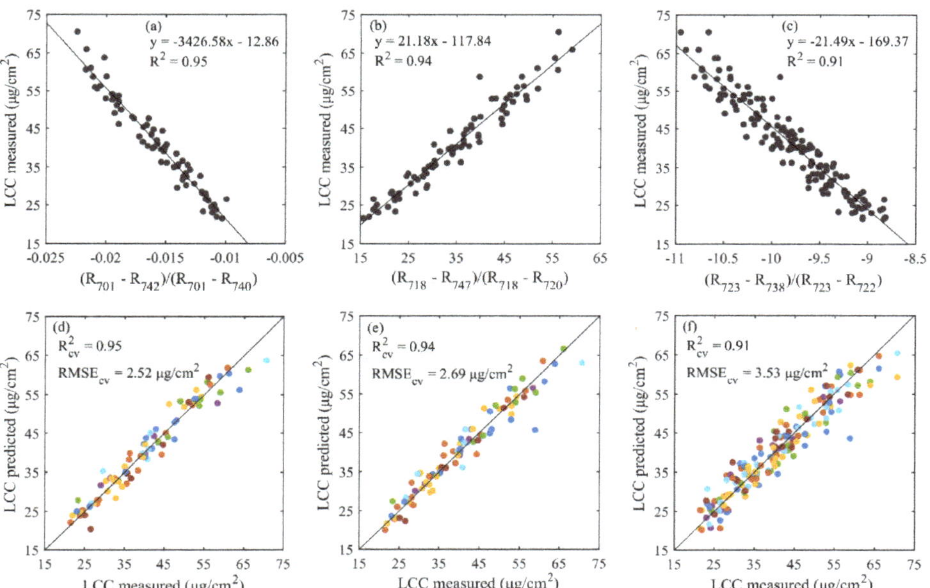

Figure 6. Relationships between MDATTs and LCC (**a–c**) and scatter plots between observed LCC and LCC predicted by the associated linear models (**d–f**). The different colors indicate the 10 fold cross-validation subsets. The left column is for the adaxial dataset, the middle column is for the abaxial dataset, and the right column is for the bifacial dataset.

Differences in adaxial and abaxial reflectance properties resulted in different MDATT indices and associated retrieval accuracies. The reliability of applying the optimal index for the adaxial surface to estimate LCC from bifacial reflectance measurements was investigated (Table 3). It showed that compared to the bifacial MDATT, applying the adaxial MDATT to estimate LCC from bifacial reflectance produced considerable errors (R^2_{cv} = 0.87, RMSE$_{cv}$ = 4.14). In light of the errors, it was necessary to consider the influence of spectral differences between adaxial and abaxial sides when estimating LCC from reflectance mixed by the two sides.

Table 3. LCC retrieval accuracy of the adaxial and bifacial MDATT indices when applied to bifacial dataset.

Dataset	Adaxial MDATT		Bifacial MDATT	
	R^2_{cv}	$RMSE_{cv}$ ($\mu g/cm^2$)	R^2_{cv}	$RMSE_{cv}$ ($\mu g/cm^2$)
Bifacial reflectance	0.87	4.14	0.91	3.53

3.3. Relationships Between DLARI and Peanut LCC

3.3.1. Performance of DLARIs Incorporating Wavelengths Between 660 and 750 nm

Because the most robust wavelength region was between 660 nm and 750 nm in the case of the MDATT, all possible DLARI combinations formed by Equation (3) using wavelengths from 660 nm to 750 nm were calculated. According to the principle of selecting the index that demonstrated the highest R^2, optimal DLARIs were established for each dataset (Table 4). The results demonstrated that the wavelengths used in the DLARIs were similar to those used in the MDATTs (Table 2). When compared with the MDATTs for the adaxial surface, the DLARI did not improve retrieval accuracy ($RMSE_{cv}$ = 2.53), the DLARI for the abaxial surface demonstrated a marginal advantage over the abaxial MDATT ($RMSE_{cv}$ = 2.62). In the case of the bifacial dataset, the DLARI achieved some improvement over the bifacial MDATT ($RMSE_{cv}$ = 3.34).

Table 4. Cross-validation results for the optimal dorsiventral leaf adjusted ratio indices (DLARIs) derived using wavelengths between 660 and 750 nm in the case of the adaxial, abaxial, and bifacial datasets.

Index	Dataset	Wavelength Region Considered (nm)	Optimal Wavelengths (nm)	R^2	R^2_{cv}	$RMSE_{cv}$ ($\mu g/cm^2$)
	Adaxial reflectance	660–750	λ_1: 740; λ_2: 742; λ_3: 703; λ_4: 731	0.95	0.95	2.53
DLARI	Abaxial reflectance	660–750	λ_1: 714; λ_2: 746; λ_3: 718; λ_4: 720	0.94	0.94	2.62
	Bifacial reflectance	660–750	λ_1: 731; λ_2: 741; λ_3: 722; λ_4: 750	0.91	0.92	3.34

3.3.2. Performance of DLARIs Incorporating Wavelengths Over 750 nm

In the DLARIs for the bifacial dataset, the wavelength selected for λ_4 was at the limit of the considered region (i.e., 750 nm), indicating that relevant information might be contained at longer wavelengths. When evaluated, DLARIs incorporating longer wavelengths (around 820 nm) achieved higher retrieval accuracies than those described in Section 3.3.1. (Table 5). It showed that for the three datasets, the optimal wavelengths of λ_1 and λ_3 moved to approximately 730 nm and 720 nm, where the differences in adaxial and abaxial reflectance were less than 5% (Figure 2). The optimal location of λ_2 moved to the red-edge shoulder, which means less sensitivity to leaf structure [42], while the optimal location of λ_4 moved to the NIR, where there is less absorption by leaf pigments [12]. The new adaxial DLARI and abaxial DLARI demonstrated advantages over the DLARIs derived from reflectance over 660 and 750 nm ($RMSE_{cv}$ = 2.37; $RMSE_{cv}$ = 2.58). The new bifacial DLARI not only substantially improved the retrieval accuracy ($RMSE_{cv}$ = 2.81), but also enhanced its sensitivity to LCC (R^2_{cv} = 0.94).

Table 5. Cross-validation results for the optimal DLARIs derived using wavelengths between 660 and 820 nm in the case of the adaxial, abaxial, and bifacial datasets.

Index	Dataset	Wavelength Region Considered (nm)	Optimal Wavelengths (nm)	R^2	R^2_{cv}	RMSE$_{cv}$ (μg/cm^2)
	Adaxial reflectance	660–820	λ_1: 735; λ_2: 753; λ_3: 715; λ_4: 819	0.96	0.96	2.37
DLARI	Abaxial reflectance	660–820	λ_1: 731; λ_2: 755; λ_3: 722; λ_4: 774	0.95	0.95	2.58
	Bifacial reflectance	660–820	λ_1: 732; λ_2: 754; λ_3: 724; λ_4: 773	0.94	0.94	2.81

Relationships between LCC and the optimal DLARIs established are shown in Figure 7, as are scatter plots of the associated retrievals and observed values. For the adaxial and the abaxial datasets (Figure 7a,b,d,e), the indices $(R_{735} - R_{753})/(R_{715} - R_{819})$ and $(R_{731} - R_{755})/(R_{722} - R_{774})$ attained higher retrieval accuracies ($R^2_{cv} = 0.96$, RMSE$_{cv} = 2.37$; $R^2_{cv} = 0.95$, RMSE$_{cv} = 2.58$) than the MDATT indices ($R^2_{cv} = 0.95$, RMSE$_{cv} = 2.52$; $R^2_{cv} = 0.94$, RMSE$_{cv} = 2.69$). For the bifacial dataset (Figure 7e–f), the index $(R_{732} - R_{754})/(R_{724} - R_{773})$ achieved an R^2_{cv} of 0.94 and RMSE$_{cv}$ of 2.81, demonstrating a substantial advantage over the bifacial MDATT index ($R^2_{cv} = 0.91$, RMSE$_{cv} = 3.53$) and the DLARI derived using wavelengths shorter than 750 nm. The results revealed that the DLARIs incorporating longer wavelengths efficiently improved LCC estimation accuracy, whether for the adaxial, abaxial or bifacial datasets.

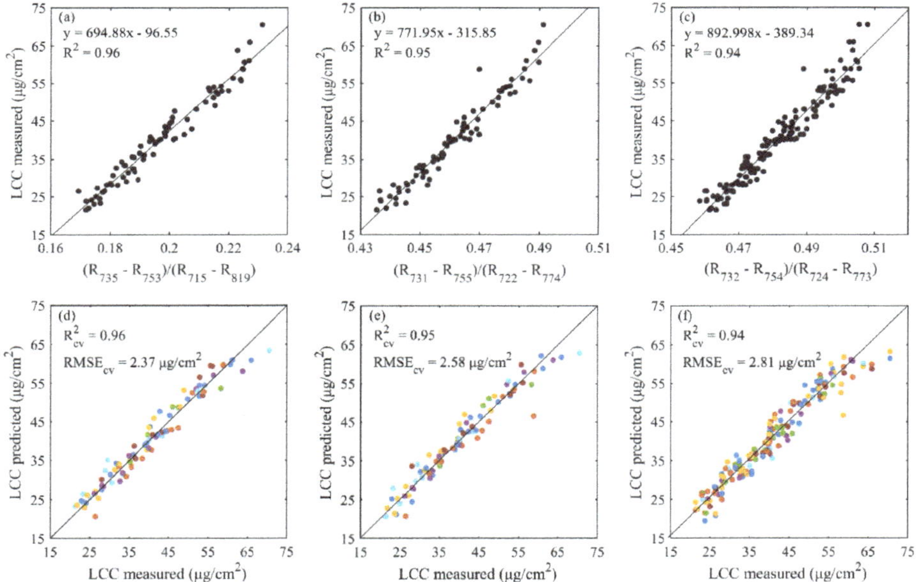

Figure 7. Relationships between optimal DLARI indices and LCC (**a–c**) and scatter plots between observed LCC and LCC predicted by the associated linear models (**d–f**). The different colors indicate the 10 fold cross-validation subsets. The left column is for the adaxial dataset, the middle column is for the abaxial dataset, and the right column is for the bifacial dataset.

3.4. Comparing Developed Indices with Those of Previous Studies

The performance of the published vegetation indices for LCC retrieval using adaxial and bifacial reflectance measurements is shown in Figure 8, as is the performance of the indices developed in this study. The published indices, which ranged from single- to four-band formulae, produced reliable retrievals of LCC when applied to adaxial reflectance measurements. In general, the three-band indices performed better than the two-band indices and the two-band indices performed better than Gitelson's index. The best performing indices were MTCI, DATT, and Maccioni followed by the four-band index VOG2. The red edge position index performed worse than all the three-band indices. However, when applied to the bifacial dataset, much lower R^2_{cv} and higher RMSE$_{cv}$ values were obtained. The VOG1, MTCI, mSR705, mND705, and VOG2 indices yielded RMSE$_{cv}$ values of approximately 3.5 from the adaxial dataset, while the RMSE$_{cv}$ values increased to 7.5 when applied to the bifacial dataset. Although the MTCI, mSR705, Maccioni, and DATT share the same format, the Maccioni and DATT indices (which employ two of the same wavelengths) performed better on the bifacial dataset. This was possibly because one of the bands used by Maccioni and Datt is located within the NIR region (780 nm and 850 nm), where there is little absorption by any leaf pigments ($\Sigma k_i C_i = 0$) [12]. This partly reduces spectral differences caused by the different absorption properties of pigments at the adaxial and abaxial surfaces.

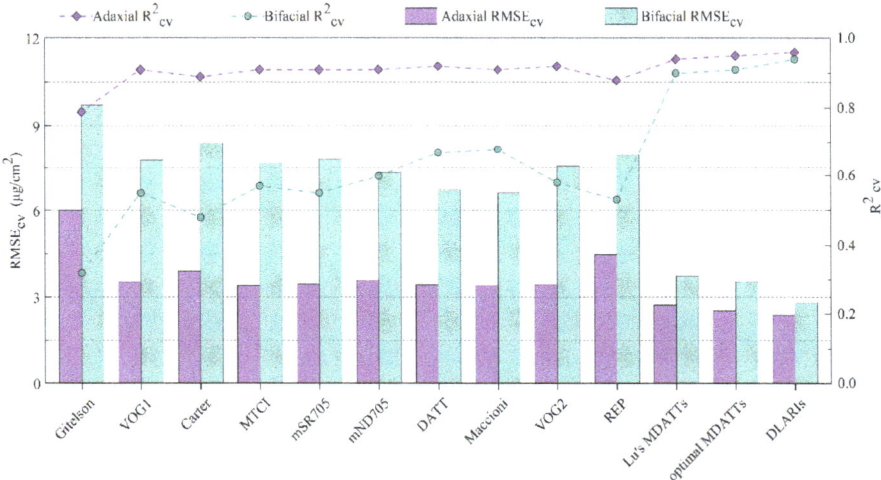

Figure 8. Comparison of published vegetation indices and the indices developed in this study for LCC estimation using adaxial and bifacial reflectance measurements.

Among the published indices, Lu's MDATTs, which uses different wavelengths for the adaxial and bifacial surfaces, provided the most accurate LCC retrievals (RMSE$_{cv}$ = 2.72; RMSE$_{cv}$ = 3.73), although they did not perform as well as the indices developed in this study. The two MDATTs were proposed for estimating the LCC of woody plants, such as white poplar (*Populus alba*) and grapevine (*Vitis* L.) [21]. The difference in wavelength combinations and retrieval accuracy between Lu's MDATTs and the MDATT optimized in this study can be attributed to the differences in phenotypic expressions (such as leaf hair, wax, palisade tissues, spongy tissues, etc.) between woody plant leaves and peanut leaves. By adding an additional band to the MDATT, the DLARI substantially improved retrieval accuracy, especially for bifacial reflectance measurements. When compared with the published vegetation indices, the indices developed in this study achieved the highest retrieval accuracies for estimating peanut LCC, whether for the adaxial or mixed surfaces.

3.5. Comparison of the DLARI and MDATT

The difference between the MDATT and DLARI was the substitution of λ_1 with an additional wavelength (λ_4). We evaluated the improvement of incorporating this additional wavelength by calculating the maximum R^2 of all band combinations for the DLARI formula based on the three datasets.

The maximum R^2 was derived from combinations by fixing λ_4 while changing λ_1, λ_2, and λ_3 from 700 nm to 760 nm. The wavelength regions for λ_4 were from 700 nm to 900 nm. The results are plotted in Figure 9. It shows that when λ_4 was at 700 nm to 760 nm, the maximum R^2 derived from adaxial DLARIs and abaxial DLARIs were similar to those of the adaxial MDATT (0.95) and abaxial MDATT (0.94), while bifacial DLARIs achieved higher correlations with LCC than the bifacial MDATT (0.91) since λ_4 was higher than 709 nm. When λ_4 was higher than 760 nm, all three DLARIs were much more correlated to LCC than the three MDATTs. The highest R^2 of the adaxial DLARI and bifacial DLARI were obtained at λ_4 equal to 819 nm and 773 nm, respectively, and then rapidly became lower than those of the MDATT when λ_4 was located above 890 nm. For the abaxial dataset, the highest R^2 appeared at λ_4 equal to 774 nm and then became lower than the abaxial MDATT when λ_4 was above 840 nm. Compared to the MDATT, the effective regions of λ_4 for the adaxial DLARI were from 760 nm to 890 nm and for the abaxial DLARI when they were from 760 nm to 840 nm. For the bifacial DLARI, the robust wavelength regions of λ_4 were from 709 nm to 890 nm.

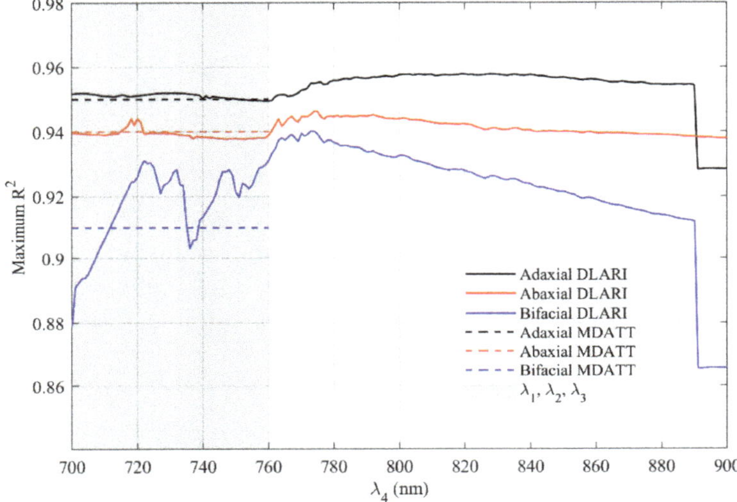

Figure 9. Maximum R^2 between LCC and MDATTs, DLARIs with λ_1, λ_2, and λ_3 from 700 nm to 760 nm and λ_4 from 700 nm to 900 nm.

The above bifacial DLARI was derived from a dataset composed of the same quantity of adaxial and abaxial reflectance measurements. In order to evaluate the impact of the abaxial reflectance on the performance of the two dorsiventral leaf adjusted indices, we divided the 84 abaxial reflectance samples into 6 parts and accumulated them into the adaxial dataset, then we calculated the optimal DLARIs and MDATTs and their accuracies, estimating LCC for each sub-dataset. The results are shown in Figure 10, which shows that the RMSE$_{cv}$ of MDATT dramatically increased from 2.52 to 3.35 when adding one-sixth of the abaxial samples into the adaxial dataset and then linearly increased to 3.55. With the increase of abaxial reflectance, the RMSE$_{cv}$ of DLARI stably increased from 2.37 to 2.82. Compared to the MDATT, the DLARI possessed a linear response to the impact of abaxial reflectance. The optimal wavelengths for the DLARI and MDATT derived from each sub-dataset showed unobvious changes with the addition of the abaxial reflectance. It can be concluded that the presence of abaxial leaves decreased the accuracy of DLARI and MDATT for LCC retrieval, but had no obvious influence on the optimal wavelengths for DLARI and MDATT.

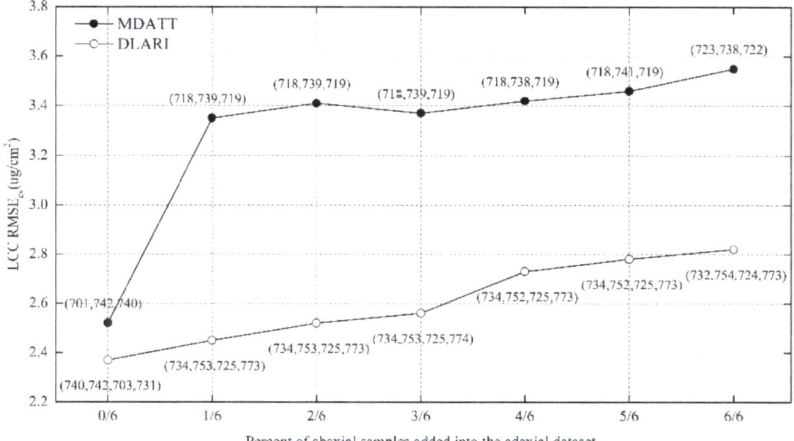

Figure 10. The impact of the dorsiventral leaf structure on DLARI and MDATT in terms of the estimation accuracy and optimal wavelengths. The numbers in the brackets are the optimal wavelengths for DLARI and MDATT with a unit of nm.

4. Discussion

In the case of narrow band indices such as the MDATT, different wavelength combinations can provide the best performance for different vegetation types. Lu et al. [21] suggested the robust wavelength region for λ_1 is from 723 to 885 nm and for λ_2 and λ_3 from 697 to 771 nm for woody plants. We optimized the wavelengths used in the MDATT to increase its suitability for peanut LCC estimation, but the effects from the dorsiventral leaf structure remained. The optimal wavelengths for the bifacial MDATT were 723 nm, 738 nm, and 722 nm. Using the MDATT as a basis, we constructed a DLARI to further decrease the impact of abaxial leaves. Compared with the MDATT and the published indices without considering the dorsiventral leaf structure, the three DLARIs performed best. The adaxial DLARI improved the estimation accuracy to 2.37 with a high R^2 of 0.96. The abaxial DLARI achieved a performance with $R^2 = 0.95$ and RMSE = 2.58. The bifacial DLARI showed an R^2 value of 0.94 and RMSE of 2.81. The results showed that the DLARIs not only improved the retrieval of LCC from the adaxial side of the leaf, but also further reduced the impact of differences in the adaxial and abaxial leaf reflectance thus increasing LCC estimation from the bifacial reflectance.

The measured leaf reflectance can be composed of the external (surface) reflectance (R_s) of the leaf and internal reflectance (R_i) of the leaf [42]. The reflectance of the adaxial and abaxial leaf side differ both in the R_s and R_i. The MDATT and DLARI formulae both successfully removed R_s according to Equations (2) and (3), respectively. The left R_i is influenced by pigments concentrations and absorption properties which are different at the two sides of the leaf [16]. For the MDATT, the optimal wavelengths for λ_1 and λ_3 changed from 701 nm and 740 nm to 723 nm and 722 nm (Figure 10). The reflectance of the adaxial surface and abaxial surface at 723 nm and 722 nm showed minimum difference (Figure 2). The optimal wavelengths λ_2 for bifacial MDATT was located at 738 nm where the reflectance of both sides showed similar sensitivity to LCC (Figure 3). The ability of MDATT to decrease the R_i effect contributed to the combination of these three wavelengths. For the DLARI, with the addition of abaxial samples into the adaxial dataset, the four wavelengths gradually changed to approximately 732 nm, 754 nm, 724 nm, and 773 nm (Figure 10). At 732 nm and 724 nm, the adaxial reflectance showed higher sensitivity to LCC than the abaxial reflectance. In contrast, at 754 nm and 773 nm, the abaxial reflectance showed stronger correlation to LCC than the adaxial reflectance (Figure 3). In addition, the optimal λ_3 was located at the region where spectral differences among the two sides of the leaf were negligible (Figure 2). The wavelength near 754 nm is known as the red-edge shoulder and has shown considerable potential in suppressing the influence of leaf structure [43]. These factors contributed to DLARI being optimal for LCC estimation when bifacial reflectance measurements were used.

Compared to the published vegetation indices, the DLARI not only decreased the effect of dorsiventral leaf structure but also significantly improved LCC estimation from the adaxial reflectance measurements. In fact, when multiple bands are available, there is no reason to limit to two-band or three-band indices. For instance, the first three wavelengths used in the DLARI were similar to that used in the MDATT for the adaxial dataset, but the accuracy (RMSE$_{cv}$) was improved from 2.52 to 2.37 by adding the fourth wavelength. Our results proved that indices based on four bands led to further improvements compared to two-band and three-band indices.

As previously mentioned, the reliability of narrow-band indices can be influenced by a range of phenotypic characteristics. Further work is required to assess the application of DLARI to estimate LCC for other crop species. The robust wavelength regions proposed should provide a good starting point for optimizing the index for other crop species.

The potential of satellites, such as Sentinel-2, to map crop biophysical variables has been shown by many studies [44,45]. The Sentinel-2 multispectral instrument includes three bands in the red-edge region centered at 705, 740, and 775 nm, which were found to be of great interest for crop monitoring [46]. Unmanned aerial vehicle (UAV) platforms coupled with imaging sensors are able to collect multispectral or hyperspectral imagery and offer great possibilities in the precision farming [47,48]. When using these remote sensing techniques to investigate peanut canopy information, the spectral information collected by the sensors may not only come from the adaxial leaf surfaces but also the abaxial leaf surfaces. Our results provide evidence that ignoring the spectral difference among the two faces introduces significant errors in LCC estimation. Further work should consider this effect when estimating peanut chlorophyll content or other biochemistry parameters at the canopy scale. The application of DLARI on remote sensing sensors to estimate canopy chlorophyll content is yet to be tested.

5. Conclusions

In this study, we focused on the development and optimization of dorsiventral leaf structure adjusted indices to minimize the impact of spectral differences between adaxial and abaxial leaf surfaces when retrieving peanut LCC. The wavelengths used by the MDATT were optimized for peanut, while a new dorsiventral leaf adjusted index was proposed to improve the LCC retrieval accuracy. The optimal MDATT index for retrieving LCC from bifacial reflectance measurements was $(R_{723} - R_{738})/(R_{723} - R_{722})$ with an $R^2{}_{cv}$ of 0.91 (RMSE$_{cv}$ = 3.53). The DLARI incorporated an additional wavelength in the NIR and exhibited the best retrieval accuracy when compared to the MDATT and other previously published indices. The DLARIs of $(R_{735} - R_{753})/(R_{715} - R_{819})$ and $(R_{732} - R_{754})/(R_{724} - R_{773})$ are recommended for retrieval of LCC using adaxial and bifacial reflectance, respectively. These two DLARIs delivered excellent cross-validation accuracies ($R^2{}_{cv}$ = 0.96, RMSE$_{cv}$ = 2.37; $R^2{}_{cv}$ = 0.94, RMSE$_{cv}$ = 2.81). The effective wavelength regions for DLARI were from the red edge to the NIR. Compared to the MDATT, the DLARI showed stronger correlation to LCC and less sensitivity to abaxial surface structure. This research provided new insights into the impact of spectral differences between adaxial and abaxial leaf surfaces on LCC estimation and proposed DLARI to improve LCC retrieval accuracy. The spectral differences between adaxial and abaxial leaf surfaces should be considered when estimating peanut canopy parameters. Further studies should be carried out to verify the applicability of DLARI to other plant species which have similar physiological response to solar radiation and drought stress as peanut.

Supplementary Materials: The following are available online at http://www.mdpi.com/2072-4292/11/18/2148/s1, Figure S1: Photographs of peanut canopies in the field.

Author Contributions: Conceptualization, Y.D., M.X. and Z.W.; methodology, Y.D. and M.X.; formal analysis, M.X. and Y.D.; data curation, H.W.; writing—original draft preparation, M.X. and Y.D.; writing—review and editing, A.H., L.A.B., Q.X. and X.X.; funding acquisition, Z.W.

Funding: This research was funded by the National Key Research and Development Program of China, grant number 2016YFD0200102, the Technology Development Program of Jilin Province, China, grant number 20180201012GX and "The 13th Five-Year plan" Science and Technology Project of the Department of Education, Jilin Province, grant number JJKH20170915KJ.

Acknowledgments: The support provided by the China Scholarship Council (CSC) during a visit by Yanling Ding (No.201806625001) to the University of Technology Sydney is acknowledged. The authors would like to thank the editors and reviewers who handled our paper.

Conflicts of Interest: The authors declare no conflict of interest.

References

1. FAOSTAT. 2017. Available online: http://www.fao.org (accessed on 21 May 2019).
2. Akram, N.A.; Shafiq, F.; Ashraf, M. Peanut (*Arachis hypogaea* L.): A prospective legume crop to offer multiple health benefits under changing climate. *Compr. Rev. Food Sci. Food Saf.* **2018**, *17*, 1325–1338. [CrossRef]
3. Wu, C.; Niu, Z.; Tang, Q.; Huang, W. Estimating chlorophyll content from hyperspectral vegetation indices: Modeling and validation. *Agric. For. Meteorol.* **2008**, *148*, 1230–1241. [CrossRef]
4. Daughtry, C.S.T.; Walthall, C.L.; Kim, M.S.; de Colstoun, E.B.; McMurtrey, J.E. Estimating corn leaf chlorophyll concentration from leaf and canopy reflectance. *Remote Sens. Environ.* **2000**, *74*, 229–239. [CrossRef]
5. Anjum, S.A.; Xie, X.Y.; Wang, L.C.; Saleem, M.F.; Man, C.; Lei, W. Morphological, physiological and biochemical responses of plants to drought stress. *Afr. J. Agric. Res.* **2011**, *6*, 2026–2032.
6. Mulla, D.J. Twenty five years of remote sensing in precision agriculture: Key advances and remaining knowledge gaps. *Biosyst. Eng.* **2013**, *114*, 358–371. [CrossRef]
7. Hunt, E.R.; Doraiswamy, P.C.; McMurtrey, J.E.; Daughtry, C.S.T.; Perry, E.M.; Akhmedov, B. A visible band index for remote sensing leaf chlorophyll content at the canopy scale. *Int. J. Appl. Earth Obs. Geoinf.* **2013**, *21*, 103–112. [CrossRef]
8. Verrelst, J.; Malenovský, Z.; Van der Tol, C.; Camps-Valls, G.; Gastellu-Etchegorry, J.-P.; Lewis, P.; North, P.; Moreno, J. Quantifying vegetation biophysical variables from imaging spectroscopy data: A review on retrieval methods. *Surv. Geophys.* **2019**, *40*, 589–629. [CrossRef]
9. Féret, J.-B.; François, C.; Gitelson, A.; Asner, G.P.; Barry, K.M.; Panigada, C.; Richardson, A.D.; Jacquemoud, S. Optimizing spectral indices and chemometric analysis of leaf chemical properties using radiative transfer modeling. *Remote Sens. Environ.* **2011**, *115*, 2742–2750. [CrossRef]
10. Malenovský, Z.; Homolová, L.; Zurita-Milla, R.; Lukeš, P.; Kaplan, V.; Hanuš, J.; Gastellu-Etchegorry, J.-P.; Schaepman, M.E. Retrieval of spruce leaf chlorophyll content from airborne image data using continuum removal and radiative transfer. *Remote Sens. Environ.* **2013**, *131*, 85–102. [CrossRef]
11. Ustin, S.L.; Gitelson, A.A.; Jacquemoud, S.; Schaepman, M.; Asner, G.P.; Gamon, J.A.; Zarco-Tejada, P. Retrieval of foliar information about plant pigment systems from high resolution spectroscopy. *Remote Sens. Environ.* **2009**, *113*, S67–S77. [CrossRef]
12. Datt, B. A new reflectance index for remote sensing of chlorophyll content in higher plants: Tests using eucalyptus leaves. *J. Plant Physiol.* **1999**, *154*, 30–36. [CrossRef]
13. Sims, D.A.; Gamon, J.A. Relationships between leaf pigment content and spectral reflectance across a wide range of species, leaf structures and developmental stages. *Remote Sens. Environ.* **2002**, *81*, 337–354. [CrossRef]
14. Gitelson, A.A.; Gritz †, Y.; Merzlyak, M.N. Relationships between leaf chlorophyll content and spectral reflectance and algorithms for non-destructive chlorophyll assessment in higher plant leaves. *J. Plant Physiol.* **2003**, *160*, 271–282. [CrossRef] [PubMed]
15. Kalariya, K.A.; Singh, A.L.; Goswami, N.; Mehta, D.; Mahatma, M.K.; Ajay, B.C.; Chakraborty, K.; Zala, P.V.; Chaudhary, V.; Patel, C.B. Photosynthetic characteristics of peanut genotypes under excess and deficit irrigation during summer. *Physiol. Mol. Biol. Plants* **2015**, *21*, 317–327. [CrossRef] [PubMed]
16. Hlavinka, J.; Nauš, J.; Špundová, M. Anthocyanin contribution to chlorophyll meter readings and its correction. *Photosynth. Res.* **2013**, *118*, 277–295. [CrossRef] [PubMed]
17. Stuckens, J.; Verstraeten, W.W.; Delalieux, S.; Swennen, R.; Coppin, P. A dorsiventral leaf radiative transfer model: Development, validation and improved model inversion techniques. *Remote Sens. Environ.* **2009**, *113*, 2560–2573. [CrossRef]

18. Baránková, B.; Lazár, D.; Nauš, J. Analysis of the effect of chloroplast arrangement on optical properties of green tobacco leaves. *Remote Sens. Environ.* **2016**, *174*, 181–196. [CrossRef]

19. Lu, X.; Lu, S. Effects of adaxial and abaxial surface on the estimation of leaf chlorophyll content using hyperspectral vegetation indices. *Int. J. Remote Sens.* **2015**, *36*, 1447–1469. [CrossRef]

20. Lu, S.; Lu, X.; Zhao, W.; Liu, Y.; Wang, Z.; Omasa, K. Comparing vegetation indices for remote chlorophyll measurement of white poplar and Chinese elm leaves with different adaxial and abaxial surfaces. *J. Exp. Bot.* **2015**, *66*, 5625–5637. [CrossRef]

21. Lu, S.; Lu, F.; You, W.; Wang, Z.; Liu, Y.; Omasa, K. A robust vegetation index for remotely assessing chlorophyll content of dorsiventral leaves across several species in different seasons. *Plant Methods* **2018**, *14*, 15. [CrossRef]

22. Dash, J.; Curran, P.J. The MERIS terrestrial chlorophyll index. *Int. J. Remote Sens.* **2004**, *25*, 5403–5413. [CrossRef]

23. Huete, A.; Didan, K.; Miura, T.; Rodriguez, E.P.; Gao, X.; Ferreira, L.G. Overview of the radiometric and biophysical performance of the MODIS vegetation indices. *Remote Sens. Environ.* **2002**, *83*, 195–213. [CrossRef]

24. Verrelst, J.; Rivera, J.P.; Veroustraete, F.; Muñoz-Marí, J.; Clevers, J.G.P.W.; Camps-Valls, G.; Moreno, J. Experimental Sentinel-2 LAI estimation using parametric, non-parametric and physical retrieval methods—A comparison. *ISPRS J. Photogramm. Remote Sens.* **2015**, *108*, 260–272. [CrossRef]

25. Wang, W.; Yao, X.; Yao, X.; Tian, Y.; Liu, X.; Ni, J.; Cao, W.; Zhu, Y. Estimating leaf nitrogen concentration with three-band vegetation indices in rice and wheat. *Field Crop. Res.* **2012**, *129*, 90–98. [CrossRef]

26. Xie, Q.; Dash, J.; Huang, W.; Peng, D.; Qin, Q.; Mortimer, H.; Casa, R.; Pignatti, S.; Laneve, G.; Pascucci, S.; et al. Vegetation indices combining the red and red-edge spectral information for leaf area index retrieval. *IEEE J. Sel. Top. Appl. Earth Obs. Remote Sens.* **2018**, *11*, 1482–1493. [CrossRef]

27. Brown, L.A.; Dash, J.; Lidón, A.L.; Lopez-Baeza, E.; Dransfeld, S. Synergetic exploitation of the Sentinel-2 missions for validating the Sentinel-3 ocean and land color instrument terrestrial chlorophyll index over a vineyard dominated mediterranean environment. *IEEE J. Sel. Top. Appl. Earth Obs. Remote Sens.* **2019**, *12*, 2244–2251. [CrossRef]

28. Dash, J.; Curran, P.J.; Tallis, M.J.; Llewellyn, G.M.; Taylor, G.; Snoeij, P. Validating the MERIS Terrestrial Chlorophyll Index (MTCI) with ground chlorophyll content data at MERIS spatial resolution. *Int. J. Remote Sens.* **2010**, *31*, 5513–5532. [CrossRef]

29. Vuolo, F.; Dash, J.; Curran, P.J.; Lajas, D.; Kwiatkowska, E. Methodologies and uncertainties in the use of the terrestrial chlorophyll index for the Sentinel-3 mission. *Remote Sens.* **2012**, *4*, 1112–1133. [CrossRef]

30. Kong, W.; Huang, W.; Liu, J.; Chen, P.; Qin, Q.; Ye, H.; Peng, D.; Dong, Y.; Mortimer, A.H. Estimation of canopy carotenoid content of winter wheat using multi-angle hyperspectral data. *Adv. Space Res.* **2017**, *60*, 1988–2000. [CrossRef]

31. Sartory, D.P.; Grobbelaar, J.U. Extraction of chlorophyll a from freshwater phytoplankton for spectrophotometric analysis. *Hydrobiologia* **1984**, *114*, 177–187. [CrossRef]

32. Arnon, D.I. Copper enzymes in isolated chloroplasts. polyphenol oxidase in beta vulgaris. *Plant Physiol.* **1949**, *24*, 1–15. [CrossRef] [PubMed]

33. Baret, F.; Andrieu, B.; Guyot, G. A simple model for leaf optical properties in visible and near-infrared: Application to the analysis of spectral shifts determinism. In *Applications of Chlorophyll Fluorescence in Photosynthesis Research, Stress Physiology, Hydrobiology and Remote Sensing*; Springer: Dordrecht, The Netherlands, 1988; pp. 345–351.

34. Main, R.; Cho, M.A.; Mathieu, R.; O'Kennedy, M.M.; Ramoelo, A.; Koch, S. An investigation into robust spectral indices for leaf chlorophyll estimation. *ISPRS J. Photogramm. Remote Sens.* **2011**, *66*, 751–761. [CrossRef]

35. Drusch, M.; Del Bello, U.; Carlier, S.; Colin, O.; Fernandez, V.; Gascon, F.; Hoersch, B.; Isola, C.; Laberinti, P.; Martimort, P.; et al. Sentinel-2: ESA's Optical High-Resolution Mission for GMES Operational Services. *Remote Sens. Environ.* **2012**, *120*, 25–36. [CrossRef]

36. Vogelmann, J.E.; Rock, B.N.; Moss, D.M. Red edge spectral measurements from sugar maple leaves. *Int. J. Remote Sens.* **1993**, *14*, 1563–1575. [CrossRef]

37. Carter, G.A. Ratios of leaf reflectances in narrow wavebands as indicators of plant stress. *Int. J. Remote Sens.* **1994**, *15*, 697–703. [CrossRef]

38. Maccioni, A.; Agati, G.; Mazzinghi, P. New vegetation indices for remote measurement of chlorophylls based on leaf directional reflectance spectra. *J. Photochem. Photobiol. B Biol.* **2001**, *61*, 52–61. [CrossRef]
39. Guyot, G.; Baret, F. Utilisation de la haute resolution spectrale pour suivre l'etat des couverts vegetaux. In Proceedings of the Spectral Signatures of Objects in Remote Sensing, Modane, France, 18–22 January 1988; p. 279.
40. Meyer, H.; Lehnert, L.W.; Wang, Y.; Reudenbach, C.; Nauss, T.; Bendix, J. From local spectral measurements to maps of vegetation cover and biomass on the Qinghai-Tibet-Plateau: Do we need hyperspectral information? *Int. J. Appl. Earth Obs. Geoinf.* **2017**, *55*, 21–31. [CrossRef]
41. Reicosky, D.A.; Hanover, J.W. Physiological effects of surface waxes. *Plant Physiol.* **1978**, *62*, 101–104. [CrossRef]
42. McClendon, J.H.; Fukshansky, L. On the interpretation of absorption spectra of leaves–I. Introduction and the correction of leaf spectra for surface reflection. *Photochem. Photobiol.* **1990**, *51*, 203–210. [CrossRef]
43. Croft, H.; Chen, J.M.; Zhang, Y. The applicability of empirical vegetation indices for determining leaf chlorophyll content over different leaf and canopy structures. *Ecol. Complex.* **2014**, *17*, 119–130. [CrossRef]
44. Kross, A.; McNairn, H.; Lapen, D.; Sunohara, M.; Champagne, C. Assessment of RapidEye vegetation indices for estimation of leaf area index and biomass in corn and soybean crops. *Int. J. Appl. Earth Obs. Geoinf.* **2015**, *34*, 235–248. [CrossRef]
45. Xie, Q.; Dash, J.; Huete, A.; Jiang, A.; Yin, G.; Ding, Y.; Peng, D.; Hall, C.C.; Brown, L.; Shi, Y.; et al. Retrieval of crop biophysical parameters from Sentinel-2 remote sensing imagery. *Int. J. Appl. Earth Obs. Geoinf.* **2019**, *80*, 187–195. [CrossRef]
46. Delloye, C.; Weiss, M.; Defourny, P. Retrieval of the canopy chlorophyll content from Sentinel-2 spectral bands to estimate nitrogen uptake in intensive winter wheat cropping systems. *Remote Sens. Environ.* **2018**, *216*, 245–261. [CrossRef]
47. Kanning, M.; Kühling, I.; Trautz, D.; Jarmer, T. High-Resolution UAV-Based Hyperspectral Imagery for LAI and Chlorophyll Estimations from Wheat for Yield Prediction. *Remote Sens.* **2018**, *10*, 2000. [CrossRef]
48. Tahir, N.; Naqvi, S.M.; Lan, Y.; Zhang, Y.; Wang, Y.; Afzal, M.; Cheema, M.J. Real time monitoring chlorophyll content based on vegetation indices derived from multispectral UAVs in the kinnow orchard. *Int. J. Precis. Agric. Aviat.* **2018**, *1*, 24–31. [CrossRef]

Article

Eco-Friendly Estimation of Heavy Metal Contents in Grapevine Foliage Using In-Field Hyperspectral Data and Multivariate Analysis

Mohsen Mirzaei [1], Jochem Verrelst [2], Safar Marofi [3,*], Mozhgan Abbasi [4] and Hossein Azadi [5,6,7]

[1] Environmental Pollutions, Grape Environmental Science Department, Research Institute for Grapes and Raisin (RIGR), Malayer University, Malayer 65719-95863, Iran; mohsen.mirzaei@stu.malayeru.ac.ir

[2] Image Processing Laboratory (IPL), Parc Científic, Universitat de València, 46980 Paterna, València, Spain; jochem.verrelst@uv.es

[3] Grape Environmental Science Department, Research Institute for Grapes and Raisin (RIGR), Malayer University & Water Science Engineering Department, Bu-Ali Sina University, Hamedan 65178, Iran

[4] Faculty of Natural Resource and Earth Science, Shahrekord University, Shahrekord 8815648456, Iran; mozhgan.abbasi@sku.ac.ir

[5] Department of Geography, Ghent University, 9000 Ghent, Belgium; hossein.azadi@ugent.be

[6] Research Group Climate Change and Security, Institute of Geography, University of Hamburg, 20146 Hamburg, Germany

[7] ISUMADECIP, Faculty of Environmental Science and Engineering, Babeş-Bolyai University, 400084 Cluj-Napoca, Romania

[*] Correspondence: marofi@basu.ac.ir; Tel.: +98-9183143686

Received: 22 October 2019; Accepted: 16 November 2019; Published: 20 November 2019

Abstract: Heavy metal monitoring in food-producing ecosystems can play an important role in human health safety. Since they are able to interfere with plants' physiochemical characteristics, which influence the optical properties of leaves, they can be measured by in-field spectroscopy. In this study, the predictive power of spectroscopic data is examined. Five treatments of heavy metal stress (Cu, Zn, Pb, Cr, and Cd) were applied to grapevine seedlings and hyperspectral data (350–2500 nm), and heavy metal contents were collected based on in-field and laboratory experiments. The partial least squares (PLS) method was used as a feature selection technique, and multiple linear regressions (MLR) and support vector machine (SVM) regression methods were applied for modelling purposes. Based on the PLS results, the wavelengths in the vicinity of 2431, 809, 489, and 616 nm; 2032, 883, 665, 564, 688, and 437 nm; 1865, 728, 692, 683, and 356 nm; 863, 2044, 415, 652, 713, and 1036 nm; and 1373, 631, 744, and 438 nm were found most sensitive for the estimation of Cu, Zn, Pb, Cr, and Cd contents in the grapevine leaves, respectively. Therefore, visible and red-edge regions were found most suitable for estimating heavy metal contents in the present study. Heavy metals played a significant role in reforming the spectral pattern of stressed grapevine compared to healthy samples, meaning that in the best structures of the SVM regression models, the concentrations of Cu, Zn, Pb, Cr, and Cd were estimated with R^2 rates of 0.56, 0.85, 0.71, 0.80, and 0.86 in the testing set, respectively. The results confirm the efficiency of in-field spectroscopy in estimating heavy metals content in grapevine foliage.

Keywords: field spectroscopy; hyperspectral; heavy metals; grapevine; PLS; SVM; MLR

---

1. Introduction

In-field spectroscopy provides a time and cost-efficient and accurate way to monitor plant stress [1–3]. These hyperspectral data are sensitive to small differences in plant features; i.e., plant disease [4–6], water content [7,8], biomass assessment [8,9], crops quantity and quality [10,11], species and varieties discrimination [12–14], and heavy metal stress [1,2,15].

Heavy metal contamination in food-producing ecosystems is considered to be a major environmental problem due to its potential hazard to humans and other organisms and due to the intention to protect the safety of food chains [16,17]. Within the selection of human food, grapes and their secondary products (wine, jam, juice, jelly, vinegar, grape seed oil, and raisins) play an important role. Therefore, the safety of vineyards in terms of heavy metals is a key factor in grape production and wine industries [17,18]. In viticulture areas, the excessive and prolonged usage of fertilizers and pesticides releases heavy metals (i.e., Cu, Zn, Cd, Pb, Cr, Ni, Hg, and As), which has been considered in many studies [16–20]. According to Milićević et al. [18] and Sun et al. [17], significant correlations occur between heavy metal concentration in soil, grapevine parts (leaf, skin, pulp, and seed), and wine. Alagić et al. [21] also concluded that the grapevine has some highly effective strategies involved in tolerance to heavy metal stress, which makes it an excellent plant species for phytostabilization purposes. Therefore, grapevine foliage monitoring can potentially demonstrate heavy metal concentration states in other parts of the plant and is also acknowledged to be a bio-indicator of heavy metals in the enclosing environment.

Heavy metal stress can produce some changes in plant morphological and biochemical characteristics [15]. This is because the leaf spectral response is mainly affected by plant structural and morphological characteristics; i.e., the leaf's intracellular and extracellular structure, and biochemical parameters such as nitrogen, pigments, and water contents [22–27].

Usually, heavy metal concentrations are detected in plant samples by acid digestion–solvent extraction followed by hydride generation atomic absorption spectrometry [28,29]. This tedious approach is expensive and destructive. Alternatively, by modeling the relationships between the heavy metal concentrations and foliar spectral characteristics, these concentrations can be efficiently estimated without using any chemical solvents. Therefore, by analyzing leaf spectral data, it becomes possible to investigate the biochemical and morphological changes caused by heavy metal stress [15,30]. It should be noted that in-field spectroscopy is one of the most attractive fields in remote sensing studies and can record specific spectral data to any object such as fingerprints [31,32]. Hyperspectral sensors can be used in the in-field spectroscopy process and so provide a framework for spectral reflectance acquisition in hundreds of narrow and contiguous bands/wavelengths [24,26]. Accordingly, it is expected that a plant being exposed to heavy metal stress will lead to subtle differences in the spectral curve as opposed to a healthy plant. These differences mainly occur in the visible and near-infrared regions of the electromagnetic spectrum [33].

Several studies have made specific use of the application of crop spectral characteristics through in-field spectroscopy data and multivariate statistical analysis to promote the prediction of heavy metal content in plant samples. For instance, Font et al. [28] and Font et al. [29] applied visible and near-infrared spectroscopy and the modified partial least squares (PLS) method to forecast metal content in prostrate amaranth and rice, with determination coefficients of 0.63 and 0.65, respectively. In another study, Rosso et al. [34] examined the spectral and physiological responses of *Salicornia virginica* to heavy metal (Cd and V) stress in laboratory conditions. The potential of in-field spectroscopy to detect heavy metal contents was also investigated by Ni et al. [35], Gu et al. [36], Liu et al. [37], Liu et al. [38], and Li et al. [39] in the case of dominant plants in the Poyang lake wetlands, *Brassica rapa chinesis*, rice, *Phragmites australis*, and vegetables, respectively.

It is worth noting that in-field spectroscopy delivers a large amount of spectral data, whereby each of the wavelengths may be associated with one of the plant parameters [40]. Therefore, identifying optimal wavelengths to monitor any parameter—e.g., heavy metal concentrations—is an important step in applying these data [41]. In this regard, the usage of multivariate statistical techniques such as the PLS method [14,40,42,43], multiple linear regression (MLR) [41,44,45], and support vector machines (SVM) [12,40,46] can help with feature selection, data reduction, and modelling the existing relationships between hyperspectral data and plant characteristics. Many studies have also taken advantage of spectral indices to minimize atmospheric and background disturbances and illustrate plant characteristics [3,15,30,45,47]. These indices are mathematical spectral transformations of two or more bands designed to enhance the spectral response of vegetation properties [12,40,46]. Hence, spectral indices calculated from foliar reflectance data may reveal the biochemical and physiological properties of leaves, which may be responsible for monitoring plant characteristics [46]. Despite the proven performance of in-field spectroscopy in estimating heavy metal contents in plants, to the best of our knowledge, such a study has never been employed on grapevines leaves.

Altogether, this study was designed with the following goals: (i) developing hyperspectral libraries of healthy and heavy metal-stressed grapevine leaves (Vitis vinifera cv. Askari, as a common grapevine variety in Iran) by using full range in-situ spectroscopy (350-2500 nm), (ii) evaluating the potential of in-field spectroscopy for estimating heavy metals (Cu, Zn, Pb, Cr, and Cd) concentrations in grapevine foliage, (iii) investigating two types of hyperspectral data (wavelengths vs. spectral indices) and identifying the most appropriate features to estimate each studied metal in grapevine foliage, and (iv) comparing the performance of SVM and MLR algorithms in modeling the relationships between the foliar spectral response and heavy metal concentrations.

2. Materials and Methods

2.1. Pollutant Exposure Experiments

Since experience in the evaluation of in-field spectroscopy when estimating heavy metal contents in grapevine leaves is lacking, we chose to conduct this research in a laboratory-controlled environment. Therefore, treatments for heavy metal stress were applied to grapevine seedlings. For this purpose, five treatments at varying levels of Cu, Zn, Pb, Cr, and Cd were considered, and in each treatment, four repetitions were carried out (a total of 84 grapevine seedlings were examined).

It should be noted that the objective of this experiment was not to determine the sensitivity of grapevine to pollutants. We only intended to add heavy metal contents to the grapevine to compare its spectral differences with healthy leaf samples. The common grapevine variety in the study area is *Vitis vinifera cv. Askari*; all seedlings belonged to this variety to eliminate the effect of variety change on spectral characteristics [43,48]. Experiments were conducted outdoors in full sun between March and September 2018. Each grapevine seedling sample was placed in an individual plastic pot (length and width 25 cm × 10 cm) and was randomly divided amongst the studied treatments. The seedlings were two years old, and their height at the beginning of the experiment was between 20 and 30 cm. All seedlings were in the same conditions in terms of soil, pot size, sunlight exposure, watering, temperature, and humidity. In Figure 1, a schematic of the applied treatments is displayed. The first treatment served as a control to monitor the potential effects of soil, water, and air on the transfer of heavy metals to grapevine seedlings. In the second treatment, the maximum allowed level (MAL) of Cu, Zn, Pb, Cr, and Cd in irrigation water provided stress to the seedlings. All contamination levels were increased in the third, fourth, and fifth treatments as two, three, and four times the metal MALs in irrigation water, respectively. A stress program was applied to treatments 2–5 by dissolving the metal salts (nitrate form) in irrigation water. Salt metals have a high solubility, resulting in the absorption of the metals by plant organs [34]. According to the Iranian Water Quality Standard (IWQS), the MALs for Cu, Zn, Pb, Cr, and Cd in irrigation water are 200, 2000, 100, and 10 mcg/l, respectively. Seedlings were examined for a period of seven months, and they were stressed during each month (a total of seven stresses were applied). At the end of the stress period and before the beginning of the fall season (September 2018), a spectrophotometric analysis of grapevine seedlings leaves was applied.

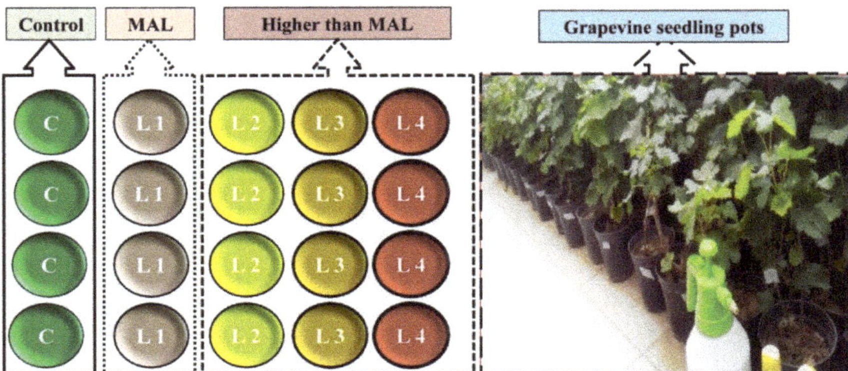

Figure 1. Schematic design of the treatment for each studied metal (C: control, L1 to L4: level 1 to level 4 stressed, MAL: maximum allowed level) and image of applied grapevine seedling pots.

2.2. Spectra Acquisition

At least five leaves of each seedling pot were collected for spectroscopy measurements (a total of 420 spectra samples were taken), and afterwards, individual reflectance spectra were measured by pot. In this study, the grapevine foliar spectral reflectance was measured using the ASD FieldSpec 3 spectroradiometer in the full range (350–2500 nm). This instrument is supported by three separate spectrometers (first: 350–975 nm, second: 976–1770, and third: 1771–2500 nm). The ASD spectral resolutions in the range of 350–1000 nm and 1000–2500 nm are 3 and 10 nm with sampling intervals of 1.4 and 2 nm, respectively. In accordance with Kumar et al. [49], the electromagnetic spectrum in the range of 350 to 2500 nm can be classified into four regions including visible (VIS), red-edge region (RDE), near-infrared (NIR), and mid-infrared (MIR), with ranges of 350~700, 680~750, 700~1300, and 1300–2500 nm, respectively. We performed the spectroscopy experiment in a fully dark room in order to reduce the effect of wind, water vapor, temperature, and other environmental disturbance [12]. In this study, each spectral sample was recorded in 100 automatic replicates. Then, we applied the ViewSpect version 6.0 in order to convert spectral curves into test files and analyze them by statistical software.

For each sample, the reflectance spectrum was recorded at 2151 wavelengths (350–2500 nm), which gave a large amount of data, not all of which may be useful for the study purpose. Therefore, in this study, 32 spectral indices were calculated to evaluate their ability to estimate heavy metal contents. The spectral indices which are used in this study were calculated based on the method indicated by Mirzaei et al. [12], although no specific spectral indices exist to detect heavy metal contamination [1]. Table 1 shows the indices that have demonstrated sensitivity in previous studies to small differences in plant characteristics [12,46,50].

Table 1. Characteristics of studied hyperspectral indices [13,46].

Indices	Equation	Indices	Equation
Cellulose Absorption Index, CAI	$0.5(R2000 + R2200) - R2100$	Gitelson and Merzlyak Chlorophyll, GM1 and 2	$GM1 = (R750)/(R550)$
Moisture Stress Index, MSI	$(R1600)/(R820)$		$GM2 = (R750)/(R700)$
Normalized Difference Water Index, NDWI	$(R860 - R1240)/(R860 + R1240)$		$Lic1 = (R800 - R680)/ (R800 + R680)$
Disease Water Stress Index, DWSI	$(R802 + R547)/(R1657 + R682)$	Lichtenthaler Indices, Lic1 to 3	$Lic2 = (R440)/(R690)$
Band ratio at 975 nm, RATIO975	$2 \times R960 - 990/(R920 - 940 + R1090 - 1110)$		$Lic3 = (R440)/(R740)$
Band ratio at 1200 nm, RATIO975-2	$2 \times R1180 - 1220/(R1090 - 1110 + R1265 - 1285)$	Simple Ratio Pigment Index, SRPI	$(R430)/(R680)$
Leaf Chlorophyll Index, LCI	$(R850 - R710)/(R850 + R680)$	Normalized Phaepophytiniz Index, NPQI	$(R415 - R435)/(R415 + R435)$
DattA	$(R780 - R710)/(R780 - R680)$	Normalized Pigment Chlorophyll Ratio Index, NPCI	$(R680 - R430)/(R680 + R430)$
Modified Red Edge Normalized Difference VegetationIndex, mNDVI705	$(R750 + R705)/(R750 + R705 - 2 \times R445)$	Greenness Index, GI	$(R554)/(R677)$
Chlorophyll Index, SGB	$(R750 - R445)/(R705 - R445)$	Water Index at 1180nm, WI1180	$(R900)/(R1180)$
Structure Intensive Pigment Index, SIPI	$(R445 - R800)/(R680 - R800)$	Normalized Difference Vegetation Index, NDVI	$(R831 - R667)/(R831 + R667)$
Simple Ratio, SR	$(R774)/(R677)$	Carter Index, CI	$(R760/R695)$
Reflectance at 550 nm, R550	$(R550)$	Vogelman Index, VOG	$(R740/R720)$
Reflectance at 680 nm, R680	$(R680)$	Carotenoid Reflectance Index, CRI	$R800(1/R520 - 1/R550)$
Water Index, WI	$(R900)/(R970)$	Photochemical Reflectance Index, PRI	$PRI1 = (R531 - R570)/(R531 + R570)PRI2 = 1.5(R830 - R660)/(R830 - R660 + 0.5)PRI3 = (R539 - R570)/(R539 + R570)$

R: Reflectance.

2.3. Heavy Metal Laboratory Analysis

The leaves of each pot were placed in polyethylene bags and converted separately in the laboratory after obtaining the foliar reflectance spectra. The leaf samples were dried for 24 h in an oven at 45 °C to achieve a constant weight [16]. The samples were powdered and stored for further analysis with a stainless-steel mill. We then digested one gram of each grapevine sample with HNO3 + HClO4 (3:1 v/v) for about 4 h at a low temperature (about 40 °C) [51]. All digested samples were then diluted and filtered to 25 mL. Finally, a Graphite-Furnace Atomic Absorption Spectrophotometer (GA-AAS, Model: Analytik Jena, Germany) was used to analyze all samples in triplicate. The concentrations of heavy metal samples were expressed as dry weight (DW) mg/kg. The device detection limits for Zn, Cu, Pb, Cr, and Cd were 0.008, 0.025, 0.01, 0.04, and 0.009 mg/kg, respectively. Based on the analysis, the relative standard deviation accuracy was less than 9%. To evaluate the accuracy of analytical techniques, a spike-and-recovery analysis was performed. Post-analyzed samples were accentuated and homogenized with varying amounts of standard metal solutions. The recovery ranged from 90% to 108% of the spiked sample [52].

2.4. Feature Selection/Partial Least Squares (PLS)

In summary, the dependent variables were the contents of Cd, Cr, Cu, Pb, and Zn in grapevine leaves, while the independent variables were wavelengths (count: 2151) and spectral indices (count: 32). However, a large number of independent variables can reduce the performance of the relationship modelling between spectral data and metal contents. To mitigate this, we needed a feature selection process to identify optimal features (wavelengths and spectral indices) to forecast the concentration of each metal, individually. Also, before applying statistical operations, it is recommended to scale each variable linearly to the same standard range, especially in the machine learning methods [40,53].

The values of wavelengths, spectral indices, and heavy metal concentrations were therefore scaled to the range between 0 and 1, as follows:

$$N_i = \frac{x_i - x_{\min}}{x_{\max} - x_{\min}} \tag{1}$$

where N_i is the normalized value, x_i is the original data, and $x_{\min}$ and $x_{\max}$ are the minimum and maximum of each variable's percentages, respectively.

Given the high-dimensional spectral dataset, the use of multivariate statistical analysis is an appropriate solution for achieving optimal features to estimate each metal. PLS is a robust and well-known statistical analysis in relation to hyperspectral data that has shown acceptable performance in many studies [12,40]. This statistical analysis method generates some new components instead of using existing inputs, based on the least square regression. Unlike principal components analysis (PCA), PLS considers response variables in the data reduction process [54]. Fitting a regression model between input and output variables, high collinear spectral data, and the high processing speed are the other advantages of the PLS method. The PLS-developed components are capable of explaining community variance by a simpler structural mechanism. Accordingly, the importance of each input variable is realized by its factor load in each component [12]. We therefore selected optimal independent variables (wavelengths or spectral indices) based on the maximum factor load in each developed PLS component. These variables were considered to be the most representative of the related components. Based on the PLS results, the optimal wavelengths and indices were identified and introduced to the next step (modelling). Wold et al. [55] has provided more information about the assumptions and applications of the PLS.

2.5. Modelling the Relationship Between Spectral Data and Heavy Metal Contents

After the identification of the optimal wavelengths and relevant indices by the PLS, two types of modelling algorithms (SVM and MLR) were applied to estimate heavy metal concentrations based on hyperspectral data. To assess the estimation performance of each model, two goodness-of-fit indicators—specifically, the coefficient of determination (R^2) and root mean squared error (RMSE)—were used [40]. All achieved data in this study were randomly separated into two sections: 70% as training data and 30% as testing data. As such, the performance of each developed model was individually reported for training and testing sets.

2.5.1. Support Vector Machine (SVM)

SVM is a nonparametric learning algorithm for regression and classification goals and for hyperspectral data mining [56–58]. In the SVM procedure, the n-dimensional input vectors are conveyed into a high-dimensional feature space, and consequently, the optimal separating hyper-planes are developed [59]. Here, the SVM regression algorithm was used in multiple scenarios and designs to gain the best performance for modelling the relationship between the in-field hyperspectral data and the measured heavy metal concentration in grapevine leaves. To this end, the input vectors were linked to the outputs with a kernel function [12]. Regression SVM-type 1 with different kernel functions—i.e., radial basis functions (RBF), polynomials, and a sigmoid shape—was applied. In order to achieve an optimal training constant, V-fold cross validation was used, and kernel function parameters (coefficient, gamma, and degree) were altered to give a high-performance score [60]. More details about the assumptions and structure of SVM are provided by Stitson et al. [59] and Cristianini and Shawe-Taylor [61].

2.5.2. Multiple Linear Regressions (MLR)

MLR is a parametric regression algorithm that attempts a relationship model between two or more independent variables and a response variable with a linear fitting. It has the capacity to select appropriate input data. In this study, the forward selection method of MLR was applied to increases the R^2 value by adding an independent variable [40]. The Durbin–Watson statistic was applied to test autocorrelation in the residuals from statistical regression analysis. Durbin–Watson values close to 2 (1.5–2.5) indicate that there is no autocorrelation detected in the samples. Additionally, in order to detect multicollinearity in regression analysis, thw variance inflation factor (VIF) was considered (VIFs exceeding 10 are signs of serious multicollinearity) [62,63]. The general form of the MLR equation is as follows:

$$\text{HMC} = a_0 + a_1 LM_1 + a_2 LM_2 + \ldots + a_n LM_n \tag{2}$$

where HMC is the heavy metal concentration in grapevine leaves, a $_{(i=0,1,\ldots,n)}$ are the parameters generally estimated by least squares, and X $_{(i=1,2,\ldots,n)}$ are the independent variables (i.e., wavelengths and spectral indices).

3. Results and Discussion

3.1. Reflectance Spectra of Healthy and Stressed Leaves

The average reflectance spectrum of a healthy grapevine vs. stressed grapevine leaves due to heavy metal stress is shown in Figure 2. In the VIS region, the light absorption rate of the stressed grapevine was drastically decreased. This is due to the fact that the spectral characteristics of plants in this region are regularly motivated by pigments [64,65]. Accordingly, this suggests that heavy metal stress reduced pigment contents. Various spectral characteristics between healthy and stressed leaves can also be observed in the RDE, NIR, and MIR regicns (Figure 2). As Vogelmann [66], Slaton et al. [23], and Strever [67] stated, plant pigments do not absorb the light in the NIR and MIR regions; therefore, the plant leaf reflectance is significantly increasing in these regions. Additionally, the spectral characteristics of plant leaf in the NIR and MIR regions were changed by structure/morphology and water contents, respectively [54]. According to Figure 2, in the NIR and MIR regions, a lower reflectance was observed in healthy grapevine leaves as opposed to the stressed grapevine. Although other driving variables such as structural parameters and water contents were not measured in this study, it can be concluded that the stress caused by heavy metals had a significant effect on the leaf optical properties.

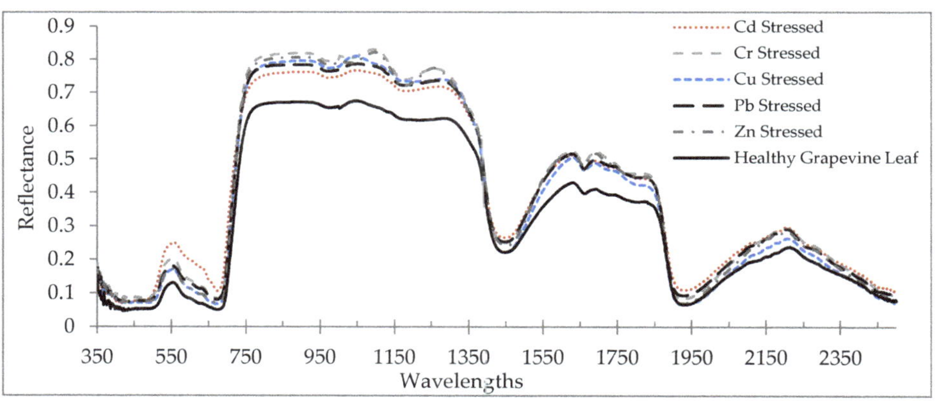

Figure 2. Average reflectance spectrum of healthy grapevine leaves vs. the heavy metal-stressed grapevine leaves (from 350 to 2500 nm).

3.2. Correlation Coefficient

Figure 3 displays the correlation coefficient between grapevine leaves' reflectance (350–2500 nm) and their heavy metal concentrations (Cu, Zn, Pb, Cr, and Cd). The correlation coefficients were noisiest in the range from 350 to 400 nm due to atmospheric effects. Of particular interest is that the highest absolute correlation coefficient took place in the range of 350 to 400 nm in relation to Cr, Pb, and Zn. Cd showed the best correlation with the wavelengths in the VIS region (400–680 nm), while it dropped sharply in the RDE region (680–750 nm) (Figure 3). This suggests that the RDE region is one of the best options for introducing optimal wavelengths to estimate Cd concentrations in grapevine leaves. Also, the other heavy metals caused subtle fluctuations in the RDE region, and their correlation coefficients tended to be positive. This finding indicates the potential of this region to forecast metal contents in the grapevine leaves. Similar correlation coefficients were observed for Cu, Zn, Cr, and Pb in the NIR spectrum region (750–1300 nm), but Cd had a varied correlation curve in this range. In the MIR region (2500–1300 nm), the heavy metal correlation coefficients were closer together (Figure 3). With the exception of Pb, the remaining metals were negatively correlated with most wavelengths of this region.

In comparison to a related study by Zhuang [41], a similar correlation graph between spectral response (400–2500 nm) and heavy metal contents (Cu, Zn, Pb, Cd, As, and Fe) was obtained. A comparison of Figure 3 with the study results of Zhuang [41] shows that the correlation pattern between the heavy metal contents and the spectral response is not alike. Therefore, the structural and biochemical differences between the studied species (grapevine and rice) and the level of spectroscopy (leaf or canopy level) can be considered as the most important drivers justifying these differences.

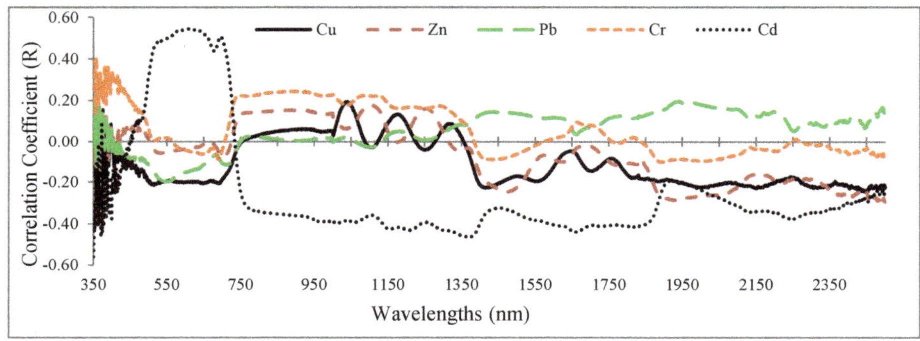

Figure 3. Correlation coefficient between the heavy metal concentration (determined by laboratory analysis) and spectral response of grapevine leaf samples (350 to 2500 nm).

3.3. Optimal Feature Selection

Determining the optimal wavelengths to monitor the desired plant parameters within the vast hyperspectral bands is one of the most critical operations in spectroscopy [43,46,56]. Commonly, a small number of wavelengths/spectral indices are selected with maximum performance for the study purpose, while missing data should be minimal [46,68]. Thus, we chose the PLS method because of its high adaptability with hyperspectral data to recognize optimal predictive variables (wavelengths and spectral indices) for estimating heavy metals in grapevine leaves [42]. Identifying the fit number of components is one of the most imperative factors in applying the PLS results because the number of components can directly determine the number of model input variables. Accordingly, the cross-validation algorithm was applied to optimize the number of PLS components [43], and then the optimum variable for each of the components was identified. Figure 4 shows the number of optimal components and the wavelength factor loads of the metals studied. This figure shows that the numbers of developed fit components were 4, 6, 5, 6, and 4, for Cu, Zn, Pb, Cr, and Cd, respectively. Therefore, based on the introduced components, the wavelengths and spectral indices which had the highest correlation with the components were identified. They can be subsequently used as optimal spectral wavelengths and indices in the relevant modelling process, especially for estimating metal concentrations in the grape leaves [43].

As shown in Figure 4, the wavelengths in the vicinity of 2431, 809, 489, and 616 nm can be recognized as an optimal rate for estimating Cu content in grapevine leaves. In the same method, wavelengths in the vicinity of 2032, 883, 665, 564, 688, and 437 nm; 1865, 728, 692, 683, and 356 nm; 863, 2044, 415, 652, 713, and 1036 nm; and 1373, 631, 744, and 438 nm were the optimal wavelengths for estimating Zn, Pb, Cr, and Cd, respectively. Based on these results, the VIS, RDE, NIR, and MIR regions introduced eight, eight, three, and five wavelengths for estimating the studied heavy metals, respectively. The most delicate regions to estimate the studied heavy metals in the grapevine leaves were RDE and VIS (particularly the blue region). Consistent with this finding, Liu et al. [38] and Zhuang [41] also reported that VIS and RDE delivered the most optimal wavelengths for estimating heavy metal contents. Moreover, according to the results, the RDE was one of the most influential regions in introducing optimal wavelengths for estimating the contents of Zn, Pb, Cr, and Cd. In confirmation with this finding, Gu et al. [36] noted the RDE region as being sensitive to estimate the variances of metal contents (especially Cd). They suggested the wavelength of 782 nm as an optimal wavelength for estimating Cd concentration in *Brassica rapa* leaves.

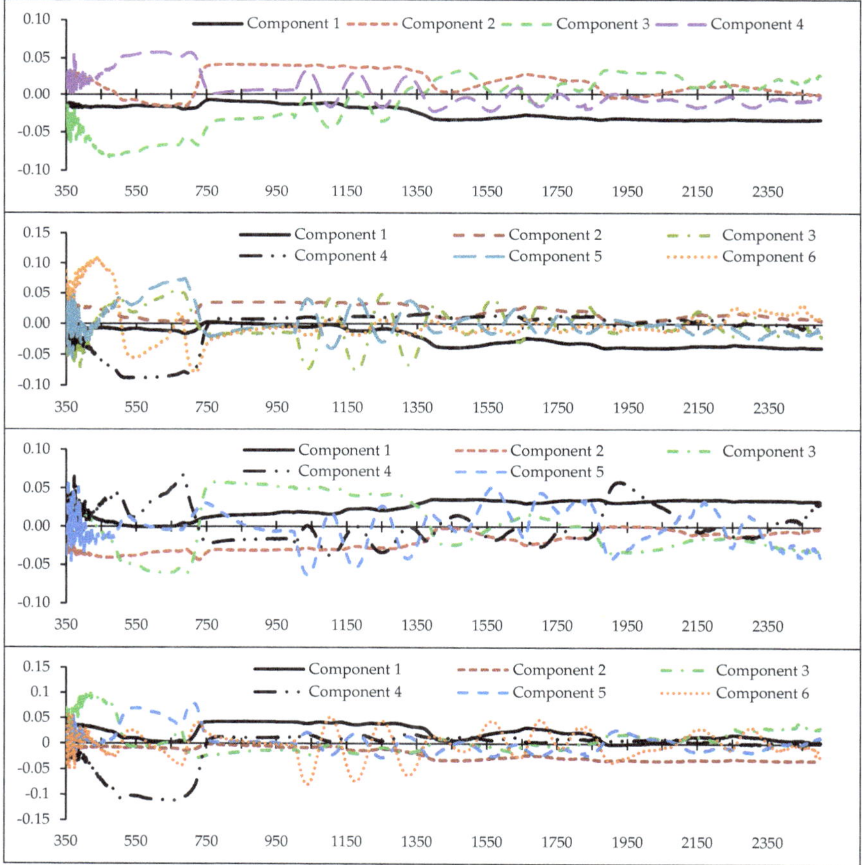

Figure 4. *Cont.*

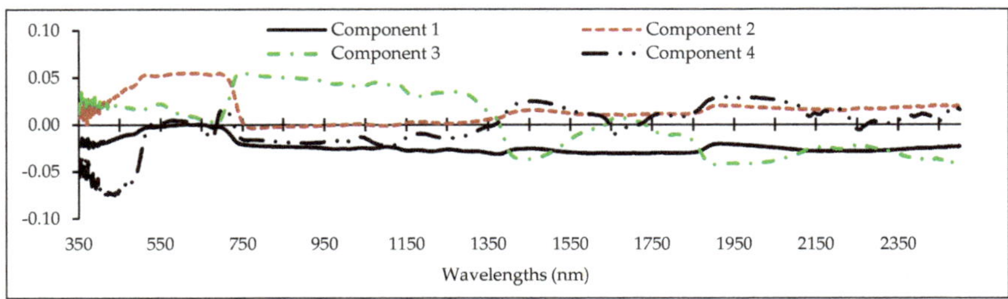

Figure 4. The factor load of wavelengths (350–2500 nm) in the optimal components extracted by the the partial least squares (PLS) method for estimating heavy metal concentrations (from top to bottom) in the grapevine leaves (vertical axis is the factor load).

In the same way, the optimal spectral indices for estimating contents of Cu, Zn, Pb, Cr, and Cd were also determined based on the interpretation of the PLS results. In Table 2, a summary of the PLS results is presented, which is used to determine the optimal indices to estimate the heavy metal concentrations. As an optimal index for the estimation of Pb, Cr, and Cd concentrations, the Structure Intensive Pigment Index (SIPI) (proposed by Penuelas et al. [69]), which represents the ratio of carotenoids to chlorophyll, was the most frequent index among the studied indices. Furthermore, the Disease Water Stress Index (DWSI) and Moisture Stress Index (MSI) indices, which are sensitive to water levels in vegetation (water stress), were identified as optimal indices for estimating Zn–Pb and Cu–Cd, respectively. It is worth remarking that the Normalized Difference Vegetation Index (NDVI) was not chosen as the optimal index to predict the studied metal contents. On the other hand, according to Zhuang [41], the NDVI band ratios were extremely useful in monitoring the contents of metals in the paddy canopy. Therefore, it can be argued that, in addition to the structural and biochemical differences between grapevine and paddy species, the differences in studied spectral indices are another reason for differences in the optimal spectral indices.

Table 2. Summary of the PLS results on the number of components and optimal indices for estimating heavy metal contents in grapevine leaves.

Heavy Metal	No. of Optimal Components	Cumulative Variance (%)	Optimal Indices in Components
Cu	4	82	SR, CAI, RATIO9752, and DWSI
Zn	5	84	R680, WI, Lic1, MSI, and PRI2
Pb	4	88	VOG, MSI, SIPI, and R550
Cr	4	92	mNDVI705, GI, RATIO975, and SIPI
Cd	2	81	SIPI and DWSI

3.4. Modelling and Accuracy Assessment

After determining the optimal spectral wavelengths and indices, two regression approaches—i.e., MLR and SVM—were applied to model the relationships between spectral data and heavy metal concentrations. Table 3 illustrates the best-developed models and validation results using the SVM algorithm. Based on this table, the RBF function was selected as the optimal central function in 60% of the developed models, followed by the linear function (30%). These two functions were therefore considered as the optimal functions for relevant modelling in the studied grapevine leaves.

Table 3. Modelling and validation results of the best support vector machine (SVM) models based on optimal wavelengths and spectral indices for estimating heavy metal concentrations in grapevine leaves in training and testing sets. RBF: radial basis function.

Hyperspectral Data Type	Heavy Metal	Model Structure						Train		Test	
		Kernel Function	No. of Vectors	Coefficient	Degree	Gamma	R^2	RMSE *		R^2	RMSE *
Wavelengths	Cu	RBF	13	-	-	0.25	0.97	7.46		0.54	25.06
	Zn	Linear	25	-	-	-	0.67	22.50		0.42	29.65
	Pb	RBF	21	-	-	0.20	0.89	22.28		0.71	24.09
	Cr	Linear	30	-	-	-	0.84	5.61		0.71	7.82
	Cd	RBF	34	-	-	0.25	0.78	98.16		0.77	103.09
Spectral Indices	Cu	Linear	32	-	-	-	0.88	13.01		0.50	25.46
	Zn	RBF	23	-	-	0.8	0.92	13.42		0.85	15.94
	Pb	RBF	24	-	-	0.4	0.85	22.49		0.67	24.51
	Cr	RBF	43	-	-	0.32	0.80	7.27		0.79	6.11
	Cd	Polynomial	19	1	11	0.7	0.88	91.94		0.86	102.85

* mg/kg: dry weight.

Table 4 shows the modelling results using the MLR method. In cases where the Durbin–Watson coefficient ranged from 1.5 to 2.5, there was a lack of self-correlation between error terms in the regression model that included 60% of the presented models [40]. However, in relation to the presented models for Pb–Cd (based on wavelengths) and Cr–Cd (based on spectral indices), the Durbin–Watson coefficient was less than 1.5 and lacked one of the most important conditions for using regression modelling. VIF was also considered for the multicollinearity checking between the predictor variables in the regression models. According to Table 4, there was serious multicollinearity (some predictor VIFs exceeded the critical threshold of 10) in the Pb-based-wavelength and Zn-based-spectral index models. Therefore, these models violate the key assumption of multiple linear regression, making these models invalid.

Table 4. The results of modelling and validation of the best multiple linear regression (MLR) models based on optimal wavelengths and spectral indices for estimating heavy metals concentrations in grapevine leaves in training and testing sets.

Hyperspectral Data Type	Heavy Metal	Predictor Variable VIF	Sig. of Regression Confident	Durbin–Watson	Model Structure	Train		Test	
						R^2	RMSE *	R^2	RMSE
Wavelengths	Cu	All < 10	<0.05	2.17	$C_{Cu} = -1.27 - (0.28 \times R2431) + (4.08 \times R809) - (5.32 \times R489) - (8.73 \times R616)$	0.94	9.35	0.56	25.60
	Zn	All < 10	<0.05	2.18	$C_{Zn} = -1.11 - (5.77 \times R2032) - (1.83 \times R665) + (2.38 \times R564) + (13.85 \times R688) - (7.7 \times R437)$	0.73	20.46	0.47	399.13
	Pb	Some cases >10	>0.05	1.39	$C_{Pb} = 0.46 - (5.1 \times R692) + (6.24 \times R683)$	0.32	25.29	0.13	27.28
	Cr	All < 10	<0.05	1.59	$C_{Cr} = 0.61 + (18.08 \times R415) - (1.41 \times R2044) - (4.01 \times R652) - (1.99 \times R1036) + (1.11 \times R713)$	0.84	5.58	0.78	6.79
	Cd	All < 10	<0.05	1.38	$C_{Cd} = 0.98 + (2.76 \times R1373) + (3.15 \times R631) + (1.04 \times R744) - (5.09 \times R438)$	0.63	132.79	0.64	117.26
Spectral Indices	Cu	All < 10	<0.05	1.74	$C_{Cu} = -2.95 + (3.38 \times SR) - (0.01 \times CAI) + (6.76 \times RATIO9752) - (0.77 \times DWSI)$	0.89	12.63	0.52	25.33
	Zn	Some cases >10	<0.05	1.81	$C_{Zn} = -2.26 - (11.34 \times R680) + (41.89 \times WI) + (20.68 \times Lic1) - (3.63 \times MSI) - (4.14 \times PRI2)$	0.87	15.73	0.70	20.38
	Pb	All < 10	<0.05	1.55	$C_{Pb} = 2.53 - (1.33 \times VOG) + (1.93 \times MSI) + (0.85 \times SIPI)$	0.50	24.45	0.15	27.03
	Cr	All < 10	<0.05	1.06	$C_{Cr} = -4.97 + (5.23 \times mNDVI705) + (0.17 \times GI) - (1.28 \times RATIO975)$	0.59	8.48	0.60	8.78
	Cd	All < 10	<0.05	1.27	$C_{Cd} = -6.66 + (4.70 \times SIPI) + (1.13 \times DWSI)$	0.66	121.77	0.67	112.17

* mg/kg: dry weight, Rn: reflections at a certain wavelength, Cn: concentration of a certain heavy metal.

3.4.1. Modelling of Cu Concentration

Figure 5 illustrates the distribution of the observed vs. predicted concentration of Cu in the test set. In some cases, the predicted values were significantly lower than the observed values, which led to a sharp decrease in their accuracy. The optimal wavelengths in the SVM and MLR approaches can predict test samples with 54 and 56% accuracy, respectively. Hence, as a general finding, using wavelengths has a more acceptable performance as opposed to using spectral indices for estimating Cu concentration in the grapevine leaves. In relation to the modelling approaches, it should be noted that, although MLR yielded a slightly superior R^2 than SVM (at the test set), the SVM–RMSE (25.06) was lower than the MLR–RSME (25.65 mg/kg); therefore, the SVM's performance seems more acceptable (see also Tables 3 and 4).

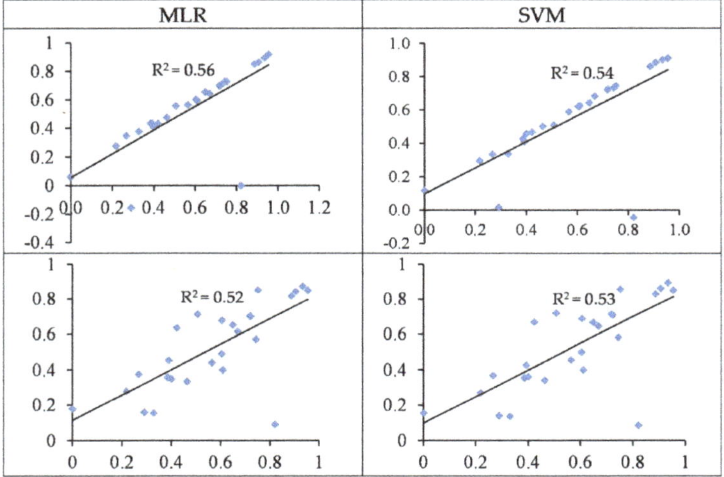

Figure 5. Standardized values (between 0 and 1) of the observed (horizontal axis) and the predicted (vertical axis) concentration of Cu based on wavelengths (top) and spectral indices (bottom) in the testing sets of the SVM and MLR methods.

3.4.2. Modelling of Zn Concentration

The SVM and MLR approaches based on wavelengths were able to predict the Zn contents with accuracies of 42–47% and based on spectral indices with accuracies of 70–85% in the testing set, respectively (Tables 3 and 4). As shown in Figure 6, the predicted values overestimated the observed values in most cases of wavelength-based models. However, a more uniform distribution was found between the observed and predicted values in spectral indices-based models. Therefore, spectral indices-based models tend to be preferred for predicting Zn contents in the grapevine leaves.

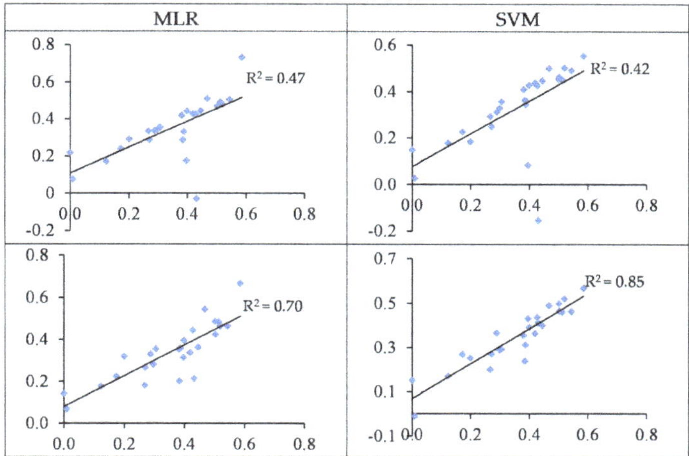

Figure 6. Standardized values (between 0 and1) of the observed (horizontal axis) and the predicted (vertical axis) concentration of Zn based on wavelengths (top) and spectral indices (bottom) in the testing sets of the SVM and MLR methods.

3.4.3. Modelling of Pb Concentration

The MLR models based on wavelengths and spectral indices yielded a low performance in testing sets with accuracies of 13% and 15%, respectively (Table 4 and Figure 7). Conversely, the SVM model performed more reasonably and predicted Pb contents based on wavelengths and spectral indices in the testing set with accuracies of 71% and 67% (RMSE: 22.49 and 24.51 mg/kg), respectively (Table 3 and Figure 7). It can thus be deduced that SVM is better at estimating Pb contents in the grapevine leaves as opposed to MLR. It should also be noted that the wavelength–SVM model had a more acceptable performance as compared to spectral indices. The obtained results therefore suggest that the wavelength–SVM model is an optimal scenario for estimating Pb contents in grapevine leaves.

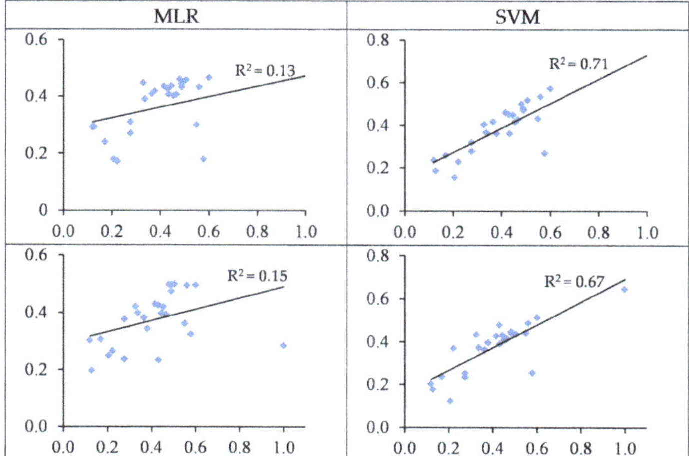

Figure 7. The standardized values (between 0–1) of the observed (horizontal axis) and the predicted (vertical axis) concentration of Pb based on wavelengths (top) and spectral indices (bottom) in the testing sets of the SVM and MLR methods.

3.4.4. Modelling of Cr Concentrations

Figure 8 shows the distribution pattern of the observed vs. predicted values of Cr in the test set. As shown, the predicted values were overestimated in most cases of MLR, but the predictions of SVM were closer to the observed contents. Overall, the usage of the spectral indices–SVM model was an optimal scenario for estimating Cr contents in the studied grapevine leaves (see also Tables 3 and 4).

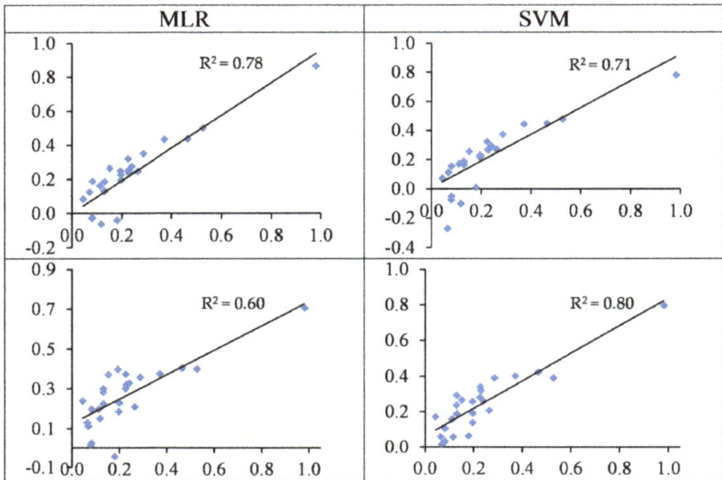

Figure 8. The standardized values (between 0–1) of the observed (horizontal axis) and the predicted (vertical axis) concentration of Cr based on wavelengths (top) and spectral indices (bottom) in the testing sets of the SVM and MLR methods.

3.4.5. Modelling of Cd Concentrations

The wavelengths-based and spectral indices-based MLR models can estimate Cd contents with accuracies of 64% and 67% in the testing set, respectively (Table 4 and Figure 9). On the other hand, the accuracies of SVM were 77% and 86% in the testing set, respectively. Thus, the SVM outperformed the MLR method at estimating Cd concentrations in the grapevine leaves. It must be admitted, however, that the majority of observed values were around zero, leading to biased estimations. This is also reflected in the RMSE values, which were higher than the other studied metals. Overall, the best model presented by the SVM approach (based on spectral indices) had an RMSE value of 102.85 mg/kg dry weight in the testing set (see also Table 3).

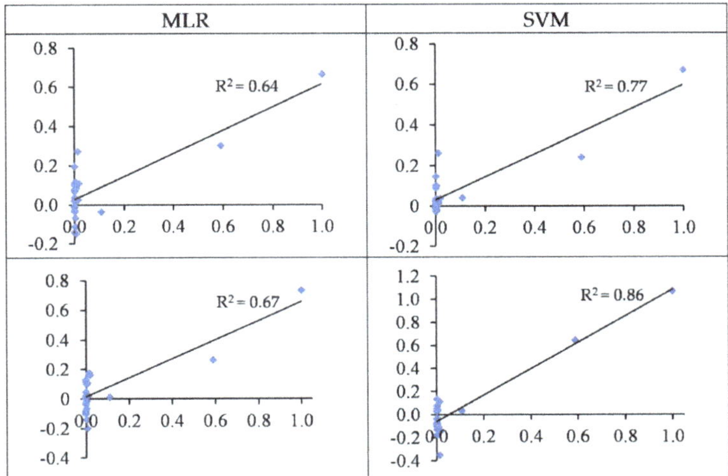

Figure 9. The standardized values (between 0–1) of the observed (horizontal axis) and the predicted (vertical axis) concentration of Cd based on wavelengths (top) and spectral indices (bottom) in the testing sets of the SVM and MLR methods.

3.5. Summarizing Heavy Metal Modelling

Grapevine leaves are a suitable option for the study of the absorption and accumulation of heavy metals [21]. Therefore, the monitoring of heavy metal concentration can ensure food security as well as the reduction of health and ecological risks [16]. In this study, a stress–stroke method was employed to ensure the appearance of heavy metals in grapevine foliage. This method was also used in similar studies [34,38]. It is important to note that expanding heavy metal masses in plant foliage leads to an increase in the number of reactive oxygen species [70]. Reactive oxygen species are produced in the course of electron transfer activities—mainly in chloroplasts and mitochondria. They also have an important role in consequences such as plant growth retardation, chlorophyll content reduction, inhibition of enzymatic activity, damage to biological molecules (such as lipids, proteins, and nucleic acids, especially DNA), cell membrane peroxidation, and damage to important cellular organelles such as chloroplasts and mitochondria [71,72]. Heavy metal stress, like other non-biotic stresses, leads to changes in the pathways of synthesis of secondary plant metabolites and increases or decreases these compounds [73,74]. It was also observed that heavy metal stress leads to changes in the cuticle position of the leaves and the openings of leaves' stomas [73]. Considering the effect of heavy metals on the physico-chemical changes in the plant, the spectral pattern of the plant can change, which leads to the spectral pattern differentiation of stressed leaves from healthy leaves. These differentiations can be determined by field-based spectrometry.

According to our results, SVM and MLR prediction methods performed similarly in estimating Cu contents, but in relation to Zn, Pb, Cr, and Cd, the SVM models outperformed the MLR models (Tables 3 and 4). Therefore, the SVM regression method tends to be preferred. Although, in related studies, MLR was the most-used model due to its clarity and structure simplicity [30,41,45], the results of this study recommend SVM for future investigations. The most important reason for the superiority of SVM as opposed to MLR can be attributed to the nature of the relationships between independent and dependent variables. SVM regression was able to perform more accurately in estimating heavy metals due to its high flexibility in training by using both linear and nonlinear functions in the kernel equation [75]. Similarly, a comparison between MLR and artificial neural network (ANN) methods was performed to estimate heavy metals in rice leaves [38]; the results also showed the superior performance of ANN as opposed to MLR.

A comparison between the results obtained for the testing set and the optimal spectral indices and wavelengths in estimating heavy metal contents in various studies was conducted and is shown in Table 5. Based on the R^2 rate of the test set, the performance order of the presented models was Cd > Zn > Cr > Pb > Cu (Table 5). Therefore, the predictive accuracies for Cd, Zn, and Cu were 86, 85, and 56%, respectively. Li (2011) listed a prediction order accuracy of heavy metals in vegetation as Cr > Pb > Cu > Zn. Furthermore, Zhuang [41] ranked the prediction accuracy of heavy metals in rice as Cu > Pb > Zn, which is different from the findings of the present study (Table 5). The rate of prediction accuracy of Pb in this study is close to the findings of Li [44] and Zhuang [41]. The accuracy of Cr prediction content is also comparable to the results of Li et al. [39]. According to Li [44], Zhuang [41] and Ping et al. [30], Cu predictions were, respectively, 60, 76, and 69%, higher than the present study's result (56%) (Table 5). However, the present study was able to estimate Zn contents with a higher accuracy compared with the results of Li [44], Zhuang [41], and Kooistra et al. [45], as well as Cd contents as compared to the findings of Ping et al. [30] and Liu et al. [37].

As a final remark, in many studies, RDE and VIS regions were reported to be sensitive to the stress caused by heavy metals [36,38,41]. The comparison of the optimal spectral indices and wavelengths selected for the heavy metal rate predicted in the present study and other related studies show discrepancies (Table 5). The number of spectral samples, spectroscopy acquisition level, spectral range, calculated spectral indices, as well as statistical analyses for data reduction and relationship modelling can all play a role in explaining these differences. Finally, it should also be pointed out that each heavy metal has a special effect, leading to distinct responses depending on the plant species (including leaf colour changes, chlorosis, necrosis, dwarfism, giant, leaf and root spreading, etc.), which can justify this finding [76].

Table 5. Comparison results of the best models presented in this study and other similar studies in relation to the estimation of heavy metal contents in plant species using field-based spectrometry.

Metal	Reference	Plant/Species	Approach	Optimal Spectral Indices/Wavelengths	R^2
	Present study	Grape	MLR	R616, R489, R809, R2431	0.56
Cu	Li [44]	Vegetation	MLR		0.60
	Zhuang [41]	Paddy/Rice	MLR		0.76
	Ping et al. [30]	Maize	MLR	NI15, NI11	0.69
	Present study	Grape	SVM	R680 WI, Lic1, MSI, PRI2	0.85
	Li [44]	Vegetation	MLR		0.48
Zn	Zhuang [41]	Paddy/Rice	MLR	661.96×R2210-136.26	0.34
	Kooistra et al. [45]	Grass	MLR	MSAVI2	0.64
	Liu et al. [37]	Rice	ANN		0.95
	Present study	Grape	SVM	R1865, R728 R692, R683, R356	0.71
Pb	Li [44]	Vegetation	MLR		0.77
	Zhuang [41]	Paddy/Rice	MLR		0.70
	Ping et al. [30]	Maize	MLR	NI15, NI17	0.87
	Present study	Grape	SVM	mNDVI705, GI, RATIO975, SIPI	0.80
Cr	Ping et al. [30]	Maize	MLR	NI5, R553	0.49
	Li et al. [39]	Vegetation	MLR	R688, R672, R874, R677, R678, R679, R 680, R566	0.81
	Present study	Grape	SVM	SIPI, DWSI	0.86
Cd	Ping et al. [30]	Maize	MLR	NI11, NI17	0.63
	Liu et al. [37]	Rice	MLR		0.70

NI11: (R700–R690)/(R700+R690), NI15: (R760–850–R350–400)/(R760–850+R350–400), NI17:(R1220–R510)/(R1220+R510), R*n*: Reflections at a certain wavelength.

4. Conclusions

In this study, we examined the suitability of in-field hyperspectral data (wavelengths from 350 to 2500 nm and 32 spectral indices) in the estimation of heavy metal contents (Cu, Zn, Pb, Cr, and Cd) in vine leaves. Our most important findings are listed as follows:

(i) The grapevine's foliar spectral signatures (reflectance characteristics) altered when applying heavy metal stress due to their effects on the biochemical components and the leaves' structure. Considerable changes are observed in the VIS, RDE, NIR, and MIR regions of the electromagnetic spectrum.

(ii)   Significant correlations are found between the heavy metal contents and the grapevine's foliar spectral response, especially in VIS and RDE regions.

(iii)   From the reflectance data, 32 spectral indices were formulated using two or more bands. In PLS analysis, it was found that the Simple Ratio (SR), Cellulose Absorption Index (CAI), RATIO9752, and DWSI; R680, Water Index (WI), Lic1, MSI, and Photochemical Reflectance Index (PRI)2; Vogelman Index (VOG), MSI, SIPI, and R550; mNDVI705, Greenness Index (GI), RATIO975, and SIPI; and SIPI and DWSI are more responsive to heavy metal contents compared with the other indices. They are considered to be optimal indices to estimate Cu, Zn, Pb, Cr, and Cd concentrations, respectively.

(iv)   Also based on the PLS results, the wavelengths in the vicinity of 2431, 809, 489, and 616 nm; 2032, 883, 665, 564, 688, and 437 nm; 1865, 728, 692, 683, and 356 nm; 863, 2044, 415, 652, 713, and 1036 nm; and 1373, 631, 744, and 438 nm are optimal for estimating Cu, Zn, Pb, Cr, and Cd contents in the grapevine leaves, respectively. Accordingly, VIS and RDE emerged as the most sensitive regions for monitoring heavy metal contents in grapevine leaves.

(v)   In most cases, the SVM regression models yielded more accurate performances when estimating heavy metal contents as opposed to the MLR models. For the best SVM structures, the concentrations of Cu, Zn, Pb, Cr, and Cd are estimated with R^2 values of 0.56, 0.85, 0.71, 0.80, and 0.86 in the testing set, respectively.

(vi)   As a general finding, spectral indices yielded more acceptable performance as opposed to wavelengths in forecasting heavy metal contents in the grapevine leaves.

Altogether, the scenario of joining spectral indices with SVM regression is suggested as the most appropriate method for predicting heavy metal contents in the grapevine leaves. At the same time, this conclusion underpins the usage of in-field spectroscopy data and multivariate statistical analysis for the rapid and eco-friendly monitoring of heavy metals in food-producing ecosystems. This study further revealed that the spectral responses of foliar grapevine and other agriculture/horticulture species to heavy metal stress need to be better understood. Similar studies are required to investigate heavy metal spectral signatures in other plant species. Eventually, the ultimate goal of this research line is to integrate field data with spectral data from overpassing aerial and satellite sensors to up-scale and automate the monitoring strategy to the field scale.

Author Contributions: M.M. performed the calculus, designed and conducted the experiments and wrote the paper, J.V. edited and supervised the form of the paper, S.M. managed and supported the heavy metal measurements, M.A. and H.A. observed the scientific content of the paper.

Funding: This work was supported by the European Research Council (ERC) under the ERC-2017-STG SENTIFLEX project (grant agreement 755617) and Jochem Verrelst was supported by a Ramón y Cajal Contract (Spanish Ministry of Science, Innovation and Universities). Mohsen Mirzaei was also supported by the Research Institute for Grapes and Raisin (RIGR), Malayer University.

Acknowledgments: We thank the scientific editor and the reviewers for their valuable suggestions.

Conflicts of Interest: The authors declare no conflict of interest.

References

1.   Wang, P.; Huang, F.; Liu, X. Assessment of heavy metal stress using hyperspectral data. In Proceedings of the 2017 IEEE International Geoscience and Remote Sensing Symposium (IGARSS), Fox Worth, TX, USA, 23–28 July 2017; pp. 6178–6181.

2.   Wang, F.; Gao, J.; Zha, Y. Hyperspectral sensing of heavy metals in soil and vegetation: Feasibility and challenges. *ISPRS J. Photogramm. Remote Sens.* **2018**, *136*, 73–84. [CrossRef]

3.   Zarco-Tejada, P.J.; González-Dugo, V.; Berni, J.A. Fluorescence, temperature and narrow-band indices acquired from a UAV platform for water stress detection using a micro-hyperspectral imager and a thermal camera. *Remote Sens. Environ.* **2012**, *117*, 322–337. [CrossRef]

4.   Bock, C.; Poole, G.; Parker, P.; Gottwald, T. Plant disease severity estimated visually, by digital photography and image analysis, and by hyperspectral imaging. *Crit. Rev. Plant Sci.* **2010**, *29*, 59–107. [CrossRef]

5. Golhani, K.; Balasundram, S.K.; Vadamalai, G.; Pradhan, B. A review of neural networks in plant disease detection using hyperspectral data. *Inf. Process. Agric.* **2018**, *5*, 354–371. [CrossRef]

6. Thomas, S.; Kuska, M.T.; Bohnenkamp, D.; Brugger, A.; Alisaac, E.; Wahabzada, M.; Behmann, J.; Mahlein, A.-K. Benefits of hyperspectral imaging for plant disease detection and plant protection: A technical perspective. *J. Plant Dis. Prot.* **2018**, *125*, 5–20. [CrossRef]

7. Clevers, J.G.; Kooistra, L.; Schaepman, M.E. Estimating canopy water content using hyperspectral remote sensing data. *Int. J. Appl. Earth Obs. Geoinf.* **2010**, *12*, 119–125. [CrossRef]

8. Murphy, R.J.; Whelan, B.; Chlingaryan, A.; Sukkarieh, S. Quantifying leaf-scale variations in water absorption in lettuce from hyperspectral imagery: A laboratory study with implications for measuring leaf water content in the context of precision agriculture. *Precis. Agric.* **2019**, *20*, 767–787. [CrossRef]

9. Ma, Y.; Fang, S.; Peng, Y.; Gong, Y.; Wang, D. Remote Estimation of Biomass in Winter Oilseed Rape (*Brassica napus* L.) Using Canopy Hyperspectral Data at Different Growth Stages. *Appl. Sci.* **2019**, *9*, 545. [CrossRef]

10. Beeri, O.; Phillips, R.; Hendrickson, J.; Frank, A.B.; Kronberg, S. Estimating forage quantity and quality using aerial hyperspectral imagery for northern mixed-grass prairie. *Remote Sens. Environ.* **2007**, *110*, 216–225. [CrossRef]

11. Ryu, C.; Suguri, M.; Umeda, M. Estimation of the quantity and quality of green tea using hyperspectral sensing. *J. Jpn. Soc. Agric. Mach.* **2010**, *72*, 46–53.

12. Mirzaei, M.; Marofi, S.; Abbasi, M.; Solgi, E.; Karimi, R.; Verrelst, J. Scenario-based discrimination of common grapevine varieties using in-field hyperspectral data in the western of Iran. *Int. J. Appl. Earth Obs. Geoinf.* **2019**, *80*, 26–37. [CrossRef]

13. Cho, M.A.; Sobhan, I.; Skidmore, A.K.; De Leeuw, J. Discriminating species using hyperspectral indices at leaf and canopy scales. *Int. Arch. Spat. Inf. Sci.* **2008**, 369–376.

14. Manevski, K.; Manakos, I.; Petropoulos, G.P.; Kalaitzidis, C. Discrimination of common Mediterranean plant species using field spectroradiometry. *Int. J. Appl. Earth Obs. Geoinf.* **2011**, *13*, 922–933. [CrossRef]

15. Li, X.; Liu, X.; Liu, M.; Wang, C.; Xia, X. A hyperspectral index sensitive to subtle changes in the canopy chlorophyll content under arsenic stress. *Int. J. Appl. Earth Obs. Geoinf.* **2015**, *36*, 41–53. [CrossRef]

16. Mirzaei, M.; Marofi, S.; Solgi, E.; Abbasi, M.; Karimi, R.; Bakhtyari, H.R.R. Ecological and health risks of soil and grape heavy metals in long-term fertilized vineyards (Chaharmahal and Bakhtiari province of Iran). *Environ. Geochem. Health* **2019**, 1–17. [CrossRef]

17. Sun, X.; Ma, T.; Yu, J.; Huang, W.; Fang, Y.; Zhan, J. Investigation of the copper contents in vineyard soil, grape must and wine and the relationship among them in the Huaizhuo Basin Region, China: A preliminary study. *Food Chem.* **2018**, *241*, 40–50. [CrossRef]

18. Milićević, T.; Urošević, M.A.; Relić, D.; Vuković, G.; Škrivanj, S.; Popović, A. Bioavailability of potentially toxic elements in soil–grapevine (leaf, skin, pulp and seed) system and environmental and health risk assessment. *Sci. Total Environ.* **2018**, *626*, 528–545. [CrossRef]

19. Wei, B.; Yang, L. A review of heavy metal contaminations in urban soils, urban road dusts and agricultural soils from China. *Microchem. J.* **2010**, *94*, 99–107. [CrossRef]

20. Liang, Q.; Xue, Z.-J.; Wang, F.; Sun, Z.-M.; Yang, Z.-X.; Liu, S.-Q. Contamination and health risks from heavy metals in cultivated soil in Zhangjiakou City of Hebei Province, China. *Environ. Monit. Assess.* **2015**, *187*, 754. [CrossRef]

21. Alagić, S.Č.; Tošić, S.B.; Dimitrijević, M.D.; Antonijević, M.M.; Nujkić, M.M. Assessment of the quality of polluted areas based on the content of heavy metals in different organs of the grapevine (*Vitis vinifera*) cv Tamjanika. *Environ. Sci. Pollut. Res.* **2015**, *22*, 7155–7175. [CrossRef]

22. Guyot, G.; Baret, F.; Jacquemoud, S. Imaging spectroscopy for vegetation studies. *Imaging Spectrosc. Fundam. Prospect. Appl.* **1992**, *2*, 145–165.

23. Slaton, M.R.; Raymond Hunt, E., Jr.; Smith, W.K. Estimating near-infrared leaf reflectance from leaf structural characteristics. *Am. J. Bot.* **2001**, *88*, 278–284. [CrossRef] [PubMed]

24. Adam, E.; Mutanga, O. Spectral discrimination of papyrus vegetation (*Cyperus papyrus* L.) in swamp wetlands using field spectrometry. *ISPRS J. Photogramm. Remote Sens.* **2009**, *64*, 612–620. [CrossRef]

25. Mutanga, O.; Skidmore, A.K. Red edge shift and biochemical content in grass canopies. *ISPRS J. Photogramm. Remote Sens.* **2007**, *62*, 34–42. [CrossRef]

26. Aneece, I.; Epstein, H. Identifying invasive plant species using field spectroscopy in the VNIR region in successional systems of north-central Virginia. *Int. J. Remote Sens.* **2017**, *38*, 100–122. [CrossRef]

27. Damm, A.; Paul-Limoges, E.; Haghighi, E.; Simmer, C.; Morsdorf, F.; Schneider, F.D.; van der Tol, C.; Migliavacca, M.; Rascher, U. Remote sensing of plant-water relations: An overview and future perspectives. *J. Plant Physiol.* **2018**, *227*, 3–19. [CrossRef]

28. Font, R.; Del Rıo, M.; Vélez, D.; Montoro, R.; De Haro, A. Use of near-infrared spectroscopy for determining the total arsenic content in prostrate amaranth. *Sci. Total Environ.* **2004**, *327*, 93–104. [CrossRef]

29. Font, R.; Vélez, D.; Del Río-Celestino, M.; De Haro-Bailón, A.; Montoro, R. Screening inorganic arsenic in rice by visible and near-infrared spectroscopy. *Microchim. Acta* **2005**, *151*, 231–239. [CrossRef]

30. Ping, W.; Liu, X.; Huang, F. Retrieval model for subtle variation of contamination stressed maize chlorophyll using hyperspectral data. *Guang Pu Xue Yu Guang Pu Fen Xi = Guang Pu* **2010**, *30*, 197–201.

31. Asner, G.P.; Martin, R.E. Airborne spectranomics: Mapping canopy chemical and taxonomic diversity in tropical forests. *Front. Ecol. Environ.* **2009**, *7*, 269–276. [CrossRef]

32. Harmon, R.S.; Remus, J.; McMillan, N.J.; McManus, C.; Collins, L.; Gottfried Jr, J.L.; DeLucia, F.C.; Miziolek, A.W. LIBS analysis of geomaterials: Geochemical fingerprinting for the rapid analysis and discrimination of minerals. *Appl. Geochem.* **2009**, *24*, 1125–1141. [CrossRef]

33. Banerjee, B.P.; Raval, S.; Zhai, H.; Cullen, P.J. Health condition assessment for vegetation exposed to heavy metal pollution through airborne hyperspectral data. *Environ. Monit. Assess.* **2017**, *189*, 604. [CrossRef] [PubMed]

34. Rosso, P.H.; Pushnik, J.C.; Lay, M.; Ustin, S.L. Reflectance properties and physiological responses of Salicornia virginica to heavy metal and petroleum contamination. *Environ. Pollut.* **2005**, *137*, 241–252. [CrossRef] [PubMed]

35. Ni, C.; Zhang, D.; Song, P.; Zhao, S.; Yang, W. Hyperspectral Response of Dominant Plants in the Poyang Lake Wetlands to Heavy Metal Pollution. In *Chinese Water Systems*; Springer: Cham, Switzerland, 2019; pp. 99–112.

36. Gu, Y.; Li, S.; Gao, W.; Wei, H. Hyperspectral estimation of the cadmium content in leaves of Brassica rapa chinesis based on the spectral parameters. *Acta Ecol. Sin.* **2015**, *35*, 4445–4453.

37. Liu, M.; Liu, X.; Ding, W.; Wu, L. Monitoring stress levels on rice with heavy metal pollution from hyperspectral reflectance data using wavelet-fractal analysis. *Int. J. Appl. Earth Obs. Geoinf.* **2011**, *13*, 246–255. [CrossRef]

38. Liu, M.; Liu, X.; Li, M.; Fang, M.; Chi, W. Neural-network model for estimating leaf chlorophyll concentration in rice under stress from heavy metals using four spectral indices. *Biosyst. Eng.* **2010**, *106*, 223–233. [CrossRef]

39. Li, N.; Lue, J.; Altermann, W. Applications of spectral analysis to monitoring of heavy metal-induced contamination in vegetation. *Spectrosc. Spectr. Anal.* **2010**, *30*, 2508–2511.

40. Mirzael, M.; Abbasi, M.; Marofi, S.; Solgi, E.; Karimi, R. Spectral Discrimination of Important Orchard Species Using Hyperspectral Indices and Artificial Intelligence Approaches. *J. RS and GIS for Nat. Resour.* **2018**, *9*, 76–92.

41. ZHUANG, D.-F. Study on canopy spectral characteristics of paddy polluted by heavy metals. *Spectrosc. Spectr. Anal.* **2010**, *30*, 430–434.

42. Lehmann, J.; Große-Stoltenberg, A.; Römer, M.; Oldeland, J. Field spectroscopy in the VNIR-SWIR region to discriminate between Mediterranean native plants and exotic-invasive shrubs based on leaf tannin content. *Remote Sens.* **2015**, *7*, 1225–1241. [CrossRef]

43. Diago, M.P.; Fernandes, A.M.; Millan, B.; Tardáguila, J.; Melo-Pinto, P. Identification of grapevine varieties using leaf spectroscopy and partial least squares. *Comput. Electron. Agric.* **2013**, *99*, 7–13. [CrossRef]

44. Li, Y.M. *Spectrum Variation of Vegetation in Yanzhou Coal Mine Area and Heavy Metal Stress Characteristics*; Handong University of Science and Technology: Qingdao, China, 2011.

45. Kooistra, L.; Leuven, R.; Wehrens, R.; Nienhuis, P.; Buydens, L. A comparison of methods to relate grass reflectance to soil metal contamination. *Int. J. Remote Sens.* **2003**, *24*, 4995–5010. [CrossRef]

46. Prospere, K.; McLaren, K.; Wilson, B. Plant species discrimination in a tropical wetland using in situ hyperspectral data. *Remote Sens.* **2014**, *6*, 8494–8523. [CrossRef]

47. Zarco-Tejada, P.J.; Guillén-Climent, M.L.; Hernández-Clemente, R.; Catalina, A.; González, M.; Martín, P. Estimating leaf carotenoid content in vineyards using high resolution hyperspectral imagery acquired from an unmanned aerial vehicle (UAV). *Agric. For. Meteorol.* **2013**, *171*, 281–294. [CrossRef]

48. Gutiérrez, S.; Tardaguila, J.; Fernández-Novales, J.; Diago, M.P. Support vector machine and artificial neural network models for the classification of grapevine varieties using a portable NIR spectrophotometer. *PLoS ONE* **2015**, *10*, e0143197. [CrossRef]

49. Kumar, L.; Schmidt, K.; Dury, S.; Skidmore, A. Review of hyperspectral remote sensing and vegetation science. *Imaging Spectrom. Basic Princ. Prospect. Appl.* **2001**, 111–155.

50. Devadas, R.; Lamb, D.; Simpfendorfer, S.; Backhouse, D. Evaluating ten spectral vegetation indices for identifying rust infection in individual wheat leaves. *Precis. Agric.* **2009**, *10*, 459–470. [CrossRef]

51. Li, M.; Luo, Y.; Su, Z. Heavy metal concentrations in soils and plant accumulation in a restored manganese mineland in Guangxi, South China. *Environ. Pollut.* **2007**, *147*, 168–175. [CrossRef]

52. Orisakwe, O.E.; Nduka, J.K.; Amadi, C.N.; Dike, D.O.; Bede, O. Heavy metals health risk assessment for population via consumption of food crops and fruits in Owerri, South Eastern, Nigeria. *Chem. Cent. J.* **2012**, *6*, 77. [CrossRef]

53. Mirzaei, M.; Jafari, A.; Gholamalifard, M.; Azadi, H.; Shooshtari, S.J.; Moghaddam, S.M.; Gebrehiwot, K.; Witlox, F. Mitigating environmental risks: Modeling the interaction of water quality parameters and land use cover. *LUse Policy* **2019**. [CrossRef]

54. Xu, H.-R.; Yu, P.; Fu, X.-P.; Ying, Y.-B. On-site variety discrimination of tomato plant using visible-near infrared reflectance spectroscopy. *J. Zhejiang Univ. Sci. B* **2009**, *10*, 126–132. [CrossRef] [PubMed]

55. Wold, S.; Sjöström, M.; Eriksson, L. PLS-regression: A basic tool of chemometrics. *Chemom. Intell. Lab. Syst.* **2001**, *58*, 109–130. [CrossRef]

56. Chauchard, F.; Cogdill, R.; Roussel, S.; Roger, J.; Bellon-Maurel, V. Application of LS-SVM to non-linear phenomena in NIR spectroscopy: Development of a robust and portable sensor for acidity prediction in grapes. *Chemom. Intell. Lab. Syst.* **2004**, *71*, 141–150. [CrossRef]

57. Ferreira, M.P.; Zortea, M.; Zanotta, D.C.; Shimabukuro, Y.E.; de Souza Filho, C.R. Mapping tree species in tropical seasonal semi-deciduous forests with hyperspectral and multispectral data. *Remote Sens. Environ.* **2016**, *179*, 66–78. [CrossRef]

58. Laurin, G.V.; Puletti, N.; Hawthorne, W.; Liesenberg, V.; Corona, P.; Papale, D.; Chen, Q.; Valentini, R. Discrimination of tropical forest types, dominant species, and mapping of functional guilds by hyperspectral and simulated multispectral Sentinel-2 data. *Remote Sens. Environ.* **2016**, *176*, 163–176. [CrossRef]

59. Stitson, M.; Weston, J.; Gammerman, A.; Vovk, V.; Vapnik, V. Theory of support vector machines. *Univ. London* **1996**, *117*, 188–191.

60. Hipni, A.; El-shafie, A.; Najah, A.; Karim, O.A.; Hussain, A.; Mukhlisin, M. Daily forecasting of dam water levels: Comparing a support vector machine (SVM) model with adaptive neuro fuzzy inference system (ANFIS). *Water Resour. Manag.* **2013**, *27*, 3803–3823. [CrossRef]

61. Cristianini, N.; Shawe-Taylor, J. *An Introduction to Support Vector Machines and Other Kernel-Based Learning Methods*; Cambridge University Press: Cambridge, UK, 2000.

62. Midi, H.; Bagheri, A. Robust multicollinearity diagnostic measure in collinear data set. In Proceedings of the 4th International Conference on Applied Mathematics, Simulation, Modeling, Corfu Island, Greece, 22–25 July 2010; pp. 138–142.

63. Hair, J.F., Jr. *Multivariate Data Analysis*; Anderson Seventh Edition; Joseph, F., Hair William, C., Jr., Black Barry, J., Babin Rolph, E., Eds.; Pearson: London, UK.

64. Penuelas, J.; Gamon, J.A.; Griffin, K.L.; Field, C.B. Assessing community type, plant biomass, pigment composition, and photosynthetic efficiency of aquatic vegetation from spectral reflectance. *Remote Sens. Environ.* **1993**, *46*, 110–118. [CrossRef]

65. Boyer, M.; Miller, J.; Belanger, M.; Hare, E.; Wu, J. Senescence and spectral reflectance in leaves of northern pin oak (Quercus palustris Muenchh.). *Remote Sens. Environ.* **1988**, *25*, 71–87. [CrossRef]

66. Vogelmann, T.C. Plant tissue optics. *Ann. Rev. Plant Biol.* **1993**, *44*, 231–251. [CrossRef]

67. Strever, A.A.E. *Non-Destructive Assessment of Leaf Composition as Related to Growth of the Grapevine (Vitis vinifera L. cv. Shiraz)*; Stellenbosch University: Stellenbosch, South Africa, 2012.

68. Adam, E.; Mutanga, O.; Rugege, D. Multispectral and hyperspectral remote sensing for identification and mapping of wetland vegetation: A review. *Wetl. Ecol. Manag.* **2010**, *18*, 281–296. [CrossRef]

69. Penuelas, J.; Baret, F.; Filella, I. Semi-empirical indices to assess carotenoids/chlorophyll a ratio from leaf spectral reflectance. *Photosynthetica* **1995**, *31*, 221–230.

70. Malar, S.; Vikram, S.S.; Favas, P.J.; Perumal, V. Lead heavy metal toxicity induced changes on growth and antioxidative enzymes level in water hyacinths [*Eichhornia crassipes* (Mart.)]. *Bot. Stud.* **2016**, *55*, 54. [CrossRef] [PubMed]

71. Chaoui, A.; El Ferjani, E. Effects of cadmium and copper on antioxidant capacities, lignification and auxin degradation in leaves of pea (*Pisum sativum* L.) seedlings. *C. R. Biol.* **2005**, *328*, 23–31. [CrossRef] [PubMed]

72. Mishra, S.; Srivastava, S.; Tripathi, R.; Govindarajan, R.; Kuriakose, S.; Prasad, M. Phytochelatin synthesis and response of antioxidants during cadmium stress in *Bacopa monnieri* L. *Plant Physiol. Biochem.* **2006**, *44*, 25–37. [CrossRef] [PubMed]

73. Rai, V.; Khatoon, S.; Bisht, S.; Mehrotra, S. Effect of cadmium on growth, ultramorphology of leaf and secondary metabolites of Phyllanthus amarus Schum. and Thonn. *Chemosphere* **2005**, *61*, 1644–1650. [CrossRef] [PubMed]

74. Tirillini, B.; Ricci, A.; Pintore, G.; Chessa, M.; Sighinolfi, S. Induction of hypericins in Hypericum perforatum in response to chromium. *Fitoterapia* **2006**, *77*, 164–170. [CrossRef]

75. Gutiérrez, S.; Tardaguila, J.; Fernández-Novales, J.; Diago, M. Data mining and NIR spectroscopy in viticulture: Applications for plant phenotyping under field conditions. *Sensors* **2016**, *16*, 236. [CrossRef]

76. Yadav, S. Heavy metals toxicity in plants: An overview on the role of glutathione and phytochelatins in heavy metal stress tolerance of plants. *S. Afr. J. Bot.* **2010**, *76*, 167–179. [CrossRef]

Article

Application of UAV-Based Multi-angle Hyperspectral Remote Sensing in Fine Vegetation Classification

Yanan Yan [1,2], Lei Deng [1,3,*], XianLin Liu [1,4] and Lin Zhu [1,3]

[1] College of Resource Environment and Tourism, Capital Normal University, Beijing 100048, China; 2150901012@cnu.edu.cn (Y.Y.); liuxl@cae.cn (X.L.); 5533@cnu.edu.cn (L.Z.)
[2] China Institute of Water Resources and Hydropower Research, Beijing 100038, China
[3] Key Laboratory of 3D Information Acquisition and Application, Ministry of Education, Capital Normal University, Beijing 100048, China
[4] Chinese Academy of Surveying and Mapping, Beijing 100830, China
[*] Correspondence: denglei@cnu.edu.cn; Tel.: +86-10-6890-3472

Received: 8 October 2019; Accepted: 20 November 2019; Published: 22 November 2019

Abstract: To obtain a high-accuracy vegetation classification of high-resolution UAV images, in this paper, a multi-angle hyperspectral remote sensing system was built using a six-rotor UAV and a Cubert S185 frame hyperspectral sensor. The application of UAV-based multi-angle remote sensing in fine vegetation classification was studied by combining a bidirectional reflectance distribution function (BRDF) model for multi-angle remote sensing and object-oriented classification methods. This method can not only effectively reduce the classification phenomena that influence different objects with similar spectra, but also benefit the construction of a canopy-level BRDF. Then, the importance of the BRDF characteristic parameters are discussed in detail. The results show that the overall classification accuracy (OA) of the vertical observation reflectance based on BRDF extrapolation (BRDF_0°) (63.9%) was approximately 24% higher than that based on digital orthophoto maps (DOM) (39.8%), and kappa using BRDF_0° was 0.573, which was higher than that using DOM (0.301); a combination of the hot spot and dark spot features, as well as model features, improved the OA and kappa to around 77% and 0.720, respectively. The reflectance features near hot spots were more conducive to distinguishing maize, soybean, and weeds than features near dark spots; the classification results obtained by combining the observation principal plane (BRDF_PP) and on the cross-principal plane (BRDF_CP) features were best (OA = 89.2%, kappa = 0.870), and especially, this combination could improve the distinction among different leaf-shaped trees. BRDF_PP features performed better than BRDF_CP features. The observation angles in the backward reflection direction of the principal plane performed better than those in the forward direction. The observation angles associated with the zenith angles between −10° and −20° were most favorable for vegetation classification (solar position: zenith angle 28.86°, azimuth 169.07°) (OA was around 75%–80%, kappa was around 0.700–0.790); additionally, the most frequently selected bands in the classification included the blue band (466 nm–492 nm), green band (494 nm–570 nm), red band (642 nm–690 nm), red edge band (694 nm–774 nm), and the near-infrared band (810 nm–882 nm). Overall, the research results promote the application of multi-angle remote sensing technology in vegetation information extraction and provide important theoretical significance and application value for regional and global vegetation and ecological monitoring.

Keywords: multi-angle observation; hyperspectral remote sensing; BRDF; vegetation classification; object-oriented segmentation

1. Introduction

The vegetation ecosystem is an important foundation for ecological systems [1]. The use of remote sensing technology has become the main approach for vegetation ecological resource surveys and environmental monitoring due to the corresponding real-time, repeatability, and wide-coverage advantages [2–4]. With the development of remote sensing technology, visible light, multispectral, hyperspectral, and other sensors have been widely used in the remote sensing of vegetation [5,6], and more hyperspectral and high-resolution information has been obtained than ever before, greatly improving the accuracy of image classification [7,8].

As one of the current frontiers of remote sensing development, hyperspectral remote sensing technology has played an increasingly important role in quantitative analyses and accurate classifications of vegetation due to its ability to acquire high-resolution spectral and spatial data [9–12]. For instance, Filippi utilized an unsupervised self-organizing neural network to perform complex vegetation mapping in a coastal wetland environment [13]. Fu et al. proposed an integrated scheme for vegetation classification by simultaneously exploiting spectral and spatial image information to improve the vegetation classification accuracy [14].

From the perspective of remote sensing imaging, remote sensing vertical photography can obtain only the spectral feature projection of the target feature in one direction, and it lacks sufficient information to infer the reflection anisotropy and spatial structure [15]. Multi-angle observations of a target can provide information in multiple directions and be used to construct the bidirectional reflectance distribution function (BRDF) [16–18], which increases the abundance of target observation information; additionally, this approach can extract more detailed and reliable spatial structure parameters than a single-direction observation can [19]. Multi-angle hyperspectral remote sensing, which combines the advantages of multi-angle observation and hyperspectral imaging technology, is projected to become an effective technical method for the classification of vegetation in remote sensing images.

The UAV remote sensing platform has emerged due to its flexibility, easy operation, high efficiency, and low cost; it can efficiently acquire high-resolution spatial and spectral data on demand [20]. The UAV remote sensing platform has the ability to provide multi-angle observations and thus has become popular in multi-angle remote sensing [21–24]. Roosjen et al. studied the hyperspectral anisotropy of barley, winter wheat, and potatoes using a drone-based imaging hyperspectrometer by obtaining multi-angle observation data for hemispherical surfaces by hovering around the crops [25]. In addition, Liu and Abd-Elrahman developed an object-based image analysis (OBIA) approach by utilizing multi-view information acquired using a digital camera mounted on a UAV [26]. They also introduced a multi-view object-based classification using deep convolutional neural network (MODe) method to process UAV images for land cover classification [27]. Both methods avoided the salt and pepper phenomenon of the classified image and have achieved favorable classification results. However, it is difficult to obtain the continuous spectrum characteristics of the ground objects because of the fewer wave bands the optical sensors use. Moreover, the research does not fully mine the contribution difference of multi-angle features. Furthermore, how to use the limited multi-angle observations to construct the BRDF of ground objects to enrich the observation information of the target is also one of the difficulties in the application of multi-angle remote sensing.

In this paper, key technical issues, such as the difficulty in distinguishing complex vegetation species from a single remote sensing observation direction, the construction of the BRDF model based on UAV multi-angle observation data, and model application for vegetation classification and extraction, were studied. The purpose of this study was to discuss the role of ground object BRDF characteristic parameters in the fine classification of vegetation, thereby improving the understanding of the relationship between the BRDF and plant leaves and vegetation canopy structure parameters, as well as promoting the application of multi-angle optical remote sensing in the acquisition of vegetation information.

2. Date Sets

2.1. Study Area

The research area was in the village of Luozhuang, Changziying town, Daxing District, Beijing, as shown in Figure 1. The image was acquired on 24 August 2018. The weather was clear and cloudless. The rich vegetation types included weeds, crops, and tree species, as shown in Figure 2. The crops included soybean in the flowering and pod-bearing stages, and maize in the powder stage. The tree species included mulberry, peach, and ash trees. The vegetation grew densely, and shadows greatly affected the classification results. Therefore, shadows were recognized as a type of object in this paper. In summary, the land species were divided into eight types: weeds, soybeans, maize, mulberries, peach trees, ash trees, dirt roads, and shadows.

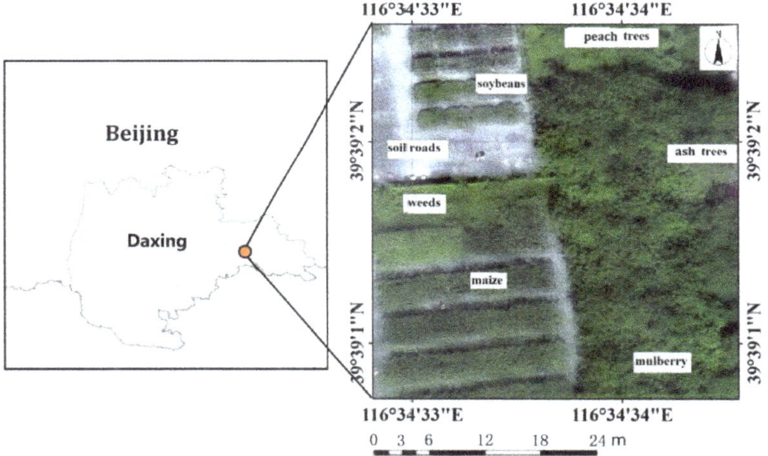

Figure 1. Study site.

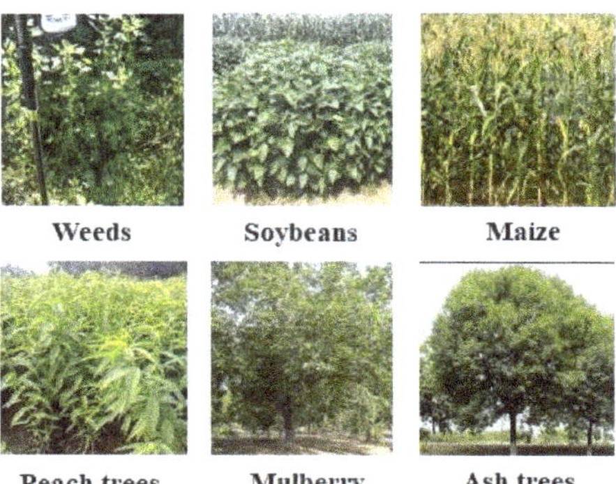

Figure 2. Schematic images of the vegetation types in the study area.

2.2. UAV Hyperspectral Remote Sensing Platform

In this paper, a Cubert S185 hyperspectral sensor mounted on a DJI Jingwei M600 PRO (Dajiang, Shenzhen, China), which is a rotary-wing vehicle with six rotors, was used to obtain research data, and is shown in Figure 3. The Cubert S185 frame-frame imaging spectrometer (Germeny) [28] simultaneously captured both low spatial resolution hyperspectral images (50 × 50 pixels) and high spatial resolution panchromatic images (1000 × 1000 pixels), and then obtained high spatial resolution hyperspectral images via data fusion using Cubert Pilot software. The sensor provides 125 spectral channels with wavelengths ranging from 450 nm to 950 nm (4-nm sampling interval). Table 1 lists the main performance parameters of the hyperspectral cameras.

(a)   (b)

Figure 3. Sensor system and UAV platform: (**a**) Cubert hyperspectral camera and (**b**) DJI Jingwei M600 PRO.

Table 1. Main parameters of the Cubert UHD 185 snapshot hyperspectral sensor (provided by the manufacturer).

Specification	Value	Specification	Value
Wavelength range	450–946 nm	Housing	28 cm, 6.5 cm, 7 cm
Sampling interval	4 nm	Digitization	12 bit
Full width at half maximum	8 nm at 532 nm, 25 nm at 850 nm	Horizontal field of view	22°
		Cube resolution	1 megapixel
Channels	125	Spectral throughput	2500 spectra/cube
Focal length	16 mm	Power	DC 12 V, 15 W
Detector	Si CCD	Weight	470 g

2.3. Flight Profile and Conditions

The drone mission was implemented from 12:10 to 12:30 on 24 August 2018. Regarding the sun's position, the zenith angle was 28.86°, and the azimuth was 169.07°. The weather was clear and cloudless, there was no wind, and the light intensity was stable. The flying height was 100 m, and the acquired hyperspectral image had a ground sample distance of 4 cm after data fusion. To ensure that the remote sensing platform obtained a sufficient observation angle for each feature and to improve the accuracy of the BRDF model construction, the flight adopted vertical photography and oblique photogrammetry (the angle of the mirror center was 30°). To obtain more abundant multi-angle observation data, the image heading and side overlap were both greater than 80%. Moreover, RTK (real-time kinematic) carrier phase difference technology was used to measure the coordinates of the ground control points with a planimetric accuracy better than 1 cm. The number of control points was 5, and they were located in areas with clear, distinguishable, and unblocked GPS signals.

2.4. Data Processing

According to the flight mission plan described above, the hyperspectral experimental dataset was successfully acquired, and the data were processed with Agisoft PhotoScan software Version 1.2.5 (St.Petersburg Russia) to generate a digital orthophoto map (DOM) and digital surface model (DSM) data for the research area. Data processing included matching according to high definition digital images and position and orientation system (POS) information at the time of image acquisition (latitude and longitude, altitude, flip, and pitch and rotation angle of the UAV flight), detecting the feature points of photos based on a dynamic structure algorithm, establishing matching feature point pairs, and arranging photos. A dense three-dimensional point cloud was generated using a dense multi-perspective stereomatching algorithm, and the ground control points were input for geometric corrections. Finally, the DSM and DOM of the experimental area were obtained, as shown in Figure 4.

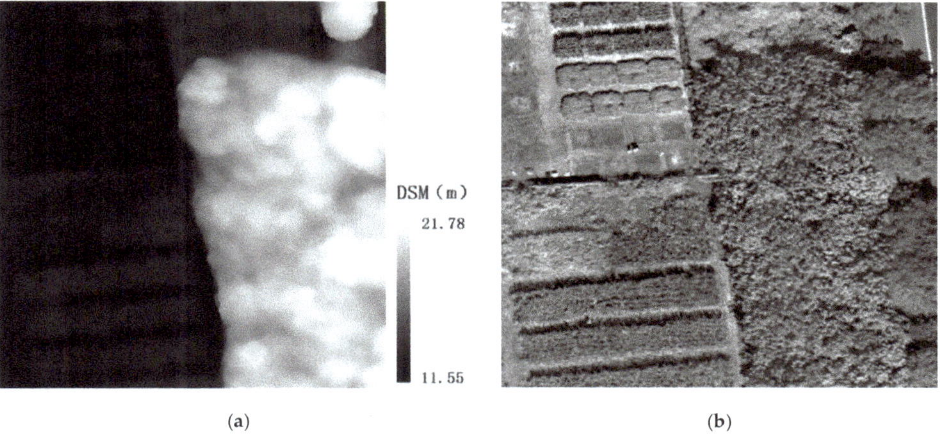

(a)　　　　　　　　　　　　　　　(b)

Figure 4. Data processing results: (**a**) DSM and (**b**) DOM.

3. Materials and Methods

This paper proposed a novel vegetation classification method combining an object-oriented classification method and BRDF. The relationship between BRDF characteristics and plant leaves and vegetation canopy structure is discussed to promote the development of multi-angle optical remote sensing in the application field of vegetation remote sensing. First, the method of image segmentation combining spectral and DSM features was studied to improve the accuracy of the object-oriented edge and the segmentation of the plaque. Second, multiple hyperspectral data sets were obtained using vertical and oblique photogrammetry, the acquired multi-angle observation data of the ground object were used, and then the semi-empirical kernel driver model was used to invert the BRDF model of each object patch. Third, according to the characteristics of BRDF for each segmentation patch, a multi-class feature set was constructed. Finally, object-oriented classification was carried out for fine vegetation classification. The specific research technology route is shown in Figure 5:

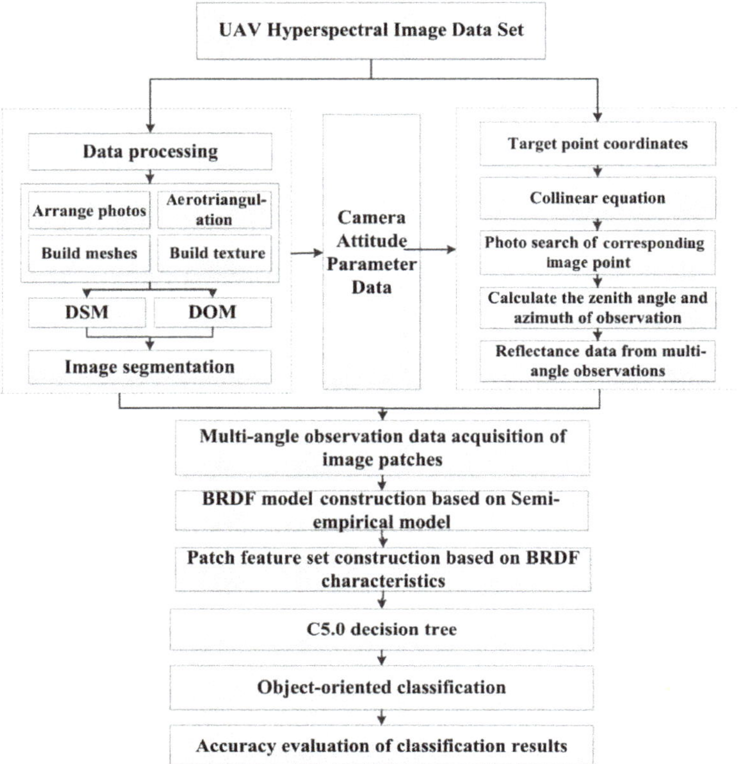

Figure 5. Flowchart of the classification method. DSM: digital surface model, DOM: digital orthophoto maps, BRDF: bidirectional reflectance distribution function.

3.1. Image Segmentation

UAV remote sensing technology can acquire DSMs through the acquisition and processing of multiple overlapping images, and this information can be used as auxiliary information to improve the image segmentation accuracy. In this study, the DSM and spectral characteristics were taken as basic information, and the object set of the UAV hyperspectral image segmentation was constructed using Definiens eCognition Developer 7.0 (München, Germany). A multi-resolution segmentation method was adopted. A segmentation scale parameter was manually adjusted using a trial and error method and finally a segmentation scale of 50 was selected, which resulted in visually correct segmentation. The shape and compactness weight parameters [29] used in the segmentation algorithm were also found using trial and error, and values of 0.05 and 0.8 were used, respectively.

Next, the extraction of the feature sets for image segmentation patches were discussed, which were used for vegetation classification.

3.2. Multi-angle Observation Data Acquisition and BRDF Model Construction

First, the maximum inscribed circle of each object patch was obtained as the attribute representative of the patch. Second, corresponding image points for each pixel inside the inscribed circle were found. Then, the average value of the reflectances in the corresponding circular area in one image was read as the reflectivity of the segmented block under different observation angles. At the same time, the observation angle of each image block with the same name was obtained. Finally, the BRDF model of the segmentation block was constructed by using the reflectance of a multi-angle observation. The specific steps were as follows:

(1) Aerotriangulation and camera attitude parameter solution:

On the basis of multi-angle image data sets with high amounts of overlap acquired through vertical photography and oblique photogrammetry, control point data obtained using synchronous field measurements were used to calculate the coordinates of pending points in the study area via aerotriangulation and then used as control points for multiple images and image correction. In this method, aerial camera stations were established for the whole network, and the acquired images were used for point transmission and network construction.

The exterior and interior orientation parameters were obtained via aerotriangulation with control point data, and the internal camera parameters were obtained via camera calibration. The camera calibration and orientation was carried out suing the Agisoft Photoscans software. Then, the coordinates of the known object in three-dimensional space, the corresponding image pixel coordinates, and the camera interior parameters were used to determine the exterior parameters of the object in a known space, namely the rotation vector and the translation vector. Finally, the rotation vector was analyzed and processed to obtain the three-dimensional altitude angle of the camera relative to the spatial coordinates of the known object by considering the pitch, rotation, and wheel angles.

(2) Search for the corresponding image points:

The corresponding image point refers to the image point of any ground object target point in different photos [30]. It was obtained by photographing the same object point multiple times at different photo points during the aerial photography. After calculating the coordinates of the pending points in the study area and the elements of the internal and external orientations of each image, a collinearity equation with digital photogrammetry was used to determine the image plane coordinates of the target point for each image; then, the characteristics of the sensor image were used to determine whether each coordinate was within the visual threshold range and to search the image for corresponding image points.

(3) Observation angle and reflectance of points with the same name:

After searching for points with the same name, the zenith angle and observation azimuth of the object point in each image and points with the same name were calculated using the orientation relationship between the camera station (projection center) and the object point. In addition, the reflectance of points with the same name was determined for the selected band image.

(4) Parameter calculation for the semi-empirical kernel driver model:

Algorithm for model bidirectional reflectance anisotropics of the land surface (AMBRALS) [31] was selected to construct the BRDF. The semi-empirical core-driven model can be expressed using Equation (1):

$$R(\theta, \partial, \sigma) = f_{iso} + f_{vol}K_{vol}(\epsilon, \partial, \sigma) + f_{geo}K_{geo}(\theta, \partial, \sigma). \tag{1}$$

The bidirectional reflectance can be decomposed into the sum of the weights of the three parts of uniform reflection, bulk reflection, and geometric optical reflection. Therefore, the value of isotropic reflection is generally equal to 1. In the core-driven model, R represents the bidirectional reflectivity, θ represents the ray zenith angle, ∂ represents the observation angle of the zenith angle, and σ represents the corresponding azimuth angle. K_{vol} and K_{geo} are the bulk nuclear reflection and geometric optical nuclear reflection, respectively. f_{iso}, f_{vol}, and f_{geo} are constant coefficients that represent the proportions of uniform reflection, bulk reflection, and geometric optical reflection, respectively. The linear regression method was used to solve for the optimal parameter values. In addition, the bulk nuclear reflection and geometric optical nuclear reflection in the formula were calculated using the ray zenith angle, the observation zenith angle, and the corresponding azimuth angle, and therein, the ray zenith angle and the azimuth angle were calculated based on the time and date the image was obtained and the coordinates of the object point.

3.3. Feature Set Construction Based on the BRDF

To evaluate the application value of the BRDF model for vegetation classification, this study extracted two types of features from the BRDF model as the basic attributes for the identification of vegetation species. The first type was bidirectional reflectance factor (BRF) predicted by the BRDF model, including the maximum (hot spot) and the minimum (dark spot) reflectance values observed in the backscattering and forward scattering regions, respectively; the multi-angle observation reflectance in the main plane of the observation (considering the maximum view zenith angle of the remote sensing sensor, which was set to 60°); and then the observations in the principal planes beginning from the 0° zenith angle in the forward and backward directions of observation with a 10° sampling interval to obtain the multi-angle observation data. The multi-angle observed reflectance of the main vertical observation plane (the angular sampling method was consistent with the main plane of observation) and joint feature set of multi-angle reflectance for the main planes (25) were also considered. Second, the BRDF model parameters f_{iso}, f_{vol}, and f_{geo} were considered [25]. Table 2 summarizes the feature sets used for vegetation species identification.

Table 2. Feature set construction using the BRDF for object-oriented classification.

		Explanatory Variable	Abbreviation
	Commonly Used	Reflectance obtained from DOM	DOM
		Vertical observation angle	BRDF_0°
BRDF Characteristics	(1) Modeled bidirectional reflectance factors (BRFs)	Hot and dark spots reflectance signatures	BRDF_HS_DS
		Observation angles on principal plane	BRDF_PP
		Observation angles on cross-principal plane	BRDF_CP
		Observation angles on principal and cross planes	BRDF_PP+CP
	(2) Model parameters	f_{iso}, f_{vol} and f_{geo}	BRDF_3f

3.4. Vegetation Classification and Accuracy Assessment

After obtaining the noise attribute information for each object according to the above scheme, the C5.0 decision tree [32] method was used to construct the vegetation species recognition model. The decision tree algorithm has a structure similar to the tree structure shown in the flow chart. This structure can intuitively display the classification rules, and the classification algorithm has a fast speed, high accuracy, and simple generation mode. This study used the SPSS Clementine V16.0 software (IBM, Chicago, USA) to achieve a fine classification of vegetation based on the C5.0 decision tree. To verify the effectiveness of the method, the image segmentation results were taken as samples, and the number of each sample was summarized, as shown in Table 3. Sixty percent of the samples were used as model training samples, and the remaining 40% were used as verification samples.

Table 3. Samples of vegetation types.

Types	Dirt Roads	Weeds	Soybeans	Maize	Mulberries	Peach Trees	Ash Trees	Shadows
Number	36	26	17	29	25	38	26	38

The quantitative evaluation of the classification results mainly included the following index factors [33]: confusion matrix (overall accuracy, producer's accuracy, and user's accuracy) and the kappa coefficient. The overall accuracy is essentially tells us out of all the reference sites, what proportion were mapped correctly. The producer's accuracy is the map accuracy from the point of view of the map maker (the producer). This is how often real features on the ground are correctly shown on the classified map or the probability that a certain land cover of an area on the ground is classified as such. The user's accuracy is the accuracy from the point of view of a map user, not the map maker. It essentially tells the user how often the class on the map will actually be present on the ground. This is referred to as the reliability. The kappa coefficient is a statistical measure of inter-rater agreement or inter-annotator agreement for qualitative (categorical) items.

4. Image Classification Results

According to the set of classification feature parameters listed in Table 2, an image classification based on C5.0 was performed. A quantitative evaluation of the classification results is shown in Table 4. The overall classification accuracy based on BRDF_0° (63.9%) was approximately 24% higher than that based on the DOM. Two principal plane reflectance feature sets (BRDF_PP+CP) were used for the fine classification of vegetation, and the best results were obtained. The overall accuracy of classification (89%) was greatly improved by 39%, and the kappa coefficient (0.870) was increased by 0.438. The classification results for the study area based on BRDF_PP+CP are shown in Figure 6.

Table 4. Classification accuracy based on a feature set construction with the BRDF (overall accuracy (OA) and kappa).

	Explanatory Variable		OA	Kappa
Commonly Used		DOM	39.8	0.301
		BRDF_0°	63.9	0.573
	Modeled bidirectional reflectance factors (BRFs)	BRDF_0°+HS+DS	77.1	0.728
BRDF Characteristics		BRDF_PP	85.5	0.828
		BRDF_CP	78.3	0.740
		BRDF_PP+CP	89.2	0.870
	BRDF model parameters	BRDF_0°_3f	78.3	0.739

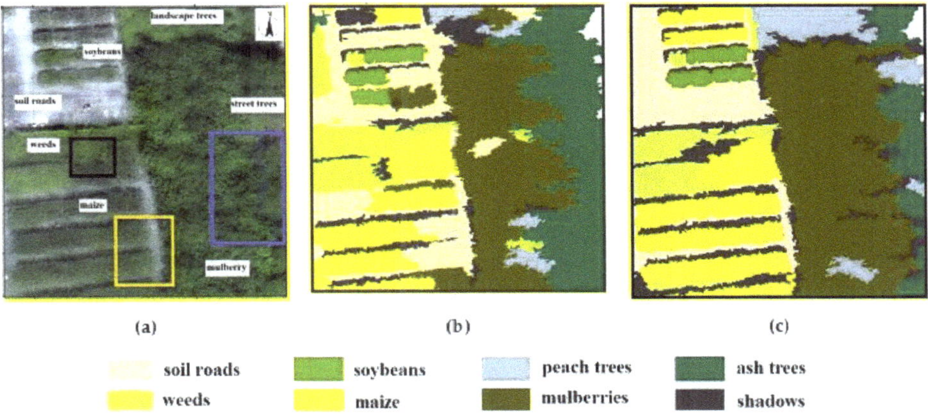

Figure 6. Classification results: (**a**) the reference map, (**b**) the map produced using DOM, and (**c**) the map produced based on multi-angle reflectance characteristics of the observed principal planes and cross-principal planes.

From Figure 6, the object-oriented vegetation classification method based on the multi-angle reflectance characteristics achieved good mapping results with clear boundaries and an accurate location distribution. BRDF_PP+CP feature sets helped to improve the recognition accuracy of the junction of different tree species, as shown in the blue rectangle in Figure 6. This was because the observation data from different angles could reflect the difference in tree structure, and the tree species could be identified well using multi-angle difference features. In addition, it could improve the accuracy of the division of corn and field roads, as shown in the yellow rectangle in Figure 6. However, although the BRDF_PP+CP greatly improved the identification accuracy for shadows, it performed poorly regarding the distinction between shadows and weeds with a low height, as shown in the black area of Figure 6. The spectral vegetation types under shadow coverage in the study area were various, and the spectral characteristics of shadow were similar to those of weeds with a low height.

5. Discussion

5.1. Applicability Assessment of BRDF Characteristic Types

For promoting the realization of the relationship between the BRDF and plant leaves and vegetation canopy structure parameters, the following subsections are given to discuss the role of the ground object BRDF characteristic parameters in the fine classification of vegetation.

(1) Performance of BRDF_0°:

Table 4 shows that compared with the classification results based on the DOM, the classification accuracy based on BRDF_0° was greatly improved. Figure 7 shows the producer's accuracy and user's accuracy of each type of land feature using two classification features.

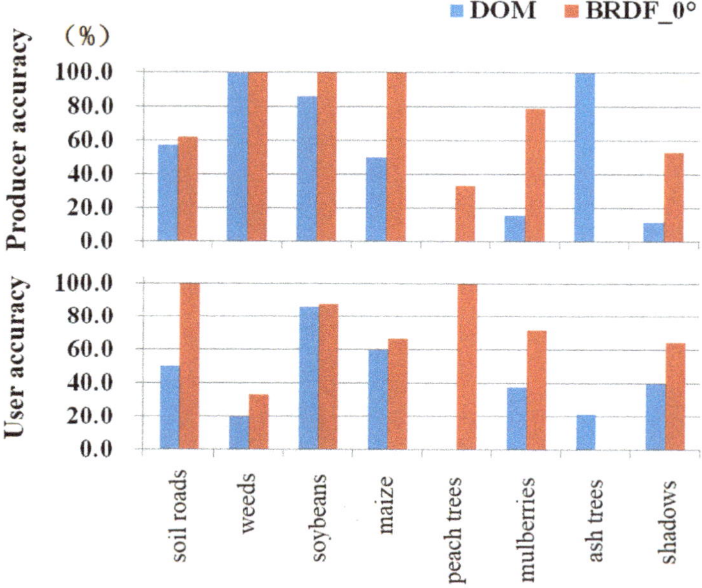

Figure 7. Vegetation classification accuracy based on DOM and BRDF_0°.

Figure 7 indicates that BRDF_0° was instrumental in the distinction among different objects. The vertical reflectance data of DOM were obtained using statistical methods. However, the method of using a semi-empirical model to construct the BRDF and invert the vertically observed reflectance combines the advantages of an empirical model and a physical model. Although the model parameters are empirical parameters, they have certain physical significance. Consequently, the observation angle of the ground objects is unified with the vertical observations through the BRDF model, which weakens the reflection characteristics of the same type of vegetation affected by the observation angle difference. Compared with the classification results obtained using DOM data, the classification accuracy obtained using BRDF_0° was greatly improved, but the recognition accuracy of dirt roads, peach trees, and ash trees was still very low. The producer's accuracy of weeds, soybeans, and maize improved to greater than 90%, but the user's accuracy improved only slightly, which indicates that the results for these three types of land features were overclassified.

(2) Hot and dark spot reflectance signatures:

Six feature sets were used to classify vegetation, namely, the vertical observation direction (BRDF_0°); hot spot observation direction (BRDF_HS); dark spot observation direction (BRDF_DS); vertical observation direction and hot spot direction (BRDF_0°+HS); vertical observation direction and dark spot direction (BRDF_0°+DS); and vertical observation direction, hot spot direction, and dark spot direction (BRDF_0°+HS+DS). The overall classification accuracy and kappa coefficients of the six feature sets are shown in Figure 8. The classification accuracy of BRDF_0°+HS+DS was the highest at about 77%. The classification effect of vegetation types using BRDF_DS was slightly worse than that using BRDF_0°, while the classification effect of vegetation types using BRDF_HS was better than that using BRDF_0°. The results show that the hot spot reflectance signature had an excellent effect in the recognition of complex vegetation types. This was because the reflection characteristics of different objects in the direction of dark spots were lower than those in the direction of hot spots, and the hot spot effects between crops and tree species were quite different. The producer and user accuracies of each type of land feature are shown in Figure 9.

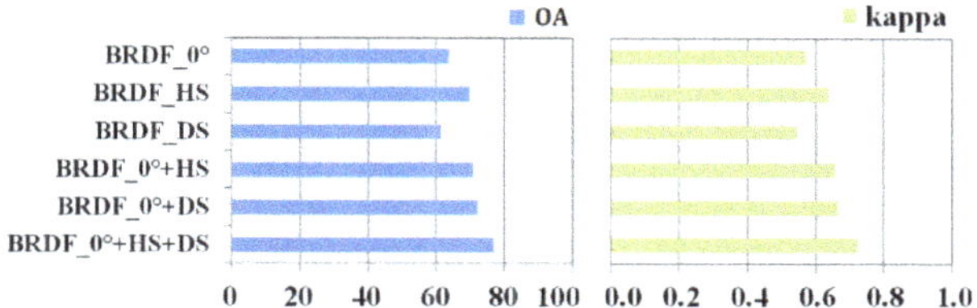

Figure 8. Overall accuracy and kappa coefficient of vegetation classification based on hot and dark spot characteristics.

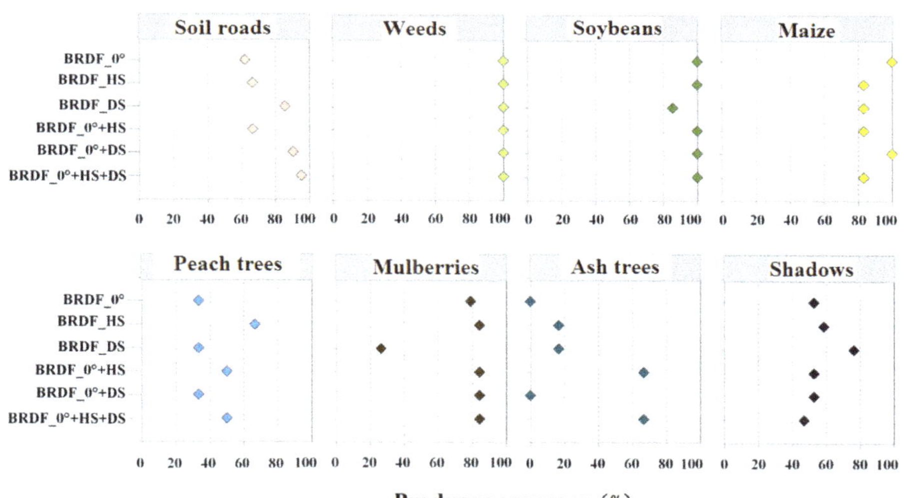

Producer accuracy（%）

Figure 9. *Cont.*

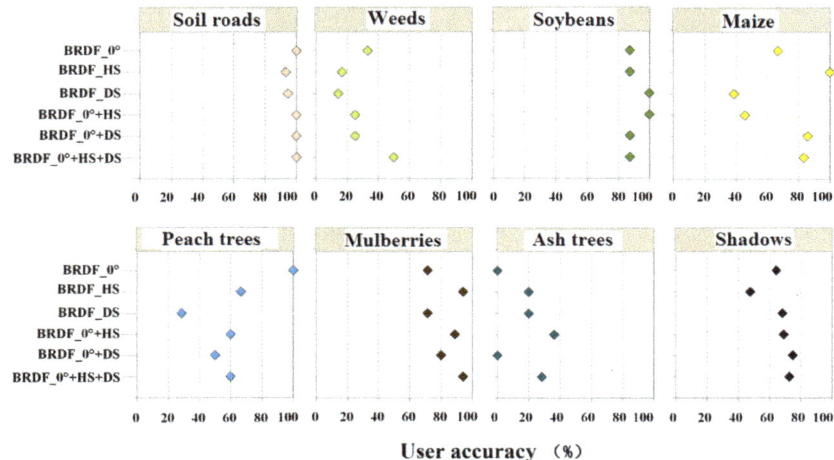

Figure 9. Producer's accuracy and user's accuracy for each vegetation type based on hot and dark spot characteristics.

From Figure 9, the combined application of dark spot and hot spot directional reflectance features improved the classification accuracy. The classification results for soybean, peach trees, mulberry trees, and ash trees using BRDF_HS were more accurate than those using BRDF_DS. In contrast, the ground objects with a high accuracy included dirt roads and shadows based on the reflection features in the dark spot direction. The research shows that the tree structure features had a high sensitivity in the hot spot direction.

(3) Multi-angle reflectance characteristics of the observed principal plane and cross-principal plane:

Four feature sets were used to classify the vegetation, namely the reflectance values from the vertical observation direction (BRDF_0°), principal plane (BRDF_PP), cross plane (BRDF_CP), principal and cross planes (BRDF_PP+CP). The corresponding classification results are shown in Figure 10. The classification accuracy using BRDF_PP+CP was the highest (OA = 88%). The reflectance characteristics from the principal plane were more conducive to the classification of complex vegetation species than those in the vertical main plane. The producer's accuracy and user's accuracy of each type of land feature are shown in Figure 11.

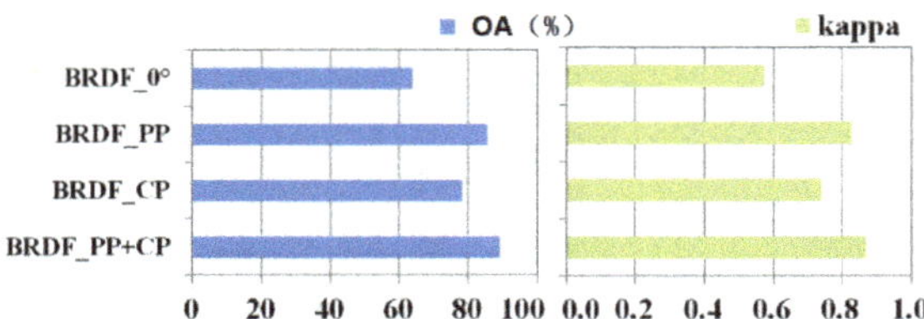

Figure 10. Overall accuracy and kappa coefficient of the vegetation classification based on the multi-angle reflectance characteristics for the observed principal plane and cross-principal plane.

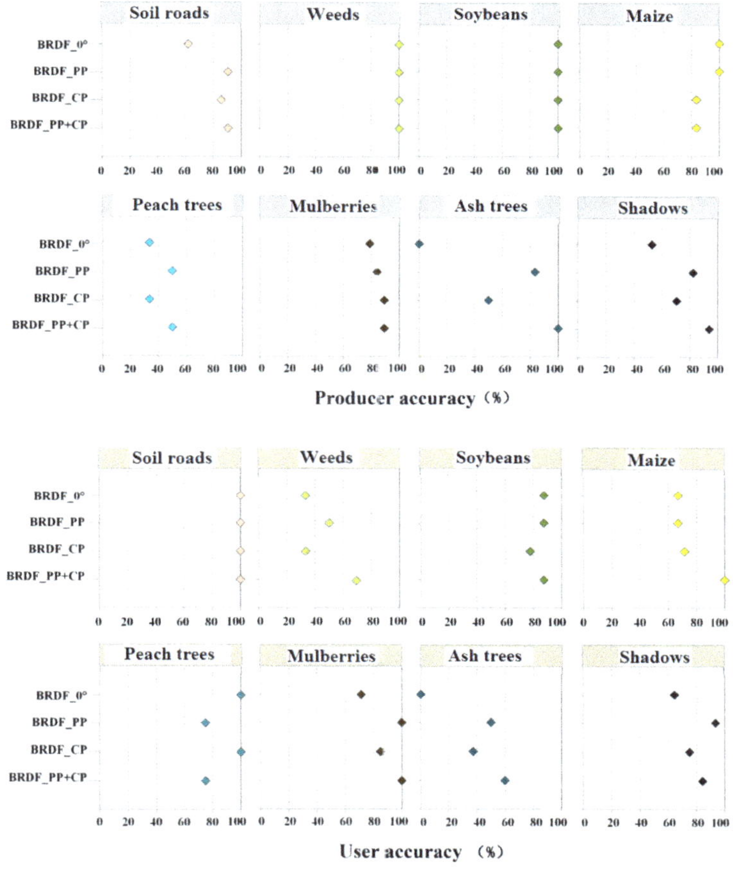

Figure 11. Producer's accuracy and user's accuracy of each vegetation type based on the multi-angle reflectance characteristics for the observed principal plane and cross-principal plane.

Figure 11 shows that the combined application of reflectance characteristics from the principal and cross planes could improve the classification accuracy. The joint classification results for the reflectance characteristics in the two main planes show that the producer's accuracy of other land features was greater than 90%, the producer's accuracy of peach seedlings was approximately 52%, and the peach seedlings were misclassified as soybean and ash trees.

(4) BRDF model parameters:

Three feature sets were used to classify vegetation, namely the reflectance from the vertical observation direction (BRDF_0°), BRDF model parameters (BRDF_3f), and reflectance from the vertical observation direction and BRDF model parameters (BRDF_0°+3f). The corresponding classification results are shown in Figure 12. The classification accuracy of BRDF_0°+3f was the highest (OA = 78%). The proportions of uniform reflection, bulk reflection, and geometric optical reflection were expressed as parameters. The addition of model parameters increased the descriptive information for the physical structure of vegetation, which contributed to the classification. The producer's and user's accuracies of each type of land feature are shown in Figure 13.

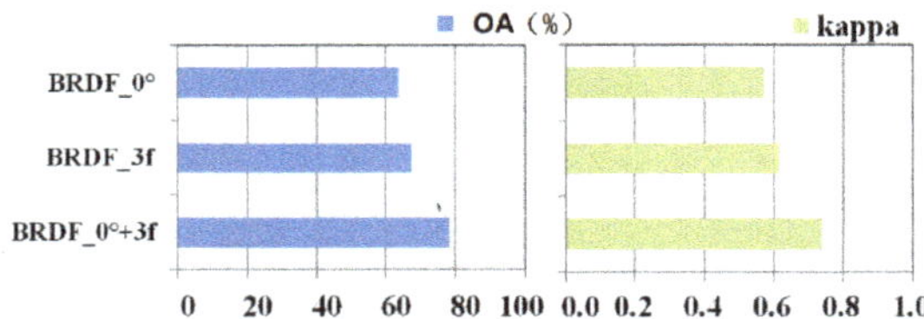

Figure 12. Overall accuracy and kappa coefficient of vegetation classification based on the BRDF model parameters.

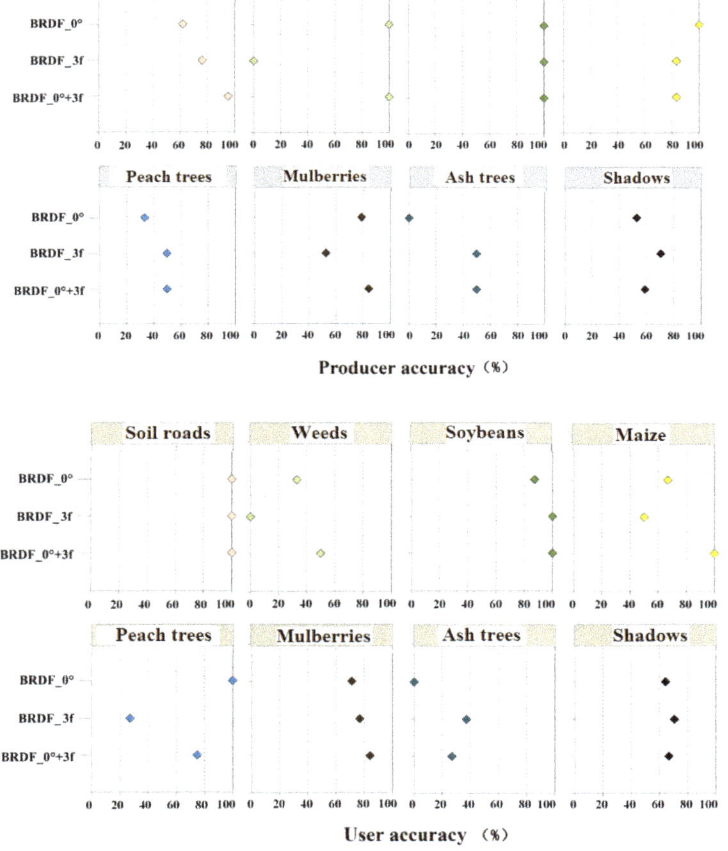

Figure 13. Producer's accuracy and user's accuracy of each vegetation type based on the BRDF model parameters.

5.2. Importance Evaluation of the Observation Angle and Band Selection

Figure 14 shows the variation in the classification accuracy (overall accuracy and kappa coefficient) based on a single observation angle feature in the main observation plane and main vertical plane. The angle feature in the main plane was observed. The angle feature located in the backward reflection direction (zenith angle between −10° and −20°) was associated with the optimal overall accuracy and kappa coefficient. In the main vertical observation plane, the classification accuracy exhibited a symmetrical phenomenon with the angle distribution, and the variation in amplitude was lower than that in the main observation plane.

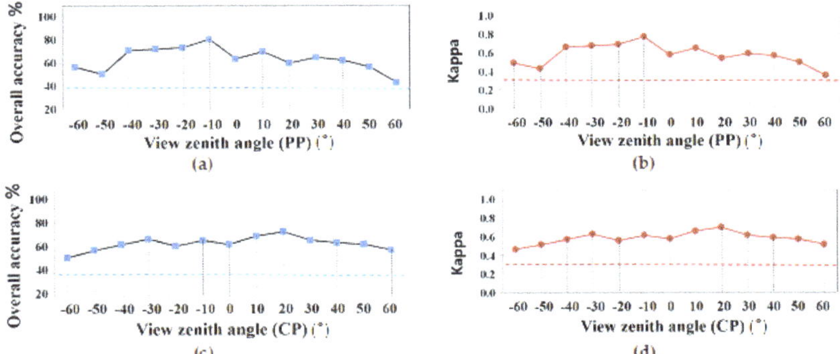

Figure 14. Observation angle importance analysis for the main plane. The blue/red dotted lines represent the classification results using DOM data.

Hyperspectral remote sensing has the advantage of providing hundreds of spectral channels of data to obtain the spectral curves to reflect the attribute differences of the object. It also provides convenience for the study of the band sensitivity of different vegetation types. Figure 15 shows the top 10 bands in terms of feature importance when only the multiband dataset for the observation zenith angle was used for classification in the main observation plane, where the diameter of the circle represents the importance degree of the band. The importance of features were calculated using SPSS Clementine software, and the indicators included the sensitivity and information gain contribution. The results show that in the main plane of observation, the blue band (466–492 nm), green band (494–570 nm), red band (642–690 nm), red edge band (694–774 nm), and near-infrared band (810–882 nm) were of high importance, among which the blue light band, red light band, and red edge band were the most important.

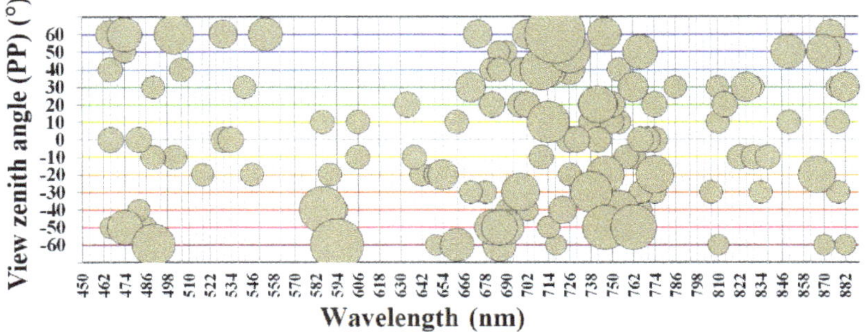

Figure 15. The importance of band selection at each angle in the principal plane. The diameter of the circle represents the importance of the band.

6. Conclusions

In this paper, the application of UAV multi-angle remote sensing in the fine classification of vegetation was studied by combining a constructed multi-angle remote sensing BRDF model with an object-oriented classification method. High-resolution image classification extraction with a UAV was the objective, and the importance of ground object BRDF characteristic parameters was discussed in detail. In addition, considering the spectral segmentation advantage of hyperspectral data and the importance of features from the two principal planes, the observation angles and band conditions of the participating classifications were further analyzed. The main conclusions are as follows.

(1) The overall classification accuracy (63.9%) based on the BRDF vertical observation reflectance characteristics was approximately 24% higher than that of traditional UAV orthophoto-based classification. The combined application of the reflection features from the main observation plane and main vertical plane yielded the best classification results, with an overall accuracy of approximately 89.2% and a kappa of 0.870.

(2) The reflectance characteristics near the hot spots were favorable for distinguishing between corn, soybean, and weeds. The combined application of the reflectance characteristics from the main observed plane could improve the classification accuracy of trees with different leaf shapes.

(3) The viewing angle characteristics in the retroreflective direction of the principal plane were better than those in the forward reflection direction. The observation angles associated with zenith angles between −10° and −20° were the most favorable for vegetation classification (sun position: zenith angle 28.86°, azimuth 169.07°).

(4) Bands of high importance for the fine classification of vegetation included the blue band (466–nm), green band (494–570 nm), red band (642–690 nm), red edge band (694–774 nm), and near-infrared band (810–882 nm), among which the blue, red, and red edge bands were the most important.

Due to the UAV hyperspectral image with a centimeter spatial resolution, when the research target size was larger than the image resolution, the introduction of an object-oriented analysis method can make the work of target recognition more accurate and efficient. Additionally, combining the construction of a multi-angle remote sensing BRDF model with an object-oriented classification method is very conducive to the study of the BRDF characteristics of canopy level vegetation. The research results provide a methodological reference and technical support for BRDF construction based on UAV multi-angle measurements, which promotes the development of multi-angle remote sensing technology in vegetation information extraction. The study provides important theoretical significance and application value for regional to global vegetation remote sensing applications. In this paper, only two classification characteristics of the reflectance and model parameters were proposed for the BRDF model. Research on the application of index characteristics, such as the vegetation index and BRDF shape index in vegetation classification, along with an evaluation of different classifiers, will be developed in future work.

Author Contributions: Conceptualization, L.D.; methodology, L.D. and Y.Y.; software, L.D. and Y.Y.; validation, Y.Y.; formal analysis, Y.Y.; investigation, L.D. and Y.Y.; resources, L.D.; data curation, L.D.; writing—original draft preparation, Y.Y.; writing—review and editing, L.D., X.L. and L.Z.; supervision, X.L. and L.Z.; project administration, L.D.

Funding: This research and APC was funded by National Key R&D Program of China, grant number no. 2018YFC0706004.

Conflicts of Interest: The authors declare no conflict of interest.

References

1. Winter, S.; Bauer, T.; Strauss, P.; Kratschmer, S.; Paredes, D.; Popescu, D.; Landa, B.; Guzmán, G.; Gómez, J.A.; Guernion, M. Effects of vegetation management intensity on biodiversity and ecosystem services in vineyards: A meta-analysis. *J. Appl. Ecol.* **2018**, *55*, 2484–2495. [CrossRef]

2. Uuemaa, E.; Antrop, M.; Roosaare, J.; Marja, R.; Mander, Ü. Landscape Metrics and Indices: An Overview of Their Use in Landscape Research. *Living Rev. Landsc. Res.* **2009**, *3*, 1–28. [CrossRef]

3. Patrick, M.J.; Ellery, W.N. Plant community and landscape patterns of a floodplain wetland in Maputaland, Northern KwaZulu-Natal, South Africa. *Afr. J. Ecol.* **2010**, *45*, 175–183. [CrossRef]

4. Liu, G.L.; Zhang, L.C.; Zhang, Q.; Zipporah, M.; Jiang, Q.H. Spatio–Temporal Dynamics of Wetland Landscape Patterns Based on Remote Sensing in Yellow River Delta, China. *Wetlands* **2014**, *34*, 787–801. [CrossRef]

5. Clark, M.L.; Buck-Diaz, J.; Evens, J. Mapping of forest alliances with simulated multi-seasonal hyperspectral satellite imagery. *Remote Sens. Envrion.* **2018**, *210*, 490–507. [CrossRef]

6. Aslan, A.; Rahman, A.F.; Warren, M.W.; Robeson, S.M. Mapping spatial distribution and biomass of coastal wetland vegetation in Indonesian Papua by combining active and passive remotely sensed data. *Remote Sens. Envrion.* **2016**, *183*, 65–81. [CrossRef]

7. Shukla, G.; Garg, R.D.; Kumar, P.; Srivastava, H.S.; Garg, P.K. Using multi-source data and decision tree classification in mapping vegetation diversity. *Spat. Inf. Res.* **2018**, 1–13. [CrossRef]

8. Shaw, J.R.; Cooper, D.J.; Sutfin, N.A. Applying a Hydrogeomorphic Channel Classification to understand Spatial Patterns in Riparian Vegetation. *J. Veg. Sci.* **2018**, *29*, 550–559. [CrossRef]

9. Melgani, F.; Bruzzone, L. Classification of hyperspectral remote sensing images with support vector machines. *IEEE Trans. Geosci. Remote Sens.* **2004**, *42*, 1778–1790. [CrossRef]

10. Goetz, A.F.H. Three decades of hyperspectral remote sensing of the Earth: A personal view. *Remote Sens. Envrion.* **2009**, *113*, S5–S16. [CrossRef]

11. Lunga, D.; Prasad, S.; Crawford, M.M.; Ersoy, O. Manifold-Learning-Based Feature Extraction for Classification of Hyperspectral Data: A Review of Advances in Manifold Learning. *IEEE Signal Proc. Mag.* **2013**, *31*, 55–66. [CrossRef]

12. Ishida, T.; Kurihara, J.; Viray, F.A.; Namuco, S.B.; Paringit, E.C.; Perez, G.J.; Takahashi, Y.; Marciano, J.J., Jr. A novel approach for vegetation classification using UAV-based hyperspectral imaging. *Comput. Elec. Agr.* **2018**, *144*, 80–85. [CrossRef]

13. Filippi, A.M.; Jensen, J.R. Fuzzy learning vector quantization for hyperspectral coastal vegetation classification. *Remote Sens. Envrion.* **2006**, *99*, 512–530. [CrossRef]

14. Fu, Y.; Zhao, C.; Wang, J.; Jia, X.; Guijun, Y.; Song, X.; Feng, H. An Improved Combination of Spectral and Spatial Features for Vegetation Classification in Hyperspectral Images. *Remote Sens.* **2017**, *9*, 261. [CrossRef]

15. Hall, F.G.; Hilker, T.; Coops, N.C.; Lyapustin, A.; Huemmrich, K.F.; Middleton, E.; Margolis, H.; Drolet, G.; Black, T.A. Multi-angle remote sensing of forest light use efficiency by observing PRI variation with canopy shadow fraction. *Remote Sens. Envrion.* **2008**, *112*, 3201–3211. [CrossRef]

16. Gatebe, C.K.; King, M.D. Airborne spectral BRDF of various surface types (ocean, vegetation, snow, desert, wetlands, cloud decks, smoke layers) for remote sensing applications. *Remote Sens. Envrion.* **2016**, *179*, 131–148. [CrossRef]

17. Xie, D.; Qin, W.; Wang, P.; Shuai, Y.; Zhou, Y.; Zhu, Q. Influences of Leaf-Specular Reflection on Canopy BRF Characteristics: A Case Study of Real Maize Canopies With a 3-D Scene BRDF Model. *IEEE Trans. Geosci. Remote Sens.* **2016**, *55*, 619–631. [CrossRef]

18. Georgiev, G.T.; Gatebe, C.K.; Butler, J.J.; King, M.D. BRDF Analysis of Savanna Vegetation and Salt-Pan Samples. *IEEE Trans. Geosci. Remote Sens.* **2009**, *47*, 2546–2556. [CrossRef]

19. Peltoniemi, J.I.; Kaasalainen, S.; Naranen, J.; Rautiainen, M.; Stenberg, P.; Smolander, H.; Smolander, S.; Voipio, P. BRDF measurement of understory vegetation in pine forests: Dwarf shrubs, lichen, and moss. *Remote Sens. Environ.* **2005**, *94*, 343–354. [CrossRef]

20. Mitchell, J.J.; Glenn, N.F.; Anderson, M.O.; Hruska, R.C.; Halford, A.; Baun, C.; Nydegger, N. 2012 4th Workshop on Hyperspectral Image and Signal Processing: Evolution in Remote Sensing (WHISPERS). In Proceedings of the Unmanned Aerial Vehicle (UAV) Hyperspectral Remote Sensing for Dryland Vegetation Monitoring, Shanghai, China, 4–7 June 2012; pp. 1–10.

21. Bareth, G.; Aasen, H.; Bendig, J.; Gnyp, M.L.; Bolten, A.; Jung, A.; Michels, R.; Soukkamäki, J. Low-weight and UAV-based Hyperspectral Full-frame Cameras for Monitoring Crops: Spectral Comparison with Portable Spectroradiometer Measurements. *Photogramm. Fernerkun. Geoinf.* **2015**, *1*, 69–79. [CrossRef]

22. Liu, H.; Zhu, H.; Wang, P. Quantitative modelling for leaf nitrogen content of winter wheat using UAV-based hyperspectral data. *Int. J. Remote Sens.* **2017**, *38*, 2117–2134. [CrossRef]

23. Zhang, C.; Kovacs, J.M. The application of small unmanned aerial systems for precision agriculture: A review. *Precis. Agric.* **2012**, *13*, 693–712. [CrossRef]

24. Baixiang, X.U.; Liu, L. A Study of the application of multi-angle remote sensing to vegetation classification. *Technol. Innov. Appl.* **2018**, *13*, 1–10.

25. Roosjen, P.; Bartholomeus, H.; Suomalainen, J.; Clevers, J. Investigating BRDF effects based on optical multi-angular laboratory and hyperspectral UAV measurements. In Proceedings of the Fourier Transform Spectroscopy, Lake Arrowhead, CA, USA, 1–4 March 2015. [CrossRef]

26. Tao, L.; Amr, A.E. Multi-view object-based classification of wetland land covers using unmanned aircraft system images. *Remote Sens. Envrion.* **2018**, *216*, 122–138. [CrossRef]

27. Tao, L.; Amr, A.E. Deep convolutional neural network training enrichment using multi-view object-based analysis of Unmanned Aerial systems imagery for wetlands classification. *ISPRS J. Photogramm. Remote Sens.* **2018**, *139*, 154–170. [CrossRef]

28. Cubert S185 Frame-Frame Imaging Spectrometer was Producted by Cubert GmbH, Science Park II, Lise-Meitner Straße 8/1, D-89081 Ulm. Available online: http://cubert-gmbh.com/ (accessed on 6 May 2019).

29. Mezaal, M.; Pradhan, B.; Rizeei, H. Improving landslide detection from airborne laser scanning data using optimized dempster–shafer. *Remote Sens.* **2018**, *10*, 1029. [CrossRef]

30. Bryson, M.; Sukkarieh, S. Building a robust implementation of bearing-only inertial SLAM for a UAV. *J. Field Robot.* **2007**, *24*, 113–143. [CrossRef]

31. Wanner, W.; Li, X.; Strahler, A.H. On the derivation of kernels for kernel-driven models of bidirectional reflectance. *J. Geophys. Res.* **1995**, *100*, 21077–21090. [CrossRef]

32. Im, J.; Jensen, R.J. A change detection model based on neighborhood correlation image analysis and decision tree classification. *Remote Sens. Envrion.* **2005**, *99*, 326–340. [CrossRef]

33. Pal, M. Random forest classifier for remote sensing classification. *Int. J. Remote Sens.* **2005**, *26*, 217–222. [CrossRef]

Letter

Modeling Hyperspectral Response of Water-Stress Induced Lettuce Plants Using Artificial Neural Networks

Lucas Prado Osco [1,*], Ana Paula Marques Ramos [2], Érika Akemi Saito Moriya [3],
Lorrayne Guimarães Bavaresco [4], Bruna Coelho de Lima [4], Nayara Estrabis [1],
Danilo Roberto Pereira [2], José Eduardo Creste [2], José Marcato Júnior [1], Wesley Nunes Gonçalves [1],
Nilton Nobuhiro Imai [3], Jonathan Li [5], Veraldo Liesenberg [6] and Fábio Fernando de Araújo [4]

[1] Faculty of Engineering, Architecture, and Urbanism and Geography, Federal University of Mato Grosso do Sul, Av. Costa e Silva, Campo Grande 79070-900, Brazil; nayara.estrabis@ufms.br (N.E.); jose.marcato@ufms.br (J.M.J.); wesley.goncalves@ufms.br (W.N.G.)

[2] Environmental and Regional Development, University of Western São Paulo, R. José Bongiovani, 700-Cidade Universitária, Presidente Prudente 19050-920, Brazil; anaramos@unoeste.br (A.P.M.R.); danilopereira@unoeste.br (D.R.P.); jcreste@unoeste.br (J.E.C.)

[3] Department of Cartographic Science, São Paulo State University, Presidente Prudente 19060-900, Brazil; eakemisaito@gmail.com (É.A.S.M.); nilton.imai@unesp.br (N.N.I.)

[4] Agronomy Development, University of Western São Paulo, R. José Bongiovani, 700-Cidade Universitária, Presidente Prudente 19050-920, Brazil; lgbavaresco@hotmail.com (L.G.B.); brunacoelhoo@outlook.com (B.C.d.L.); fabio@unoeste.br (F.F.d.A.)

[5] Department of Geography and Environmental Management and Department of Systems Design Engineering, University of Waterloo, Waterloo, ON N2L 3G1, Canada; junli@uwaterloo.ca

[6] Forest Engineering Department, State University of Santa Catarina, R. Eng. Agronômico Andrei Cristian Ferreira, Trindade, Florianópolis-SC 88040-900, Brazil; veraldo.liesenberg@udesc.br

* Correspondence: pradoosco@gmail.com; Tel.: +55-(18)-9-9798-4866

Received: 2 November 2019; Accepted: 23 November 2019; Published: 26 November 2019

Abstract: Modeling the hyperspectral response of vegetables is important for estimating water stress through a noninvasive approach. This article evaluates the hyperspectral response of water-stress induced lettuce (*Lactuca sativa* L.) using artificial neural networks (ANN). We evenly split 36 lettuce pots into three groups: control, stress, and bacteria. Hyperspectral response was measured four times, during 14 days of stress induction, with an ASD Fieldspec HandHeld spectroradiometer (325–1075 nm). Both reflectance and absorbance measurements were calculated. Different biophysical parameters were also evaluated. The performance of the ANN approach was compared against other machine learning algorithms. Our results show that the ANN approach could separate the water-stressed lettuce from the non-stressed group with up to 80% accuracy at the beginning of the experiment. Additionally, this accuracy improved at the end of the experiment, reaching an accuracy of up to 93%. Absorbance data offered better accuracy than reflectance data to model it. This study demonstrated that it is possible to detect early stages of water stress in lettuce plants with high accuracy based on an ANN approach applied to hyperspectral data. The methodology has the potential to be applied to other species and cultivars in agricultural fields.

Keywords: spectroscopy; artificial intelligence; proximal sensing data; precision agriculture

---

1. Introduction

Remote sensing is an important tool for the analysis of vegetation in agricultural fields because it allows farmers to obtain data in a faster manner than most traditional methods [1,2]. Changes in the spectral response of plants can be observed with equipment that records wavelength values, such as a spectroradiometer [3]. The analysis of the spectral signatures enables the identification of vegetation characteristics that would not be visually perceived in asymptomatic plants [4]. Because of this, many studies focusing on phytosanitary problems are based on spectroscopy. These problems include nutritional deficiencies [5–7]; diseases [8], biomass [9], and, as in the present study, water stress [10].

The spectral response of a plant differs according to the species, which motivates the creation of different approaches to model it [11]. One of the problems commonly faced by farmers is related to water stress, which limits growth and compromises its production [12]. Water stress is responsible for chlorophyll variation and impairing other biological components, such as leaf area and root size [13]. The alteration of these components results in the appearance of visual symptoms, but they are difficult to identify due to their similarity to other problems, such as diseases, malnutrition, and cold damage [14]. An alternative that can identify changes caused by water stress alone is hyperspectral analysis [10,11].

The amount of leaf water is best estimated in the near-infrared and medium-infrared spectral regions [15]. In the near-infrared region, the spectral response is associated with the structural organization of intracellular molecules located in the mesophyll, which is affected as a consequence of stress [14]. Stress may also unbalance other physiological conditions and cause changes in visible and red-edge regions [15]. Changes in these spectral regions are associated with foliar pigmentation. Studies have sought to evaluate spectral behavior in these and other regions of the spectrum [16,17]. In addition, the absorbance curve has shown a better relationship with leaf pigmentation [18], which encourages the use of absorbance data to evaluate the negative effects caused by stress in plants.

Different approaches have been adopted to model the hyperspectral response when detecting water stress in cultures. In rice, multivariate analysis models were applied to determine the spectral response of the plant under different stress levels [10]. In tomatoes, classification trees were used to separate the spectral indices that best corresponded to the induced water stress [12]. In winter wheat crops, through continuous analysis of hyperspectral data over time, it was possible to quantify water stress in relation to other variables, such as disease and nitrogen accumulation [19]. Other studies have evaluated the implications of water stress through hyperspectral data in different plants, such as vineyards [11] and citrus fruits [20].

Recently, machine learning approaches have been used in modeling the hyperspectral response of different conditions associated with vegetation [21]. The popular techniques used for analyzing data include regression analysis, vegetation indices, linear polarizations, wavelet-based filtering, and, currently, machine learning algorithms like random forest, decision tree, support vector machine (SVM), k-nearest neighbor (kNN), artificial neural networks (ANN), naïve Bayes (NB), and others [22–25]. To evaluate the hyperspectral response of plants, machine learning has already been implemented in different scenarios. A radial basis function and the kNN were used to detect citrus canker in several disease development stages [26]. ANN, NB, and kNN were also used to model pepper fusarium disease in a climate room [27]. A combination of different machine learning algorithms like SVM, ANN, and others were also evaluated to model photosynthetic variables [28].

In lettuce, water stress poses a major threat. To deal with this, commercially available seeds are being inoculated with rhizobacteria, because it mitigates the effects of the stress [29]. These effects; however, may not be visually perceptible, which makes detection by ordinary approaches difficult. Hyperspectral data have already demonstrated high potential in assessing water stress in plants in different spectral regions (350–2500 nm) [11,23,30]. However, to date, no model has evaluated the spectral response of lettuce submitted to water stress. Here we evaluate the hyperspectral response of water-stress induced lettuce with a machine learning method through ANN. The contribution of this study is twofold. Firstly, we identified the effects of water stress in lettuce and its association with their spectral response. Secondly, we evaluated the performance of the ANN algorithm to model its effects. The rest of this article is organized as follows. Section 2 presents the materials and methods adopted in this study. Sections 3 and 4 present and discuss the results obtained in the experimental analysis. Finally, Section 5 concludes the article.

2. Materials and Methods

The experiment was conducted in a growth chamber under controlled conditions in a phytotron. Twenty-eight-day-old lettuce (*Lactuca sativa* L.) plants were transplanted to pots with 0.5 kg of agricultural soil (pH (CaCl$_2$ 0.01 mol L^{-1}) 5.9; 43.9 mg dm-3 of P (Mehlich-1), 2.7 mmol dm-3 of K, 25.3 mmol dm-3 of Ca, 5.3 mmol dm-3 of Mg, 14.3 mmol dm-3 of H + Al). The experimental design was randomized with three treatments (n = 36). One treatment was carried out with B. subtilis inoculation, strain AP-3 [31]. The inoculation was performed with bacteria obtained by scraping cells multiplied in a solid medium, diluted in sterile water to a concentration of 1.0.109 cels. mL^{-1}, and 0.1 mL per inoculum. The plants were cultivated for 14 days under the same irrigation conditions, maintaining the soil at field capacity. From the 14th day on, water restriction was applied.

The treatments were conducted as follows: (i) control group, with maintenance of the field capacity and without inoculation of the plants; (ii) stress group, with maintenance of 50% of the field capacity and without inoculation of the plants, and; (iii) bacteria group, with maintenance of 50% of field capacity and inoculation of the plants. The water replacement for field capacity maintenance of only 50% was conducted by the gravimetric method. The phytotron chamber maintained the same temperature (25 °C) and lighting conditions during the experiment. Water-deficit treatment was performed for 15 days and was completed on the 30th day after the transplantation of the plants to vessels. The experimental design and analysis are summarized below (Figure 1).

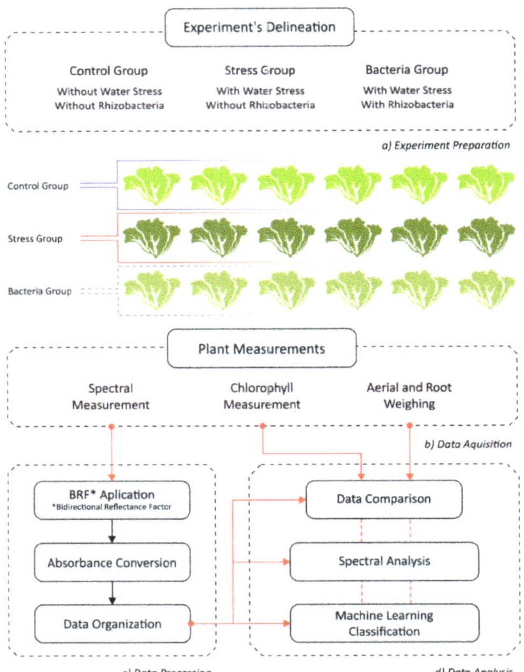

Figure 1. The workflow of the experimental analysis conducted in this study.

2.1. Spectral Data Measurements

To record the spectral response of each plant, considering the different treatments, a darkroom was prepared to avoid light interference from other materials. The spectral response of the lettuce was measured using a Fieldspec HandHeld ASD spectroradiometer, operating at a spectral range of 325–1075 nm, in 512 channels with a spectral resolution of 1.6 nm and a 1° field of view. The equipment was carefully placed close to the leaves, at a 45° inclination, in relation to the height of the plant so that its field of view (FOV) did not exceed the area of the plant and register the spectral response of the substrate. A halogenic lamp was also placed at 45° on the other side. Before each measurement, the equipment was calibrated with a Lambertian (Spectralon® plate) surface plate.

The spectroradiometer registered 10 spectral curves during the same measurement for different leaves. This resulted in 360 spectral signatures for each measured day. These signatures represent the radiance from the leaves along the electromagnetic wavelength. Because the spectroradiometer records the radiance that reaches the equipment, we needed to transform it into the reflectance factor. For that, the leaf radiance was divided by a reference radiance, which corresponds with the Lambertian plate measured previously. The spectroradiometer also has a known calibration factor (K), which must be multiplied by the values of the described operation. This factor, together with the radiance values of the reference plate of the Lambertian surface, was used to estimate the bidirectional reflectance factor (BRF), as shown in Equation (1) [32].

$$BRF(\omega_i \omega_r) = \frac{dL\,(\theta_r, \Phi_r)\,(target)}{dL\,(\theta_r, \Phi_r)\,(reference)}\,K\,(\theta_i, \Phi_i, \theta_r, \Phi_r) \tag{1}$$

where dL is the spectral radiance, ω is the solid angle, θ and Φ are in order, the zenith and azimuth angles, respectively; i is the incident flux, and r is the reflected energy flux. As mentioned, the K value is the correction factor from the equipment manufacturer. The BRF represents the spectral signature of the recorded radiometric target, also called the spectral response of the selected target.

To remove regions with a low signal-to-noise ratio, the spectral range from 380 to 1020 nm was selected to compose the spectral data, removing everything outside this range. The spectral curves were evaluated in terms of reflectance and absorbance values. Following Beer–Lambert's law, which shows that a concentration of an absorbent is proportional to the absorbance, the spectral reflectance values were converted using Equation (2).

$$A = log\left(\frac{1}{R}\right) \tag{2}$$

where A corresponds to absorbance and R corresponds to the spectral reflectance obtained with the Fieldspec HandHeld ASD spectroradiometer.

2.2. Biophysical Data Measurements

The leaf chlorophyll content ($\alpha + \beta$) was recorded using a portable chlorophyllometer (Clorofilog Falker). The measurements were taken in the leaves of the apical part, median part, and basal part of each lettuce plant. This device operates in three spectral regions, in which the first two are in the red and red-edge regions and the third one in the near-infrared region [33]. The diameter of the leaves of each plant was also measured using a millimetric tape. The plants were then detached and weighed using a digital balance. At this stage, the aerial part (leaf and stem) was removed from the root and weighed separately, obtaining the fresh mass (g). The material was then left to dry in the open air for 48 h and weighed again to find its dry mass (g).

2.3. Statistical Data Analysis

The Shapiro–Wilk test was used to verify the normality of the data related to the biophysical parameters. The ANOVA method was applied to determine the difference among the three treatments (control, stress, and bacteria groups), while the mean difference was verified using the Tukey test. A 95% confidence interval was then adopted for all statistical analyses. For spectral response curves, the correlation of single wavelengths with the selected biophysical parameters was calculated for both reflectance and absorbance values. Contour maps resulting from traditional 2D correlation analysis were also applied between the pairs of the treatments to determine the spectral intervals that presented similarity. Correlation graphs between the biophysical parameters and all the spectral wavelengths of the three groups were then plotted. The results were compared using their means, standard deviations, correlation coefficient, coefficient of regression (R^2), mean discrepancy (by bootstrapping), and root mean squared error (RMSE). The metrics evaluated here were obtained with the open-source software PAST v. 3.2 and the statistical program R v. 3.6.

2.4. ANN and Machine Learning Analysis

In a computational environment, we randomly separated 80% of the data to train the ANN algorithm and 20% to test it. The number of spectral wavelengths (n = 360) was the same for each day. The spectral wavelengths were added as input for the ANN, and one hidden layer with n neurons was considered. A linear activation function was applied in the output layer. We adopted the Adam Optimizer with regularization of a = 0.0001. We used an open-source version of the RapidMiner v. 9.4 software.

To define the best hyperparameters, we performed a cross-validation method by separating our dataset into 10 folds. This separation was stratified and we used only the training dataset (80%). In this approach, one-fold is used to validate the algorithm performance while the remaining folds are used to train the model. The test is repeated until all 10 folds are used individually as validation data. An example of the training curve being adjusted to the 1st measurement day absorbance data is plotted below (Figure 2).

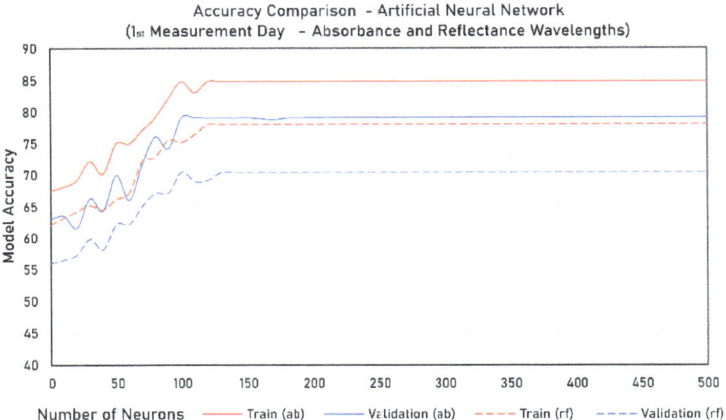

Figure 2. Example of the training curve with the difference in accuracy for the artificial neural networks (ANN) model.

We applied a hyperparametrization evaluation and detected that 100 neurons in hidden layers and a maximum number of interactions of 200 presented the ideal configuration without overfitting our model for most of the tests. Finally, we plotted an ROC (receiver operating characteristic) curve to evaluate the comparison between each classification and a confusion matrix of the ANN results. We evaluated the gain ratio and the F-score for each individual wavelength.

To test the robustness of the ANN, we compared it with other traditional machine learning algorithms, such as decision-tree; support vector machine (SVM); random forest (RF); naïve Bayes;

and logistic regression. The number of training and testing remained the same. We also performed a hyperparametrization with these algorithms. The criteria for stopping was defined once it did not return in any practical gains for the classification accuracy (%). For this, we considered the individual characteristic of each classifier, like the number of trees, nodes and leaves, number of interactions, function degree, and others.

The decision tree and random forest models provide classification trees that rely on the idea of an overall accuracy improvement by adding the predictions of combined independent predictors [34]. SVM uses a regression approach to find separation lines and can be applied in many cases where there is a distinct margin of separation. Naïve Bayes is a probabilistic classifier that applies the Bayes' theorem with independence assumptions. Lastly, logistic regression is also a regression approach that bases itself in a sigmoid function to model the predicted classes [35].

The metrics used for evaluating the performance of each algorithm was the AUC (area under the curve), overall accuracy, F1-score, precision, and recall. We compared each one of them during the four stages of the spectral response measurement: 14th, 19th, 24th, and 29th days. Both the reflectance and the absorbance values were used separately as input features. The results are presented in the following section.

3. Results

3.1. Hyperspectral and Biophysical Parameters Comparison

The water stress caused a reduction in practically all biophysical parameters. The exception was for the root development node treatment inoculated with the rhizobacteria. This indicates that the influence of the bacteria was more evident in the root system. The bacteria group had mean values similar or even higher to those found for the control group, both in the fresh mass and in the dry root (Table 1). Chlorophyll content also expressed high differences between the control group and the others.

Table 1. Results from the mean test between the biophysical parameters mean.

Treatment	Chlorophyll Content	Diameter (cm)	Weight (g)			
			Fresh (Aerial)	Fresh (Root)	Dry (Aerial)	Dry (Root)
Control Group	12.55 [b]	18.33 [a]	11.04 [a]	3.48 [bc]	1.86 [a]	0.48 [b]
Stress Group	20.02 [a]	17.17 [a]	6.87 [bc]	2.97 [c]	1.49 [ab]	0.40 [c]
Bacteria Group	19.60 [a]	16.58 [ab]	7.27 [b]	4.82 [a]	1.43 [b]	0.52 [ab]
C. V.	29.54	16.15	28.14	26.48	22.18	20.07
F value	4.677	7.942	35.56	6.952	3.293	6.236
p-value	0.0124	0.0012	<0.0001	0.0021	0.0417	0.0036

Mean values followed by equal letters in lines do not present differences at 5% in the Turkey test. C. V. = coefficient of variation.

At the end of the experiment, the lowest values of reflectance and the highest values of absorbance were observed in the control group (Figure 3). The spectral behavior of each treatment increased in difference as the stress progress continued. Stress and bacteria groups were both submitted to water stress, and their spectral response curves were distanced from the control curve. This condition can be explained by the reduction in fresh leaf mass and leaf diameter, which caused an increase in chlorophyll concentration in both groups (stress and bacteria). Furthermore, a continuous analysis over time showed that reflectance and absorbance values both increased and decreased, respectively. The stress group; however, was the one that presented a higher discrepancy over time.

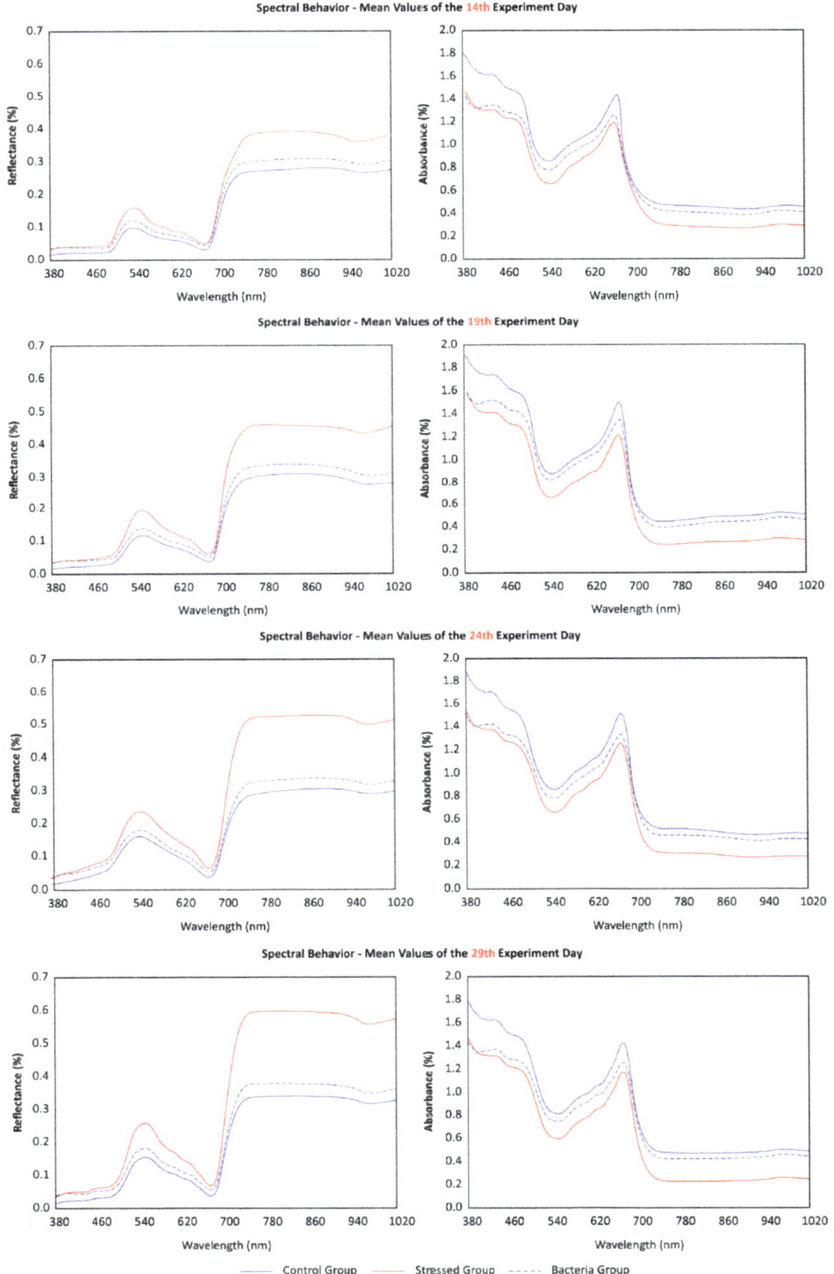

Figure 3. Comparison between the mean spectral reflectance and absorptance at selected wavelengths for each of the three treatment groups.

To evaluate the correlation between each spectral behavior in the final day of the experiment, a matrix (Figure 4) was organized with the correlation values between all groups. In general, the control group presented better correlations with the bacteria group than the stress group. The correlation was higher in the near-infrared region for all curves. The green region (from 520 to 580 nm) was more evident on the reflectance curve in the bacteria group and was better isolated when evaluated on the absorbance curve (Figure 4C,D). This could prove difficult in differentiating both groups' spectral behavior. Still, both groups returned different amplitudes in the averaged values (Figure 3), which indicates a feasible separation between them.

The relationship between the biophysical parameters and all the spectral wavelengths of the three groups is shown in Figure 5. The parameters that presented best correlations with the spectral curves were chlorophyll, fresh masses and aerial dry matter, and dry weight of roots. With the exception of the dry root weight, the aforementioned parameters presented higher negative and positive correlations in the visible region (Figure 5A,C,D), specifically in the blue region (from 380 to 460 nm) and in a smaller range in the red region (between 640 and 680 nm), which coincides with the absorption regions of chlorophyll. The weight of the root dry mass was the parameter that presented the highest correlation (r = 0.80) with the near-infrared region (700 nm onwards).

The mean values of the correlation of each parameter with the reflectance and absorbance curves were also compared (Table 2): the chlorophyll, diameter, fresh aerial weight, and dry root weight had better correlations (r = −0.803; 0.540; 0.822; and 0.709, respectively) with absorbance than with reflectance. The difference between averages ranged from 0.797 (fresh root) to 1.602 (fresh aerial), which demonstrates how much the reflectance values differ from those of absorbance. This behavior is better observed in the absorbance wavelengths of chlorophyll, fresh weight of the aerial part and in the dry-root weight, which presented values higher than 0.7, and the lowest values in root mean squared error (RMSE).

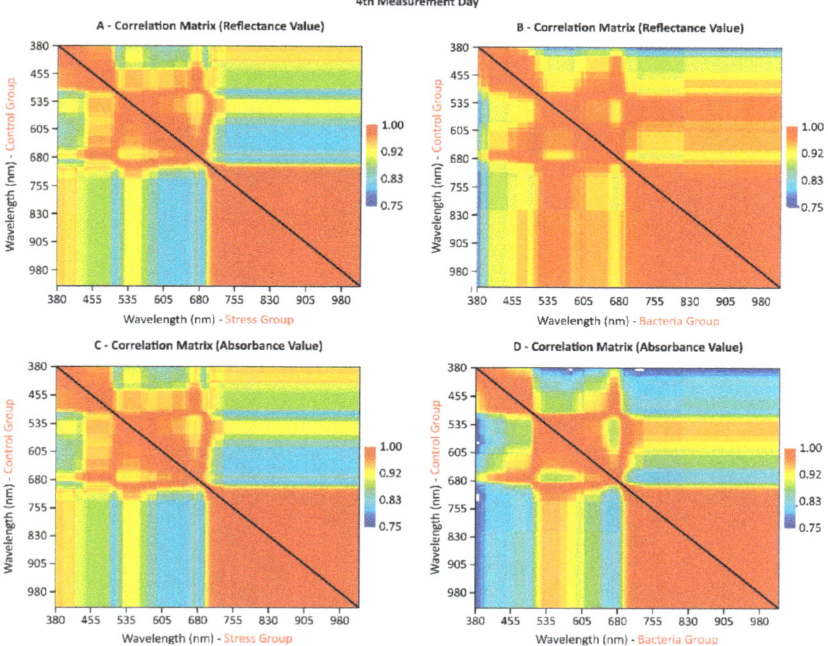

Figure 4. The correlation matrix between the mean spectral wavelengths.

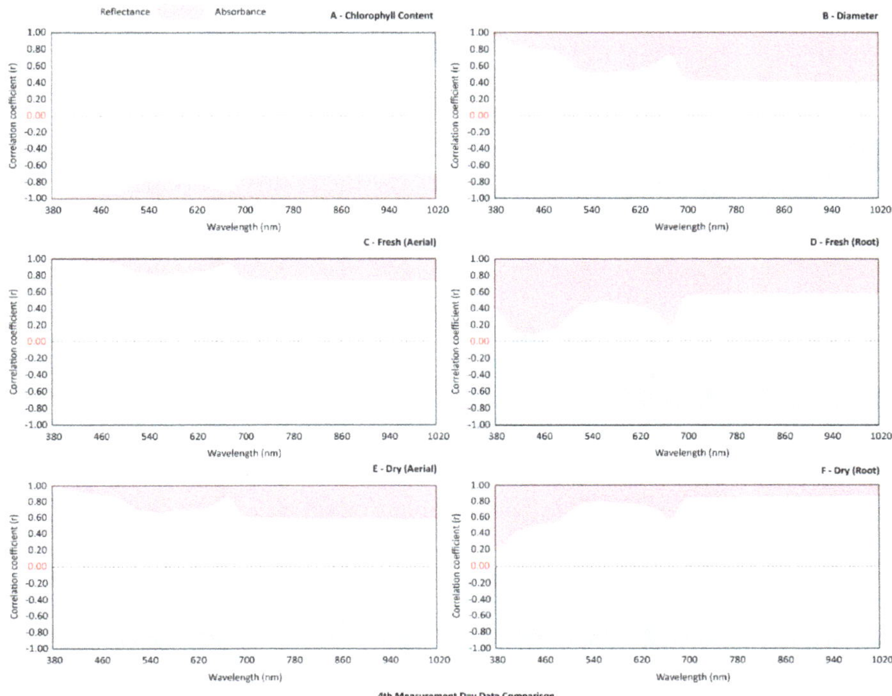

Figure 5. The correlation coefficient (r) between the spectral wavelengths (380–1020 nm) and the biophysical parameter; reflectance (blue curve) and absorbance (pink curve).

Table 2. Differences from mean values to correlations from absorbance and reflectance.

Biophysical Parameter	Mean Correlation ± Std. Dev.		Mean Difference	R^2	RMSE	R^2	RMSE
	Reflectance (r)	Absorbance (r)		Reflectance		Absorbance	
Chlorophyll Content	0.759 ± 0.131	−0.803 ± 0.104	1.562 (1.549–1.574)	0.43	0.125	0.70	0.102
Diameter	−0.486 ± 0.215	0.540 ± 0.175	1.026 (1.006–1.047)	0.11	0.128	0.38	0.180
Fresh Weight (Aerial)	−0.779 ± 0.122	0.822 ± 0.097	1.602 (1.591–1.613)	0.47	0.099	0.72	0.120
Fresh Weight (Root)	−0.421 ± 0.278	0.377 ± 0.228	0.797 (0,771–0.824)	0.36	0.108	0.17	0.209
Dry Weight (Aerial)	−0.646 ± 0.172	0.695 ± 0.139	1.341 (1.325–1.357)	0.27	0.116	0.56	0.152
Dry Weight (Root)	−0.734 ± 0.226	0.709 ± 0.186	1.443 (1.422–1.465)	0.76	0.065	0.47	0.166

3.2. Modeling the Hyperspectral Wavelengths Through Artificial Neural Network

The wavelengths were modeled by different machine learning algorithms from the 14th day of the experiment. The ANN model presented here was able to classify better than any of the remaining algorithms since the first day of measurement (Table 3). In general, the absorbance values offered better accuracy than the reflectance ones. The accuracy and other metrics were improved with each measurement, indicating an increased difference over time.

Table 3. Accuracy metrics for each of the machine learning algorithms evaluated in this study.

Model	AUC		Class. Acc. (%)		F1-Score (%)		Precision (%)		Recall (%)	
Day: 14th	**Refl.**	**Abs.**	**Refl.**	**Abs.**	**Refl.**	**Abs.**	**Refl.**	**Abs.**	**Refl.**	**Abs.**
Decision Tree	0.620	0.741	58.5	63.5	58.6	63.6	57.8	63.9	57.6	63.5
SVM	0.671	0.767	59.8	63.8	58.1	63.2	55.3	65.4	58.8	63.8
Random Forest	0.712	0.822	56.4	66.5	56.0	66.5	56.7	66.5	59.9	66.5
ANN	**0.794**	**0.924**	**70.3**	**79.6**	**68.2**	**79.7**	**71.2**	**80.1**	**70.4**	**79.6**
Naïve Bayes	0.589	0.707	44.2	54.8	46.4	54.2	49.8	55.4	50.1	54.8
Logistic Regression	0.701	0.833	66.2	71.4	60.3	71.3	65.4	71.4	61.9	71.4
Day: 19th	**Refl.**	**Abs.**	**Refl.**	**Abs.**	**Refl.**	**Abs.**	**Refl.**	**Abs.**	**Refl.**	**Abs.**
Decision Tree	0.702	0.772	60.5	69.2	59.5	69.2	64.6	69.3	60.2	69.2
SVM	0.755	0.803	58.8	60.6	59.2	58.0	58.4	60.6	58.8	60.6
Random Forest	0.811	0.891	62.3	74.5	56.5	74.5	64.0	74.5	69.2	74.5
ANN	**0.904**	**0.941**	**75.6**	**82.0**	**70.8**	**81.9**	**74.1**	**81.8**	**74.0**	**82.0**
Naïve Bayes	0.677	0.694	50.0	50.6	48.2	47.7	45.4	48.8	44.3	50.6
Logistic Regression	0.733	0.858	65.4	67.3	59.3	66.0	62.4	66.2	59.3	67.3
Day: 24th	**Refl.**	**Abs.**	**Refl.**	**Abs.**	**Refl.**	**Abs.**	**Refl.**	**Abs.**	**Refl.**	**Abs.**
Decision Tree	0.792	0.869	78.8	81.2	74.2	81.2	75.3	81.2	74.1	81.2
SVM	0.813	0.949	72.6	83.7	76.6	83.9	75.2	84.6	71.6	83.7
Random Forest	0.899	0.952	70.5	84.4	73.5	84.4	78.6	84.5	78.5	84.4
ANN	**0.922**	**0.985**	**82.5**	**90.5**	**85.4**	**90.4**	**84.7**	**90.5**	**83.6**	**90.5**
Naïve Bayes	0.684	0.869	60.2	70.4	67.9	70.7	65.8	71.4	61.4	70.4
Logistic Regression	0.868	0.971	77.9	89.4	76.1	89.3	67.8	89.4	72.1	89.4
Day: 29th	**Refl.**	**Abs.**	**Refl.**	**Abs.**	**Refl.**	**Abs.**	**Refl.**	**Abs.**	**Refl.**	**Abs.**
Decision Tree	0.818	0.845	81.2	78.0	71.3	78.0	70.2	78.0	64.9	78.0
SVM	0.879	0.902	83.7	76.6	70.2	76.8	71.7	77.5	62.3	76.6
Random Forest	0.912	0.942	84.4	83.1	74.5	83.1	75.3	83.7	70.1	83.1
ANN	**0.945**	**0.984**	**90.5**	**92.7**	**81.3**	**92.7**	**82.4**	**92.7**	**79.3**	**92.7**
Naïve Bayes	0.819	0.719	70.4	52.4	38.7	48.0	60.1	49.3	46.4	52.4
Logistic Regression	0.901	0.945	89.4	75.3	89.3	80.3	70.8	80.6	64.3	80.3

The other machine learning algorithms were also able to return similar classification accuracies. The logistic regression method presented high accuracy in the first three measurements. However, it declined over the final day. This behavior was noted for the other algorithms as well. ANN was not only able to maintain consistency over time but also presented its highest accuracy on the final day. Another observation is that, to all machine learning methods applied here, the absorbance values were more efficient in discriminating the plant groups in most of the classifications.

To visualize the differences between each group, an ROC curve of the last day of measurement was used (Figure 6). The ROCs suggest that the ANN was better to differentiate individually the three groups, while other algorithms performed worse at specific conditions. The ANN also returned a less false-positive rate than all of the other machine learning algorithms. The confusion matrix of the final measurement day also shows how the ANN had more problems in predicting the control group (89.6%) than the other groups (94.4% and 94.1%, bacteria and stress, respectively).

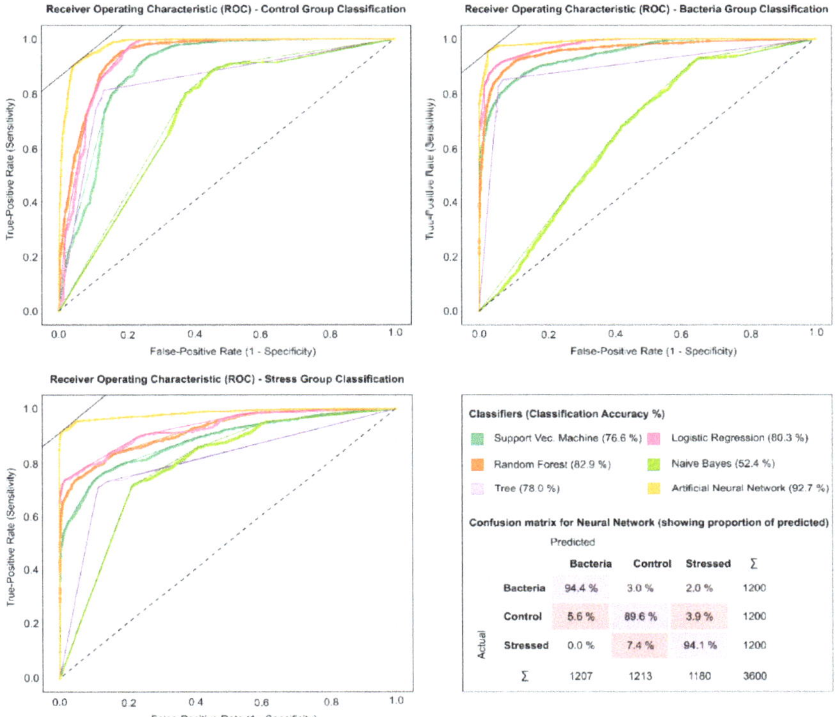

Figure 6. Receiver operating characteristic (ROC) curve comparison for each group classification and the ANN confusion matrix.

Based on this classification, the gain ratio and the relief-F were used to evaluate the contribution of individual wavelengths to the ANN model (Figure 7). These metrics suggest that the stress group presented higher differences with the control group than the bacteria group, easily distinguishable by the algorithm. However, there appears to be a higher discrepancy between the bacteria group and the control group at the blue region (380 to 440 nm). Nonetheless, the near-infrared region and the 660 to 730 nm region appears to be contributing more to the stress group response.

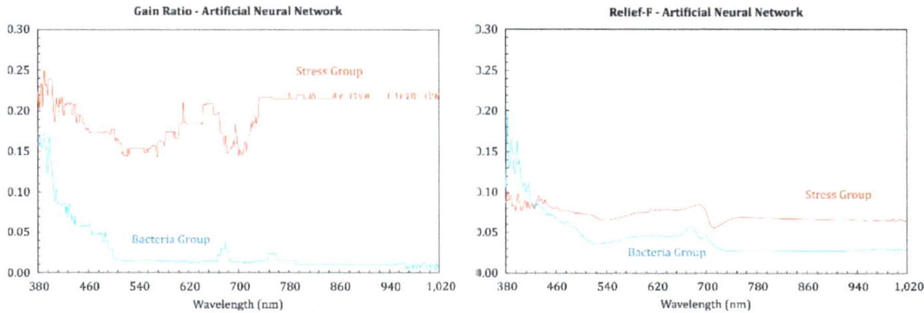

Figure 7. The individual contribution of the wavelengths to model the water-stress induced lettuces in relation to the control group.

4. Discussion

This study evaluated the spectral response of lettuce submitted to water stress while modeling its effects with an ANN and other machine learning algorithms. For that, we separated our data into three groups: control, stress, and bacteria. The reason to include the rhizobacteria in this situation is to induce a similarity with what transpires in greenhouses or horticulture models, as this bacterium is commonly present in soil and commercial seeds [29]. The addition of the bacteria group is also important to reinforce our test as it can act as a middle-ground between the stress group and the control group. We firstly evaluated the biological and physical response of the induced stress, and later compared it with the hyperspectral measurement. Lastly, we used ANN and other machine learning algorithms to classify both groups solely by their spectral response.

Our results indicate that the physiological response to water stress in early-stage lettuce is the reduction in leaf size and an increase in chlorophyll concentration. This behavior was evident both in the stress group and in the bacteria group, although with lower intensity in the latter. Changes in leaf pigmentation are noticeable in cases of plant stress [15]. However, an interesting observation is that the stress did not affect root weight in the bacteria group. This indicates that there is an effect in mitigating this stress, although this was not indicated by leaf analysis. By examining the mean spectral curves of each treatment (Figure 2) there is a small amplitude between the red-edge region wavelengths. The red-edge region is commonly known to indicate stress presence [23,34], and this may explain why the bacteria group did not differentiate much from the control group here.

Regarding the correlation between biophysical parameters and the wavelengths, it is initially perceived that absorbance wavelengths are more correlated with most biophysical parameters. An important observation to be made is the strongest correlation between chlorophyll and leaf fresh-weight with these wavelengths (Figure 4 and Table 2). This relationship between absorbance levels and the biophysical parameter was continuous throughout the experiment, particularly in regard to modeling each group's response with machine learning algorithms (Table 3). Still, the correlation was more pronounced in the green and near-infrared regions (Figure 3). This situation is also evident when observing the mean curves of each treatment (Figure 2), where the amplitude between the curves is smaller in the blue, red, and red-edge remaining regions.

The classification performed by the ANN algorithm in this study showed interesting results since the first day of measurement when lettuces had just been stressed one day before the actual measurement. This condition is important to mention as it indicates how powerful hyperspectral analysis in conjunction with machine learning algorithms can be. From the evaluation metrics used in this study, it is evident how ANN was better in distinguishing the three plant groups. By observing the phenomenon temporally, one can see how the performance of the algorithms increased (Table 3). This can be explained by the increased distinction between the wavelengths of each group. As the stress occurred, the spectral behavior of these experiments became distinct from each other. Because this study is unique in this regard, there is a lack of literature to compare with. Still, the accuracy found here is similar to or even higher than those obtained by modeling different stresses effects in plants [21,24–27].

Another contribution of this study is the evaluation of the performance of different algorithms for both reflectance and absorbance wavelengths. Absorbance curves were directly related to changes in biophysical parameters for all treatments (Figure 4 and Table 2). This persisted in the machine learning analysis, where the performance of the algorithms was superior in differentiating the three groups by using their absorbance values. Thus, it is recommended that the modeling of these effects in lettuce is preferably performed from the conversion of reflectance to absorbance data. Another observation is that, by evaluating the performance of each algorithm over time, the ANN accuracy reached its peak at the last measurement day (with 92.7%), while the other algorithms decreased in performance (from the third to the fourth day of evaluation). This indicates how feasible the ANN algorithm was in modeling the water-stress effects in comparison to the others.

Lastly, the ANN algorithm has shown high precision and recall values (Table 2) when classifying each group, as shown in Figure 5. The confusion matrix demonstrated a small decrease in performance (89.6%) for differentiating the control group from the others. Regardless, the gain ratio and relief-F metrics (Figure 6) show how each individual wavelength contributed to the ANN model. In the gain ratio analysis, there is a predominance of wavelengths in the region of blue (380 to 440 nm), red (660 to 730 nm), and near-infrared (790 nm onwards). Despite the similarity between the curves of both groups, there was a smaller amplitude difference for the blue region. This may indicate how much the blue region contributed to differentiate the effects of the stress on the bacteria group. This was similar in the assessment of relief-F curves, in which this same region presented an even higher value than the group under stress. The blue region is responsible for the absorption of chlorophyll and may be an indication of how important the stress effect was in this spectral range. Nevertheless, the model also indicated greater contributions in the red region, red-edge, and near-infrared, which corroborates with the observations made during previous results (Figure 3). Apart from other species and cultivars, future research could be conducted exploring additional spectral regions such as the shortwave infrared (SWIR) region that is unfortunately not considered in the Fieldspec HandHeld ASD spectroradiometer device.

5. Conclusions

In this study, we have applied an artificial neural network algorithm to model the hyperspectral response of water-induced stress in lettuce. The ANN algorithm detected differences since the first day of the induced stress, with an 80% classification accuracy. The algorithm continued to present an increasing performance along with time-series analysis, resulting in a final 93% accuracy. The spectral wavelengths that contributed the most for its prediction were located around 380 to 440 nm, 660 to 730 nm, and, on a lower level, 790 nm onwards. We also detected that absorbance values are more suitable to deal with this issue than reflectance. Although the rhizobacteria did mitigate the water-stress effect at some point, a spectral behavior difference was noticed by the ANN algorithm, proving its robustness. The proposed approach indicated how feasible water stress in lettuce at early stages is measurable with machine learning algorithms such as ANN in hyperspectral data. While the small number of instances (four measurement days) evaluated could provide problems for the experiment, all machine learning algorithms tested here were able to classify it appropriately. For future works, we recommend similar studies with other species and cultivars. Additionally, the method demonstrated here could be scaled up to remote sensing platforms like unmanned aerial vehicles (UAV), as currently hyperspectral sensors can be embedded in it.

Author Contributions: Conceptualization, L.P.O. and F.F.d.A.; methodology, L.P.O., É.A.S.M., L.G.B., B.C.d.L., D.R.P.; formal analysis, L.P.O.; resources, F.F.d.A., N.N I.; data curation, É.A.S.M.; writing—original draft preparation, L.P.O.; A.P.M.R. writing—review and editing, J M.J., J.E.C., J.L., V.L., W.N.G., N.E., D.R.P.; supervision, F.F.d.A., N.N.I.; project administration, F.F.d.A., N.N.I.; funding acquisition, L.P.O., F.F.d.A.

Funding: This research was partially funded by CAPES/Print (p: 88881.311850/2018-01) and FAPESP/Print (p: 2013/20328-0) V. Liesenberg is supported by FAPESC (2017TR1762) and CNPq (313887/2018-7).

Conflicts of Interest: The authors declare no conflict of interest.

References

1. Díaz-Varela, R.A.; de la Rosa, R.; León, L.; Zarco-Tejada, P.J. High-resolution airborne UAV imagery to assess olive tree crown parameters using 3D photo reconstruction: Application in breeding trials. *Remote Sens.* **2015**, *7*, 4213–4232. [CrossRef]
2. Ampatzidis, Y.; Partel, V. UAV-based high throughput phenotyping in citrus utilizing multispectral imaging and artificial intelligence. *Remote Sens.* **2019**, *11*, 410. [CrossRef]
3. Chen, J.; Li, F.; Wang, R.; Fan, Y.; Raza, M.A.; Liu, Q.; Yang, W. Estimation of nitrogen and carbon content from soybean leaf reflectance spectra using wavelet analysis under shade stress. *Comput. Electron. Agric.* **2019**, *156*, 482–489. [CrossRef]

4. Muñoz-Huerta, R.F.; Guevara-Gonzalez, R.G.; Contreras-Medina, L.M.; Torres-Pacheco, I.; Prado-Olivarez, J.; Ocampo-Velazquez, R.V. A review of methods for sensing the nitrogen status in plants: Advantages, disadvantages and recent advances. *Sensors* **2013**, *13*, 10823–10843. [CrossRef]
5. Zhai, Y.; Cui, L.; Zhou, X.; Gao, Y.; Fei, T.; Gao, W. Estimation of nitrogen, phosphorus, and potassium contents in the leaves of different plants using laboratory-based visible and near-infrared reflectance spectroscopy: Comparison of partial least-square regression and support vector machine regression methods. *Int. J. Remote Sens.* **2013**, *34*, 2502–2518. [CrossRef]
6. Moharana, S.; Dutta, S. Spatial variability of chlorophyll and nitrogen content of rice from hyperspectral imagery. *J. Photogramm. Remote Sens.* **2016**, *122*, 17–29. [CrossRef]
7. Osco, L.P.; Ramos, A.P.M.; Moriya, É.A.S.; de Souza, M.; Junior, J.M.; Matsubara, E.T.; Imai, N.N.; Creste, J.E. Improvement of leaf nitrogen content inference in Valencia-orange trees applying spectral analysis algorithms in UAV mounted-sensor images. *Int. J. Appl. Earth Obs. Geoinf.* **2019**, *83*, 101907. [CrossRef]
8. Martins, G.D.; Galo ML, B.T.; Vieira, B.S. Detecting and mapping root-knot nematode infection in coffee crop using remote sensing measurements. *IEEE J. Sel. Top. Appl. Earth Obs. Remote Sens.* **2017**, *10*, 5395–5403. [CrossRef]
9. Kross, A.; McNairn, H.; Lapen, D.; Sunohara, M.; Champagne, C. Assessment of RapidEye vegetation indices for estimation of leaf area index and biomass in corn and soybean crops. *Int. J. Appl. Earth Obs. Geoinf.* **2015**, *34*, 235–248. [CrossRef]
10. Krishna, G.; Sahoo, R.N.; Singh, P.; Bajpai, V.; Patra, H.; Kumar, S.; Sahoo, P.M. Comparison of various modeling approaches for water-deficit stress monitoring in rice crop through hyperspectral remote sensing. *Agric. Water Manag.* **2019**, *213*, 231–244. [CrossRef]
11. Loggenberg, K.; Strever, A.; Greyling, B.; Poona, N. Modeling water stress in a Shiraz vineyard using hyperspectral imaging and machine learning. *Remote Sens.* **2018**, *10*, 202. [CrossRef]
12. Elvanidi, A.; Katsoulas, N.; Ferentinos, K.; Bartzanas, T.; Kittas, C. Hyperspectral machine vision as a tool for water stress severity assessment in soilless tomato crop. *Biosyst. Eng.* **2018**, *165*, 25–35. [CrossRef]
13. Lisar, S.Y.; Motafakkerazad, R.M.M.; Rahm, I.M. Water Stress in Plants: Causes, Effects, and Responses. *Water Stress* **2012**, *10*, 1–16. [CrossRef]
14. Gerhards, M.; Schlerf, M.; Rascher, U.; Udelhoven, T.; Juszczak, R.; Alberti, G.; Inoue, Y. Analysis of airborne optical and thermal imagery for detection of water stress symptoms. *Remote Sens.* **2018**, *10*, 1139. [CrossRef]
15. Maimaitiyiming, M.; Ghulam, A.; Bozzolo, A.; Wilkins, J.L.; Kwasniewski, M.T. Early detection of plant physiological responses to different levels of water stress using reflectance spectroscopy. *Remote Sens.* **2017**, *9*, 745. [CrossRef]
16. Delloye, C.; Weiss, M.; Defourny, P. Retrieval of the canopy chlorophyll content from Sentinel-2 spectral bands to estimate nitrogen uptake in intensive winter wheat cropping systems. *Remote Sens. Environ.* **2018**, *216*, 245–261. [CrossRef]
17. Kalacska, M.; Lalonde, M.; Moore, T.R. Estimation of foliar chlorophyll and nitrogen content in an ombrotrophic bog from hyperspectral data: Scaling from leaf to image. *Remote Sens. Environ.* **2015**, *169*, 270–279. [CrossRef]
18. Min, M.; Lee, W. Determination of significant wavelengths and prediction of nitrogen content for citrus. *Am. Soc. Agric. Eng.* **2005**, *48*, 455–461. [CrossRef]
19. Huang, W.; Lu, J.; Ye, H.; Kong, W.; Mortimer, A.H.; Shi, Y. Quantitative identification of crop disease and nitrogen-water stress in winter wheat using continuous wavelet analysis. *Int. J. Agric. Biol. Eng.* **2018**, *11*, 145–152. [CrossRef]
20. Zarco-Tejada, P.; González-Dugo, V.; Berni, J. Fluorescence, temperature and narrow-band indices acquired from a UAV platform for water stress detection using a micro-hyperspectral imager and a thermal camera. *Remote Sens. Environ.* **2012**, *117*, 322–337. [CrossRef]
21. Rocha, A.D.; Groen, T.A.; Skidmore, A.K. Spatially-explicit modelling with support of hyperspectral data can improve prediction of plant traits. *Remote Sens. Environ.* **2019**, *231*, 111200. [CrossRef]
22. Ghamisi, P.; Plaza, J.; Chen, Y.; Li, J.; Plaza, A.J. Advanced spectral classifiers for hyperspectral images: A review. *IEEE Geosci. Remote Sens. Mag.* **2017**, *5*, 8–32. [CrossRef]
23. Index, S.; Xu, N.; Tian, J.; Tian, Q.; Xu, K.; Tang, S. Analysis of vegetation red edge with different illuminated/shaded canopy proportions and to construct normalized difference canopy. *Remote Sens.* **2019**, *11*, 1192. [CrossRef]

24. Gao, J.; Meng, B.; Liang, T.; Feng, Q.; Ge, J.; Yin, J.; Wu, C.; Cui, X.; Hou, M.; Liu, J.; et al. Modeling alpine grassland forage phosphorus based on hyperspectral remote sensing and a multi-factor machine learning algorithm in the east of Tibetan Plateau, China. *ISPRS J. Photogramm. Remote Sens.* **2019**, *147*, 104–117. [CrossRef]

25. Zhang, J.; Huang, Y.; Reddy, K.N.; Wang, B. Assessing crop damage from dicamba on non-dicamba-tolerant soybean by hyperspectral imaging through machine learning. *Pest Manag. Sci.* **2019**. [CrossRef]

26. Abdulridha, J.; Batuman, O.; Ampatzidis, Y. UAV-based remote sensing technique to detect citrus canker disease utilizing hyperspectral imaging and machine learning. *Remote Sens.* **2019**, *11*, 1373. [CrossRef]

27. Karadağ, K.; Tenekeci, M.E.; Taşaltın, R.A. Detection of pepper fusarium disease using machine learning algorithms based on spectral reflectance. *Sustain. Comput. Inform. Syst.* **2019**. [CrossRef]

28. Fu, P.; Meacham-Hensold, K.; Guan, K.; Bernacchi, C.J. Hyperspectral leaf reflectance as proxy for photosynthetic capacities: An ensemble approach based on multiple machine learning algorithms. *Front. Plant Sci.* **2019**. [CrossRef]

29. Enebe, M.C.; Babalola, O.O. The influence of plant growth-promoting rhizobacteria in plant tolerance to abiotic stress: A survival strategy. *Appl. Microbiol. Biotechnol.* **2018**, *102*, 7821–7835. [CrossRef]

30. Wang, J.; Xu, R.; Yang, S. Estimation of plant water content by spectral absorption features centered at 1,450 nm and 1,940 nm regions. *Environ. Monit. Assess.* **2008**, *157*, 459–469. [CrossRef]

31. Araujo, F.F.; Henning, A.A.; Hungria, M. Phytohormones and antibiotics produced by Bacillus subtilis and their effects on seed pathogenic fungi and on soybean root development. *World J. Microbiol. Biotechnol.* **2005**, *21*, 1639–1645. [CrossRef]

32. Anderson, K.; Rossini, M.; Labrador, J.P.; Balzarolo, M.; Arthur, A.; Fava, F.; Julitta, T. Vescovo. Inter-comparison of hemispherical conical reflectance factors (HCRF) measured with four fiber-based spectrometers. *Remote Sens. Sens.* **2013**, *21*, 605–617. [CrossRef]

33. FALKER. ClorofiLOGElectronic: Chlorophyll Content Meter. Available online: http://www.falker.com.br/en/product-clorofilog-chlorophyll-meter.php (accessed on 30 November 2018).

34. Liang, L.; Di, L.; Huang, T.; Wang, J.; Lin, L.; Wang, L.; Yang, M. Estimation of leaf nitrogen content in wheat using new hyperspectral indices and a random forest regression algorithm. *Remote Sens.* **2018**, *10*, 1940. [CrossRef]

35. Thomas, M. Mitchell. In *Machine Learning*, 1st ed.; McGraw-Hill, Inc.: New York, NY, USA, 1997.

 *remote sensing*

Review

Hyperspectral Classification of Plants: A Review of Waveband Selection Generalisability

Andrew Hennessy *, Kenneth Clarke and Megan Lewis

School of Biological Sciences, The University of Adelaide, Adelaide 5005, Australia;
kenneth.clarke@adelaide.edu.au (K.C.); megan.lewis@adelaide.edu.au (M.L.)
* Correspondence: andrew.hennessy@adelaide.edu.au

Received: 22 October 2019; Accepted: 24 December 2019; Published: 1 January 2020

Abstract: Hyperspectral sensing, measuring reflectance over visible to shortwave infrared wavelengths, has enabled the classification and mapping of vegetation at a range of taxonomic scales, often down to the species level. Classification with hyperspectral measurements, acquired by narrow band spectroradiometers or imaging sensors, has generally required some form of spectral feature selection to reduce the dimensionality of the data to a level suitable for the construction of a classification model. Despite the large number of hyperspectral plant classification studies, an in-depth review of feature selection methods and resultant waveband selections has not yet been performed. Here, we present a review of the last 22 years of hyperspectral vegetation classification literature that evaluates the overall waveband selection frequency, waveband selection frequency variation by taxonomic, structural, or functional group, and the influence of feature selection choice by comparing such methods as stepwise discriminant analysis (SDA), support vector machines (SVM), and random forests (RF). This review determined that all characteristics of hyperspectral plant studies influence the wavebands selected for classification. This includes the taxonomic, structural, and functional groups of the target samples, the methods, and scale at which hyperspectral measurements are recorded, as well as the feature selection method used. Furthermore, these influences do not appear to be consistent. Moreover, the considerable variability in waveband selection caused by the feature selectors effectively masks the analysis of any variability between studies related to plant groupings. Additionally, questions are raised about the suitability of SDA as a feature selection method, with it producing waveband selections at odds with the other feature selectors. Caution is recommended when choosing a feature selector for hyperspectral plant classification: We recommend multiple methods being performed. The resultant sets of selected spectral features can either be evaluated individually by multiple classification models or combined as an ensemble for evaluation by a single classifier. Additionally, we suggest caution when relying upon waveband recommendations from the literature to guide waveband selections or classifications for new plant discrimination applications, as such recommendations appear to be weakly generalizable between studies.

Keywords: hyperspectral; spectra; vegetation; plant; classification; discrimination; feature selection; waveband selection; support vector machine; random forest

---

1. Introduction

The classification of reflectance spectra to determine broad plant type or species has been explored increasingly over the past two decades. This has been driven by the increased availability of hyperspectral sensing from imaging spectrometers and field spectroradiometers, and increasing need from environmental conservation, agriculture, and forestry groups [1]. High classification accuracies, particularly at fine taxonomic units such as species, or even clones for grapevine varieties [2], has in some cases been enabled by hyperspectral observation [3]. Hyperspectral measurements have been used to classify a variety of plant types including annual gramineous weeds [4], food crops [5], arid zone shrubs [6], and montane/sub-alpine trees [7], growing in equally varied environments, including tropical wetlands [8], urban streetscapes [9], savanna plains [10], and alpine forests [11]. Due to the scale required to map and monitor the world's vegetation, fast, generalizable, and objective methods that provide results, that can be quickly and easily shared and analysed, are required. Hyperspectral imagery and data can fulfil these requirements, producing digital measurements that can be easily shared and quickly analysed with semi-automated procedures in a repeatable and objective manner. However, the potential generalisability of classification models has yet to be fully evaluated.

Hyperspectral measurements consist of numerous, finely spaced, contiguous measurements (wavebands) providing considerably more information about targets than broadband multispectral observations. These advantages come at the cost of high dimensionality and large data volumes. Hyperspectral instruments record radiance within the range of 350 to 2500 nm of the electromagnetic spectrum, with bandwidths often between 1 and 10 nm. The number of wavebands per observation varies from hundreds to thousands. Training a classification model with such large numbers of spectral features generally requires a large sample size. However, since the collection of samples for hyperspectral studies is onerous, with high costs for imagery and arduous fieldwork for gathering field measurements, sample sizes tend to be small. Data of this high dimensionality is prone to the Hughes phenomenon, also known as the curse of dimensionality, whereby an increasing number of features originally aids in improving classification, before the addition of more features decreases performance as noise and sparsity of the feature space increases [12]. This problem is exacerbated by small sample sizes [13].

In order to overcome this the ratio between sample size and data dimensionality must be improved. In this review, we focus on reducing dimensionality via feature selection, though methods of artificially increasing sample size through data augmentation, semi-supervised classification, and active learning can aid in countering the curse of dimensionality [14–16]. Hyperspectral measurements tend to include noisy or redundant features, with high levels of collinearity between wavebands. The elimination of collinearity can substantially improve classification efforts and is in fact a requirement of parametric statistical methods that assume the independence of all variables [17,18]. Additionally, feature selection inherently reveals the spectral regions that offer the greatest discriminatory power for a set of samples. Long held associations between specific spectral regions or individual wavebands and biophysical or biochemical foliar traits [19] have often guided researchers in the selection of features to differentiate species or plant types. The overall aim of this review is to assess these assumptions in light of the evidence from 22 years of hyperspectral plant studies.

Review Scope and Approach

Here, we address some important questions that motivate much hyperspectral plant research. Do the taxonomic, structural, or functional characteristics of plant types or species influence the spectral regions that are most important to classification, or are particular spectral regions consistently selected across a diversity of plant or ecological types? A review of selected features from the hyperspectral literature could identify best practices for feature selection methods, as well as detect wave-regions of high-utility, those that best generalize across taxonomic or ecological boundaries.

The search for literature spanned two decades, from January 1996 to December 2018, focusing on peer reviewed journals in the English language. Search was performed with Google Scholar using combinations of the keywords, namely *Hyperspectral, Spectra, Vegetation, Plant, Tree, Species, Identi*, Discriminat*, Classif*, Map, Feature Select*, Waveband, Band, UAV, Drone*. In order to be included, a study must have performed a feature selection technique on hyperspectral vegetation data with an aim to classify plant samples.

Many studies fulfilled the initial requirement, but did not report selected wavebands with sufficient specificity, and therefore could not be included. Here, we present waveband selections derived from 38 hyperspectral vegetation classification studies. When applicable, studies that included multiple feature selection techniques were broken into sub-studies, increasing the total number of reviewed studies to 61 (Tables 1 and 2). These included studies are from a wide variety of scales (leaf, branch, and canopy), recording methods (lab, field, aerial, satellite), taxonomic units, and bandwidths.

Additionally, a dataset was synthesised from hyperspectral measurements of 22 species of New Zealand plants collected as field spectra from four locations on the North island [20,21]. This dataset was used to examine how study design (number of classes, number of samples, included species, and feature selection method) influenced waveband selection. This was performed with the aim of determining which elements of the study design most contributed to variation seen in selected wavebands.

The remainder of this paper is structured in the following way. Section 2 provides a meta-analysis of the selected wavebands, broken down by spectral region. Section 3 identifies and describes feature selection techniques from these studies, and where possible, highlights their effects on waveband selection. Section 4 examines study design influence on waveband selection, while Sections 5 and 6 present a discussion of the results and conclusions.

Table 1. Overview of Visible/Near Infra-red (VIS/NIR) studies included in this review.

References	Wavelengths/Bandwidths	Classes	Pre-Processing	Feature Selection Method	No. Bands Selected	Accuracy %	Study Context and Spatial Scale or Resolution
[22]	350–1025 nm, 3 nm	12	Band depth	Segmented PCA	12	77.0	Successional plant communities from canopy field spectra
[23]	454–950 nm, 4 nm	8	Smoothing	SDA	14	91.4	Mangrove forest field canopy spectra
[23]	454–950 nm, 4 nm	8	Smoothing	CFS	23	92.3	Mangrove forest field canopy spectra
[23]	454–950 nm, 4 nm	8	Smoothing	SPA	23	93.1	Mangrove forest field canopy spectra
[10]	384.8–1054.3 nm, 9.23 nm	10		SAM Band Selector Addon	31	53.0	Savanna tree species from airborne imagery (1.12 m)
[11]	403–989 nm, 4.6 nm	8		Sequential Forward Floating Selection	43	74.1	Alpine tree species and 2 non-species classes, airborne imagery (1 m)
[24]	400–900 nm, 1 nm	13		Spec angle and dist., feature parameters	7	96.2	Varied plant species from lab leaf spectra
[25]	400–1000 nm, 10 nm	5		PCA, SDA, Manual selection	7	~91.4	Crop and weed species from field imagery (1.25 m)
[26]	400–900 nm, 2.6 nm	25	Smoothing	Hierarchical Clustering	13	89.0	Sub-tropical tree species from lab leaf spectra
[27]	475–900 nm, 1 nm	22		Forward Feature Selection	8	43.0	Herbaceous wetland species from field leaf spectra
[28]	325–1075 nm, 2 nm	6	Smoothing	SDA	6	92.0	Mangrove forest field canopy spectra
[28]	325–1075 nm, 2 nm	6	Smoothing, CR	SDA	17	93.6	Mangrove forest field canopy spectra
[6]	400–900 nm, 1.4 nm	8		PCA, Discriminant Analysis	13	57.0	Arid zone plant groups from field leaf spectra
[29]	384.8–1054.3, 9.23 nm	9		Random Forest, Gini Index	8	80.3	Savanna tree species from airborne imagery (1.3 m)
[29]	384.8–1054.3, 9.23 nm	9	Continuum removed	Random Forest, Gini Index	9	~79.0	Savanna tree species from airborne imagery (1.3 m)

Table 1. *Cont.*

References	Wavelengths/Bandwidths	Classes	Pre-Processing	Feature Selection Method	No. Bands Selected	Accuracy %	Study Context and Spatial Scale or Resolution
[30]	393–900 nm, 2.2 nm	6		PLSDA VIP score	78	88.8	Forestry species from airborne imagery (2.4 m)
[31]	400–800 nm, 3 nm	3		Two Sample T-test	5	69.1	Seagrass species field canopy spectra
[31]	400–800 nm, 3 nm	3	Normalized	Two Sample T-test	5	66.0	Seagrass species field canopy spectra
[31]	400–800 nm, 3 nm	3	Normalized 1st Derivative	Two Sample T-test	5	71.1	Seagrass species field canopy spectra
[31]	400–800 nm, 3 nm	3	Normalized 2nd Derivative	Two Sample T-test	5	73.2	Seagrass species field canopy spectra
[31]	400–800 nm, 3 nm	3	1st Derivative	Two Sample T-test	5	69.1	Seagrass species field canopy spectra
[31]	400–800 nm, 3 nm	3	2nd Derivative	Two Sample T-test	5	67.0	Seagrass species field canopy spectra
[7]	400–1000 nm, 3 nm	13	Normalisation	PCA, Correlation matrix, Band variance	53	77.0	European forest trees species from airborne imagery (1.6 m)

Abbreviations: PCA = Principal Component Analysis, SDA = Stepwise Discriminant Analysis, CFS = Correlation-based Feature Selection, SPA = Successive Progressions Algorithm, SAM = Spectral Angle Mapper, PLS-DA = Partial Least Squares – Discriminant Analysis (PLS-DA), VIP = Variable Importance Projection.

Table 2. Overview of Visible/Shortwave Infra-red (VIS/SWIR) studies included in this review.

References	Wavelengths/Bandwidths	Classes	Pre-processing	Feature Selection Method	Bands	Accuracy %	Study Context and Spatial Scale or Resolution
[32]	350–2500 nm @ 3, 10 nm	4		ANOVA, CART	8	97.4	Wetland species from field canopy spectra
[33]	350–2500 nm @ 3, 10 nm	4	Resampled	Random Forest	10	90.5	Wetland species from field canopy spectra
[9]	385–2450 nm @ 9.6 nm	29		Forward Feature Selection	7	79.2	Urban street tree species from airborne imagery (3.7 m)
[34]	427–2355 nm @ 10 nm	4		PCA	15	86.3	Agricultural crops, Hyperion (30 m)
[35]	350–2500 nm @ 10nm	6	Resampled	ANOVA, LDA	26	77.0	Mangrove species leaf scale

Table 2. *Cont.*

References	Wavelengths/Bandwidths	Classes	Pre-processing	Feature Selection Method	Bands	Accuracy %	Study Context and Spatial Scale or Resolution
[36]	400–2500 nm @ 16 nm	16		Best-First Search Algorithm	21	~69.5	Temperate forest ecotopes from airborne imagery (4 m)
[36]	400–2500 nm @ 16 nm	16		Random Forest	21	~69.5	Temperate forest ecotopes from airborne imagery (4 m)
[37]	350–2350 nm @ 1 nm	14		ANOVA (Tukey HSD), CART	17	98.0	Rice genotypes from canopy spectra
[38]	400–2500 nm @ 10 nm	7	Resampled	SDA	12	70.4	Eucalypt forest species from lab leaf spectra
[38]	400–2500 nm @ 10 nm	7	Resampled, 1st Derivative	SDA	13	72.4	Eucalypt forest species from lab leaf spectra
[4]	350–2500 nm @ 1.4, 2 nm	7		PCA	8	84.3	Cabbage crops and weed species from field canopy spectra
[39]	350–2450 nm @ 3, 10nm	4		Kruskal-Wallis post hoc Dunn, CART	56	~95	Giant Reed and coexisting vegetation from field canopy spectra
[40]	400–2400 nm @ 4, 6 nm	8	Smoothing	Stepwise Regression Wrapper	30	~70.0	Tropical tree species from airborne imagery (1 m)
[41]	400–2350 nm @ 10 nm	6	Continuum Removed	SDA	29	82.3	Himalayan forest species from satellite imagery (30 m)
[42]	429–2400 nm @ 2 nm	11		SDA	40	~98.0	Canadian forest tree species from lab leaf spectra
[42]	429–2400 nm @ 2 nm	11	1st Derivative	SDA	40	~98.0	Canadian forest tree species from lab leaf spectra
[42]	429–2400 nm @ 2 nm	11	2nd Derivative	SDA	40	~98.0	Canadian forest tree species from lab leaf spectra
[5]	426.5–2355 nm @ 10 nm	5		LS-means, SDA, PCA, LL-R^2	29	90.2	Crop species from satellite imagery (30 m)

Table 2. *Cont.*

References	Wavelengths/Bandwidths	Classes	Pre-processing	Feature Selection Method	Bands	Accuracy %	Study Context and Spatial Scale or Resolution
[5]	426.5–2355 nm @ 10 nm	5	Resampled	LS-means, SDA, PCA, LL-R^2	21	92.0	Crop species from canopy field spectra
[43]	415–2340 nm @ 10 nm	5		SDA	25	100	Amazon tree species from satellite imagery (30 m)
[43]	415–2340 nm @ 10 nm	5		SDA	25	100	Amazon tree species from satellite imagery (30 m)
[8]	400–2400 nm @ 5 nm	46	Smoothing, Normalization	PCA	20	82.6	Tropical wetland species from field leaf spectra
[8]	400–2400 nm @ 5 nm	46	Smoothing, Normalization	Mann-Whitney U-test	21	86.8	Tropical wetland species from field leaf spectra
[8]	400–2400 nm @ 5 nm	46	Smoothing, Normalization	ANOVA	23	83.4	Tropical wetland species from field leaf spectra
[8]	400–2400 nm @ 5 nm	46	Smoothing, Normalization	SVM	20	87.1	Tropical wetland species from field leaf spectra
[8]	400–2400 nm @ 5 nm	46	Smoothing, Normalization	Random Forest	20	86.1	Tropical wetland species from field leaf spectra
[8]	400–2400 nm @ 5 nm	46	Smoothing, Normalization	Random Forest (a)	20	84.8	Tropical wetland species from field leaf spectra
[44]	413–2440 nm @ 0.6, 11 nm	6		PCA	40	87.0	
[45]	400–2500 nm @ 2, 6, 10 nm	27	Smoothing, Continuum removed	Mann-Whitney U-test, Manual Selection	6	-	Saltmarsh vegetation types from field canopy spectra
[46]	350–2500 nm @ 3, 10 nm	7		ANOVA – post hoc Tukey-Kramer	9	94.7	Australian forest species from lab leaf spectra
[47]	390–2360 nm @ 10 nm	4	Resampled	PCA, LL-R^2, SDA, DGVI	22	97.0	Crops and savanna cover types from field canopy spectra
[48]	350–2350 nm @ 10 nm	8		SDA	20	95.0	Crop types from field canopy spectra

Table 2. *Cont.*

References	Wavelengths/Bandwidths	Classes	Pre-processing	Feature Selection Method	Bands	Accuracy %	Study Context and Spatial Scale or Resolution
[49]	350–2500 nm @ 3, 10 nm	16		Genetic Algorithm	4	~80.0	Mangrove species from lab leaf spectra
[50]	350–2500 nm @ 3, 10 nm	16		Genetic Algorithm	(30*4)	~80.0	Mangrove species from lab leaf spectra
[51]	400–2500 nm @ 10 nm	3		SDA	10	65.0	Pine tree species from airborne imagery (3.4 m)
[51]	400–2500 nm @ 10 nm	3	1st Derivative	SDA	10	77.0	Pine tree species from airborne imagery (3.4 m)
[51]	400–2500 nm @ 10 nm	3	2nd Derivative	SDA	10	72.	Pine tree species from airborne imagery (3.4 m)
[52]	350–2500 nm @ 10 nm	3	Resampled	ANOVA, LDA	15	90.0	Mangrove species from lab leaf spectra

Abbreviations: ANOVA = Analysis of Variance, CART = Classification and Regression Tree, PCA = Principal Component Analysis, LDA = Linear Discriminant Analysis, SDA = Stepwise Discriminant Analysis, LS-means = Least Squares means, LL-R^2 = Lambda-Lambda R-Squared, SVM = Support Vector Machine, DGVI = Derivative Greenness Vegetation Indices.

2. Meta-Analysis

The delineation of spectral regions in this review follows that of [3], as adapted from [53]. Imaging and non-imaging hyperspectral instruments have different sampling intervals, so a direct comparison of selected wavebands between studies is not possible. This was resolved by aggregating selected wavebands into 50 nm bins based on their band centres (Figure 1). The design and presentation of the binned wavelengths is adapted from [1] with adjustments. Additionally, the bin size of the histogram has the benefit of grouping highly correlated and often redundant wavebands together, reducing noise from the selection of correlated features from the analysis. The percentage of studies that selected wavebands within each 50 nm region is presented in the histogram, giving the selection rate for each 50-nm spectral region. The binned table and selection rate histogram (Figure 1) only give an indication of the rate with which a spectral region was selected and do not include information on the number of bands selected in each 50 nm bin, nor the determined importance of a selected band for subject discrimination.

2.1. Spectral Range

Of the studies that met the rules for inclusion in this review, 38 used hyperspectral data spanning most of the range 350–2500 nm. However, a number of studies utilised devices that recorded a more restricted wavelength range between 350 and 1100 nm, generally from 400 to 800 nm or 1000 nm (Table 2). These studies are presented separately as the absence of Shortwave Infra-red(SWIR), and much of the Near Infra-red(NIR) has shown to have an influence on waveband selection for the Visible (VIS) and partial NIR [54]. Although selection rates in the VIS/NIR studies appear similar to those from broader wavelength hyperspectral studies there are some notable differences. The initial peak in selection rates present in both sets is shifted towards shorter blue wavelengths, and a greater importance of the red edge over the red minimum is evident for the VIS/NIR studies. However, the overall pattern is the same with two peaks in the rate of selection at both the blue/green and red reflectance minima, with yellow wavelength bands having the lowest selection rate, save for the sub-400 nm bands that appear in a very limited number of studies. Although the VIS/NIR studies do not cover the full NIR region, selection rates for the red edge and shorter wavelength NIR are closely matched between both groups (Figure 1). The overall higher rates present in the VIS/NIR table results from the smaller number of studies in that group, with selection rates tending to decrease as more studies are added. Additionally, the relatively small number of studies included in the VIS/NIR group prevents the analysis of specific subsets, such as canopy and leaf. The following discussion of selection rates refers to VIS/SWIR studies (Table 1) and is generally applicable to the VIS/NIR studies, although particular discussion of the VIS/NIR studies is included when required.

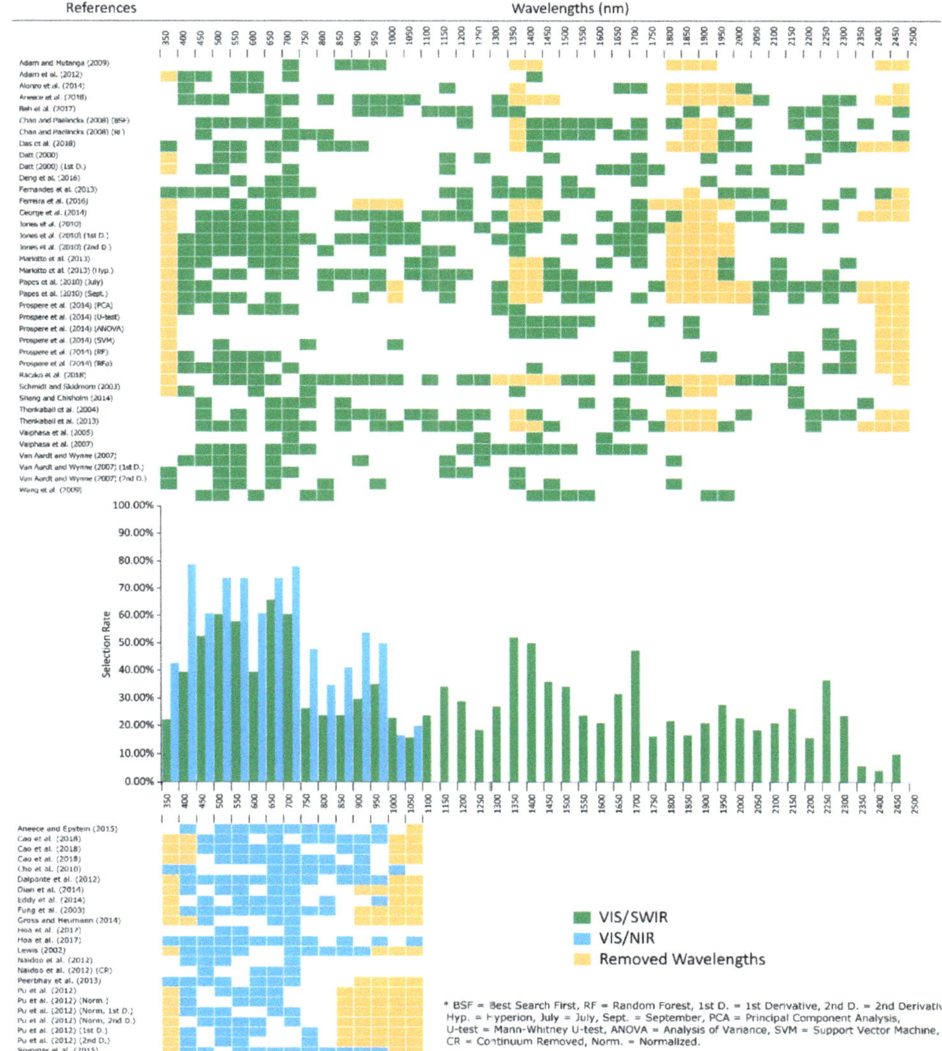

Figure 1. Waveband selection binned at 50 nm intervals for the VIS/SWIR studies (350–2500 nm) green, VIS/NIR studies (350–1100 nm) blue. Orange filled cells represent waveband regions removed from a study due to noise. Selection rate is the percentage of studies that selected a given 50 nm region for species classification. Each row of the table is an individual study, with each column being a 50 nm range bin. Green/blue shaded bins represent at least one waveband being selected from within that range, while orange shaded bins represent removed wavelength regions (e.g. major water absorption regions). Wavelength bins were only removed if the entire 50 nm region was removed due to noise/atmospheric effects in that particular study.

2.2. Visible (VIS; 400–700 nm)

Primarily a region of low reflectance in living foliage, typically as low as 5%–10% with the exception of the green peak at ~550 nm where reflectance can be more than twice that of surrounding wavelengths

(Figure 2). Reflection in the visible wavelengths is dominated by absorptions from foliar pigments. Differences in leaf pigments between species have been identified by many studies as important factors for discrimination [39], despite variability in the VIS being generally low compared to longer wavelengths [53,55]. Of the pigments, chlorophyll a and b have the strongest influence over absorption in this region, followed by those of carotenoids and anthocyanins whose effects are predominantly masked by that of chlorophyll. The visible region is one of the most influential regions for classification, with the vast majority of studies in this review selecting bands from within it. The visible wavelengths can be divided into three regions of high discriminatory value, spanning almost the entire visible range: the blue/blue-green edge (400–499 nm), the green peak centred around 550 nm, and the red reflectance minimum (650–700 nm) (Figure 2). Of these, the red reflectance minimum, specifically bands near 680 nm has previously been identified as the most commonly selected and critical band centre for crop type discrimination [56]. The continued selection of 680 nm, along with neighbouring bands in later studies has validated the importance of this region amongst agricultural crop studies [5,25,47,48,57], as well as for other vegetation types [10,17,18,26,27,29,30,46,58,59]. In addition to the obvious relationship with chlorophyll, absorption in the red region has been related to anthocyanin content, a foliar pigment responsible for the red colouration in leaves [60], particularly evident in juvenile leaves of certain species [30].

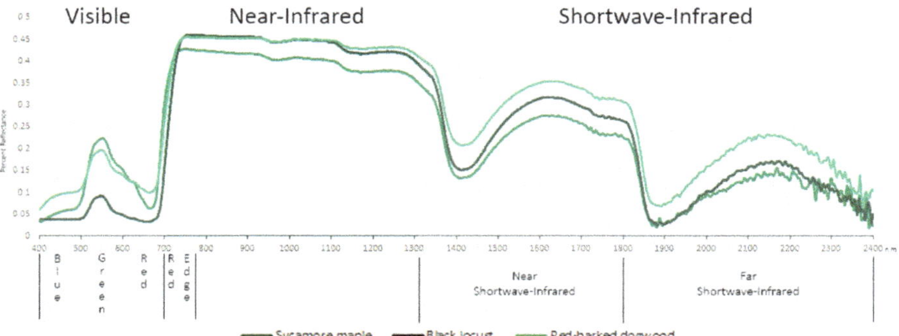

Figure 2. Example hyperspectral reflectance of 3 species of tree and key broad regions of the electromagnetic spectrum (400–2400 nm).

The green region has the second highest selection rate amongst both the VIS and entire measured spectrum (Figure 1). Wavebands selected in this region tend to be focused around the green reflectance peak at approximately 550 nm, which is strongly correlated with chlorophyll content [61]. The green peak, either manually chosen as a spectral variable as a representation of chlorophyll content or selected via feature selection, has demonstrated importance in classifying species [9,30,42,62,63]. Additionally, absorption in wavebands within the green region adjacent to the reflectance peak is associated with xanthophylls and anthocyanins. Xanthophyll pigments protect against photo-oxidation of the photosynthesis reaction centres during high light conditions [64], resulting in short term changes in reflectance at 531 nm. This band, along with 570 nm, makes up the photochemical reflectance index [65]. Anthocyanins can be estimated by an index using anthocyanin's absorption maximum near 550 nm, and a band from the red edge, usually 700 nm [66]. Although not necessarily associated with these additional pigments, studies have selected bands along the leading edge of the green reflectance peak between 500,550 nm [10].

Selection from the blue region (400–449 nm) has the third highest rate in the VIS region, though the blue-green edge (450–499 nm) has an almost equal rate of selection to the green region (55.8% and 58.8%, respectively). The importance of blue bands has been established for discriminating within groups of conifers, and between conifers and broadleaf species [67,68], though its inclusion in approximately half of the studies, many of which include non-coniferous species, indicates its importance in general for a wider range of vegetation types. Some of these non-coniferous studies focused on the savanna ecosystem, where blue bands along with the red reflectance minimum and red edge were informative [10,29]. Blue wavelengths are strongly influenced by chlorophyll absorption, along with carotenoid absorption features present in the 450–499 nm region. Carotenoids have proven important for the discrimination of senescent leaves, when the decay of chlorophyll and the diminishing of the strong chlorophyll-absorption feature reveal the carotenoid absorption feature [18].

However, studies have noted that strong similarities between the visible reflectance of different species can decrease the significance of VIS wavelengths for classification purposes. In one such study, the NIR region was more informative for distinguishing species than the VIS, with spectral differences in the VIS region being non-significant between species [69]. Additionally, in a study of tropical trees, Rivard et al. [54] performed feature selection and classification on various datasets derived from the same original spectra. One dataset included the wavelengths 350–2500 nm, another excluded the VIS, while another excluded the SWIR. Although it was found that the full spectrum produced greater overall classification accuracy, and both reduced datasets produced lower overall accuracies, individual accuracies for certain species remained high. The classification model excluding the VIS region maintained high accuracies for six out of 20 species, whereas the model excluding the SWIR maintained high accuracies for five out 20 species. Although the importance of the VIS region has been described by many authors and is clearly seen in the binned data, studies such as [54] demonstrate that wavelength importance is dependent on the species included in the study.

2.3. Red Edge (680–780 nm)

The red edge encompasses the region from the red reflectance minimum around 680 nm to the NIR shoulder at approximately 780 nm and indicates the sharp increase in reflectance from the VIS to NIR regions associated with strong chlorophyll absorptions and internal leaf structure (Figure 2). The inflection point of the slope in this region has been defined as the red edge position (REP) [70], and its strong correlation with chlorophyll concentration has seen it used as an indicator of stress and senescence in vegetation [71,72]. In the VIS-SWIR studies, the red edge region as represented by the 700–749 nm bin has the same rate of selection as the red minimum bin, whereas the VIS-NIR studies have a slightly higher red edge rate than red minimum. However, as previously stated, the delineation between the red minimum bin (650–699 nm) and the red edge bin (700–749 nm) means that bands selected from the lower point of the red edge would be included in the red minimum bin, potentially skewing red edge band selection rates.

The red edge region has been described as one of the most informative and frequently selected regions in a number of studies, where the authors have attributed its importance to its correlation with chlorophyll abundance, nitrogen concentration, water content, and structural features such as leaf area index (LAI) [3,10,11,73]. Additionally, significant variation of the red edge region between species has been documented after a first derivative transformation has been applied to the spectra [74]. The red edge has proven especially important in studies discriminating species with high levels of chlorophyll and high LAI values such as the giant reed (*Arundo donax*), in which a distinctive "red shift" is seen where the Red Edge Position (REP) is located at higher wavelengths [32,39]. This "red shift" mirrors the "blue shift" of the REP where its position is shifted towards the shorter blue wavelengths associated with a decrease in chlorophyll and used to monitor senescence or stress [75].

2.4. Near Infrared (NIR) (700–1327 nm)

The NIR is often defined to include wavelengths within the red edge region (680–780 nm) [42]: As this region has been previously discussed, this section focusses on the NIR plateau (780–1327 nm). The high reflectance of the plateau results from the scattering of photons within the leaf structure due to a change in the refractive index from liquid water to air within the inter-cellular spaces [76]. Two minor water absorption features at ~980 nm and ~1200 nm are the only major features of the plateau. Along with water content, the depth and width of these absorptions can be influenced by the spectral recording method. Canopy scale spectra tend to produce deeper and wider absorption features compared to the leaf scale, at which absorption features can vary with leaf stack thickness [3]. High levels of intraspecific variability have been identified in the NIR and related to leaf age, water, and chlorophyll concentration, as well as herbivory, necrosis, and epiphyll cover [3,38]. Wavebands selected in studies reporting these high levels of intraspecific variation have generally been limited to the water absorption features [11,38], although it has been suggested to avoid band selection from within or near water absorption features due to this high level of within-class variability, specifically for Eucalypts [46,77,78]. Despite this, [3] reported greater interspecific variability in the NIR, particularly at the canopy scale, potentially related to species-specific photon scattering caused by differences in canopy architecture, a result also reported by other studies [68,69]. However, it has been suggested that the importance of the NIR and SWIR in [3] is linked to the time delay between leaf collection and spectral measurement, causing a decrease in water content and affecting waveband importance [58].

Even when the high selection rate of the red edge is included, the average selection rate of the NIR is close to half of that of the VIS, placing it third after the near SWIR. However, there are two small peaks in the rate of selection within the NIR, in bins 950–999 and 1150–1199, both of which are associated with water absorption features near 980 and 1200 nm. Despite having one of the lowest rates, some studies have reported that bands in the NIR plateau are the most strongly discriminating [45,52].

2.5. Shortwave Infrared (SWIR) (1328–2500 nm)

Based on the binned results (Figure 1) the SWIR can be divided into two distinct regions, the near SWIR (NSWIR) from 1350–1800 nm, including the strong water absorption feature at 1350–1450 nm, and the far SWIR (FSWIR) from 1800–2500 nm, including another strong water absorption feature from 1800–2000 nm. The wavebands associated with these water absorption features, that mark the start of the SWIR and separate the near and far SWIR, are often removed from spectra due to high levels of noise, as are the bands at the far end of the SWIR above 2400 nm. Selection rates within the NSWIR are on average the second highest, primarily caused by high rates of selection at 1350–1450 and 1700–1750 nm. This initial high selection rate, spanning two consecutive bins, is associated with the water absorption feature focused around 1400 nm. However, these bins are often removed in studies, primarily when hyperspectral imagery is used due to increased noise that is not as prevalent in lab or field spectra. Selection rates then drop in the mid-NSWIR bands before peaking again for the 1700–1750 nm bin, containing wavebands often associated with lignin, cellulose, tannins, and other biochemical constituents of foliar and non-foliar plant matter [19,79]. The FSWIR has the lowest average band selection rate, with its highest selection at bin 2250–2299 nm most likely associated with the weak absorption features of cellulose and lignin present at 2270 nm [19,79].

As the selection results suggest, wavebands selected from the SWIR are reported in the literature as being associated with water absorption [17,33,38,40,46–48,58] or the weak harmonic and overtone absorptions from biochemicals such as lignin, starch, and cellulose [9,40,42,46–48,52,58,80]. However, as described in regards to the NIR, the selection of bands in or near water absorption features may not be suitable for classification in field or lab spectra, due to high levels of intraspecific variance [46,77]. Additionally, bands selected from leaf scale spectra in the two major water absorption features would not be applicable to remotely sensed imagery as they coincide with low irradiance levels resulting from atmospheric water absorption. The observation of higher selection rates in the NSWIR compared to the FSWIR has previously been made with studies noting the importance of NSWIR bands and absence of selection from the FSWIR [9,42], even when visual differences between species were apparent [52]. Possible reasons for this reduced selection of the FSWIR could be high levels of LAI or leaf water content masking the biochemical features present in this region [81], or a high correlation between the FSWIR, NSWIR, and VIS bands [9].

2.6. Canopy and Leaf Scale Spectral Selection Rates

The red edge has been demonstrated as one of the most frequently selected regions (Figure 1), though the remainder of the NIR (consisting of 12 bins from 750–1349 nm) has the second lowest mean selection rate, only slightly higher than the FSWIR. As the literature has identified an increase in importance of the NIR for canopy spectra, a comparison of band selection rates for each bin was made between canopy and leaf scale spectral studies (Figure 3). Leaf spectra were defined as only containing pure leaf reflectance, with canopy being primarily leaf spectra, though also containing non-photosynthetic vegetation and potentially background reflectance. This comparison shows a clear increase in selection rates for the NIR bins associated with water absorption features for the canopy studies, and a related decrease amongst the leaf scale spectra. Differences are also apparent in the visible regions, with a substantial increase in the selection of the leading edge of the green peak, and a decrease in selection of the trailing edge of the green peak for leaf scale studies compared to canopy level (Figure 3). This would indicate a blue-shift for green bands selected in leaf scale spectra, and a red-shift of selected bands for canopy spectra. Differences in spectral reflectance for the VIS region have been identified at different scales, with branch/canopy spectra including reflectance characteristics from non-foliar sources, shadows and uneven lighting, as well as generally displaying an increase in pigment absorption features [3,53]. Variation in selection rates is also evident in the SWIR, most notably a broad region of increased selection for canopy spectra across four bins from 1950 to 2149 nm, and a sudden peak at 1800–1850 nm. The selection peaks of the canopy spectra correspond to regions of water absorption which have demonstrated an increase in depth and width in canopy studies. However, the disparity between canopy and leaf scale spectra is potentially exaggerated by the fact that a majority of canopy studies eliminate these wavebands due to noise concerns, with the remaining few studies selecting these wavebands as being discriminatory. Increased selection of the broader region could also be related to water absorption, as well as structural components such as lignin and cellulose, particularly from non-photosynthetic material in the canopy [3]. The NSWIR however demonstrates the highest degree of conformity for a large region, covering nine bins from 1300–1750 nm.

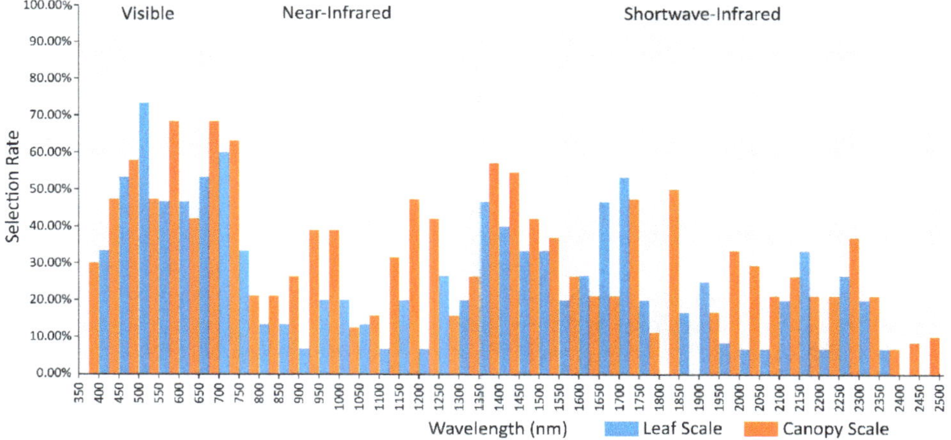

Figure 3. Feature selection rates for 350–2500 nm studies (Table 2) per 50-nm bins of both canopy and leaf scale spectra.

3. Feature Selection

Feature selection is implemented to select a subset of features to improve generalization and computation requirements while preserving or improving classification accuracy. In this review, feature and waveband selection are used interchangeably. Feature selection techniques are generally divided into three categories: filter, wrapper, and embedded methods. Filter methods are named as such as they act as a pre-processing step that filters out irrelevant features. Filter methods are known to be computationally fast and efficient, though they are generally outperformed by the other methods, as well as not able to handle nonlinear relationships [82].

3.1. Filter Methods

Analysis of variance (ANOVA) is a parametric statistical filter method to determine significant differences between group means. Related to ANOVA is the non-parametric Mann-Whitney U-test, and the Kruskal–Wallis test which extends the Mann–Whitney U-test for more than two groups [45]. Following initial dimensionality reduction by one of these methods a secondary feature selection step to further reduce the number of selected features is used, such as Linear Discriminant Analysis (LDA) [35,52], classification and regression trees (CART) [32,37,39], or the manual selection of known influential bands [8,45,46]. This secondary selection step found important bands in the VIS and SWIR, with a reduced selection of NIR bands [39,52]. However, the reverse was found by [32] where CART secondary selection was restricted to NIR wavelengths. The remainder of the studies manually selected bands that differentiated the greatest number of species pairs [8], or selected known influential bands from the wavelengths that demonstrated high levels of pairwise group variance [45,46].

3.2. Wrapper Methods

Wrapper methods search for a subset of features that gives the best classification performance, with the best performing subset being selected. Although generally considered to outperform filter methods, wrappers are known to be computationally demanding and can suffer from overfitting [82].

Two of the studies reviewed implemented genetic algorithms (GA), in which wavebands are encoded as genes that are subsequently grouped into chromosomes. These chromosomes are allowed to evolve over many generations where their fitness, as determined by a classifier, controls their likelihood to reproduce and pass their genes onto the next generation. Fitness of chromosomes is determined each generation by a chosen classifier, and with the classification accuracy of each chromosome being its fitness score, chromosomes with increased fitness are more likely to reproduce. Both studies used the same dataset of lab measured tropical mangrove leaves [49,50]. The selection of bands differed between the two studies, despite the use of the same dataset and feature selector, though methodologies did differ. The variability of selected bands with similar classification performance seen between these studies demonstrates that multiple band selections can perform classification equally well. The ensemble of chromosomes used in [50] helped to identify key regions for discriminating target species related to biophysical and biochemical aspects of the vegetation that may have been missed if a study was reliant upon the first single chromosome to reach the stopping criterion. This is apparent when comparing the bands selected in both studies, with [49] selecting no VIS bands, resulting in the authors concluding that pigments were not significant for the discrimination of the target species. However, the importance of the VIS, particularly the green region became apparent in [50] where 21 out of 120 total bands were selected from 513 ±19 nm.

Forward feature selection (FFS) is a wrapper method of feature selection that begins with a model containing a single feature that best discriminates the classes, with new features iteratively added to the model based on their ability to improve class discrimination [83]. FFS was implemented by [27] in their comparison between floral and leaf spectra, however, only the results for leaf spectra are discussed here. The leaf spectra within this study were constrained to 475–900 nm at 1 nm increments, with only eight wavebands being selected. These bands came from narrow regions of the spectra, occurring at 450–499 nm in the blue, and the red minimum and red edge from 650–749 nm. In a similar spectral range of 402.9 to 989.1 nm of airborne collected spectra, a very different feature selection trend was observed by [11] following the use of the FFS variant sequential floating feature selection (SFFS). Wavebands were selected from across the entire reduced spectrum, with a notable gap in selection occurring in the NIR between 800 and 849 nm. Selection differences exhibited between these studies could be related to the differences in target species, leaf or canopy scale spectra, or version of FFS used. The only VIS-SWIR study in this review to use FFS applied it to AVIRIS imagery of urban street trees [9]. However, feature selection was only performed to identify spectral regions responsible for species separability, with all bands used for classification. These informative spectral regions matched a number of known informative regions from the literature, such as water absorption in the NIR, cellulose and lignin features in the SWIR, and bands associated with photosynthetic pigments in the VIS. Interestingly however, the highly selected red minimum and red edge were not selected in this study, along with the majority of the NIR.

3.3. Embedded Methods

Despite being described as a wrapper method in [8], recursive feature elimination with a support vector machine (SVM-RFE) is considered to be an embedded method [84]. Embedded methods differ from wrappers, as they do not treat the classifier as a black box, rather, features are selected using information gained whilst training the classifier [85]. A claimed strength of SVM as a classifier is its reported independence of the Hughes effect, or curse of dimensionality [86,87]. However, it has been shown that SVM classifications can be affected by the Hughes effect and can benefit the from dimensionality reduction of its inputs, especially when sample sizes are small [88].

In order to be used as a feature selection method, [8] implemented recursive feature elimination (RFE) with a SVM, determining that from the original 401 bands the optimal number of features to include for classification is 20, after 1–5, 10, 15, 20, and 30 were all evaluated. The 20 bands selected demonstrated a number of trends that were not apparent in the other feature selection methods implemented in the same study. Firstly, the bands formed four distinct contiguous clusters at 520–530 nm, 745–775 nm, 1005–1030 nm, 2295–2305 nm, and then a final single band at 2345 nm. Secondly, the wavelengths of certain selected bands were also unique amongst the methods used, with SVM-RFE being the only method to select bands from the NIR plateau out of all feature selection methods implemented in [8]. Additionally, being the only method to not select bands from the NSWIR. Although not reported in a manner suitable for inclusion in Table 1, [17] also performed feature ranking with a SVM. As with [8], [17] identified the optimal number of features to be between 15 and 20, depending on the dataset, pre-processing, and feature selection methods used. Unlike [8], where the SVM selected bands from distinct contiguous regions, [17] report the SVM selecting bands evenly spread over the entire spectrum.

Random forest (RF) is an ensemble classification method, in which a number of decision tree classifiers are trained from a sub-sample of the dataset, with their results combined via a voting system. One third of samples are retained for validation purposes known as the out-of-bag (OOB) samples, with the remaining in-the-bag samples being used to construct the decision tree [89].

Of the original 72 bands in [29] between 384.8 nm and 1054.3 nm, eight were selected for classification via RF. Although no other feature selection method was implemented in this study, a previous study by [10] performed feature selection with the spectral angle mapper (SAM) add-on Selector using the same data. This resulted in the selection of a far greater 31 bands. Upon binning of the bands at 50 nm, a clear difference in the selection methods are evident (Figure 1). The RF selected bands of [29] are focused in the 400–550 nm region with a single band from the red edge at 706 nm, whereas the SAM bands are focused along the red edge and NIR plateau between 650 and 950 nm, with additional bands in the 350–450 and 1000–1050 nm regions.

As with the bands selected in [29], the RF selected bands in [36] fell within four bins in the VIS and VNIR regions. However, in [29], band selection was focused on the green region with limited selection apparent in the red and NIR plateau with the exception of a single band near the red edge inflection point. This focus was seemingly switched in [36] with bands falling into the bins along the red edge up to the NIR plateau shoulder, with the remaining bin occurring at the blue/green edge. The Chan and Paelinckx study [36] also offers a comparison to an alternative feature selection method using the best-first search (BFS) algorithm as a wrapper. The band selection techniques differ greatly in the VIS and VNIR regions with only the bins at 450–499 and 700–749 in common. However, band selection is more similar at longer wavelengths where the majority of bands were selected by both methods.

The wavebands selected via RF in [8] are in direct opposition to those selected by RF in [36]. Selected bands in [36] mainly occurred along the red edge and NIR plateau shoulder, no band was selected in this region by [8]. Instead, focus was placed on the green, yellow, and red regions of the VIS wavelengths, an area completely ignored by [36] RF selector, though significant for their BFS selection. Additionally, [8] provided the top 20 informative bands determined by a RF classifier using the full 201 waveband dataset. Although these two implementations of RF differed in selecting bands, the overall trend was very similar, with high selection rates in the VIS, low in the NIR, and similar selection throughout the SWIR.

Additionally, a study by [33] produced waveband selections similar to those in [8] with similar results in the VIS with the exception of no selection in the early green (500–549 nm), and selection of the red edge bin rather than the red minimum. The biggest difference between [33] and all other RF studies is the reduced selection at longer wavelengths, although all studies essentially ignored the NIR, [33] only selected two bands from the SWIR, both within the same NSWIR bin at the water absorption feature near 1400–1449 nm.

3.4. Comparison of Stepwise Discriminant Analysis (SDA) with non-SDA Feature Selectors

Stepwise discriminant analysis is a filter method that selects a subset of features by attempting to minimise within-class variation while simultaneously maximising between-class variation [90]. Although a number of metrics are available to determine class separability, Wilk's lambda is by far the most frequently used to enter and remove variables from the selection in a stepwise manner. Some studies reported Wilk's lambda approaching zero and becoming asymptotic, indicating near perfect separation of classes [48]. Features selected after this point can be safely removed from the model as they will not substantially increase classification accuracy. This normally resulted in the selection of 10–20 wavebands [5,38,47,48,51].

SDA in general selects wavebands more uniformly across the spectrum than other methods, though the greatest number of selected bands is still found in the VIS (Figure 4). The most significant difference for selection rates is the increased importance of the NIR beyond the red edge. The NIR demonstrates significant selection with the use of SDA in all bar a first derivative dataset from [51], and [38], with the author of the latter suggesting high levels of intraspecific variance due to differences in leaf maturity as the reason no bands were selected in this region.

Upon comparing the selection rates of SDA studies compared to non-SDA, a clear difference in selection of NIR bands is apparent. As with the difference between canopy and leaf scale spectra, the increased selection is focused around the NIR water absorption features (Figure 4). Additionally, in the VIS, there is significantly higher selection for the blue, green, and red regions in SDA studies. In order to determine if the spectral acquisition scale or feature selection technique had a greater influence on band selection, the selection rates were further subset into canopy studies using SDA and non-SDA feature selection, and leaf scale studies using SDA and non-SDA selection (Figure 5). It is apparent that the feature selection method has a greater impact on band selection rates, with SDA selecting from the NIR with far greater rates than the non-SDA methods in both canopy and leaf scale studies. The non-SDA methods demonstrated minimal selection in the NIR beyond the red edge for leaf scale spectra, with only a slight increase in selection for canopy spectra focused around the water absorption wavelengths from 1150–1250 nm. The studies that did select from the NIR with leaf scale samples via non-SDA methods stated that the selected bands represented differences in internal reflectance for leaf scale spectra [50]. The blue and red shifts around the green peak for canopy and leaf scale spectra are still evident once the data has been subset into SDA/non-SDA, although it becomes apparent that the high rates of selection in many parts of the VIS is driven by the SDA studies. However, the use of SDA does not explain the selection rates of the VIS for the reduced spectral domain

VIS/NIR studies, as only a single study used SDA for feature selection, perhaps indicating an alternate driving force. The red edge demonstrates its robustness to variations in measurement scale and band selection technique as it was frequently selected for all study subsets, although slightly less frequently for leaf scale spectra with non-SDA feature selection.

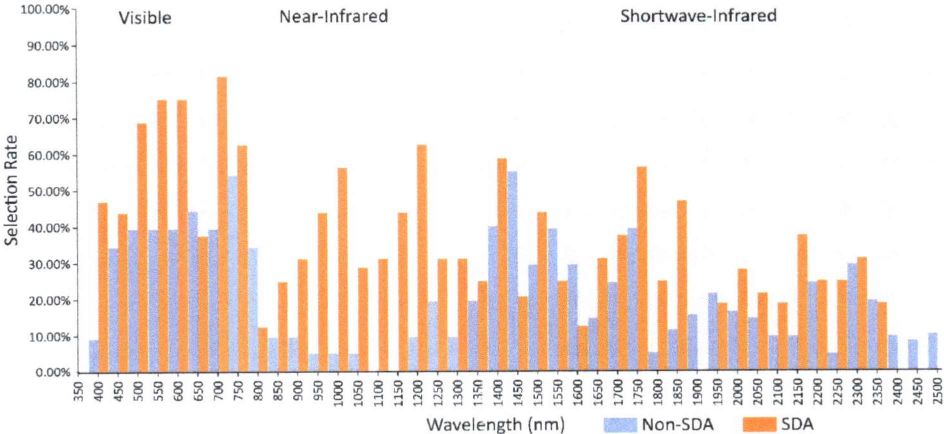

Figure 4. Feature selection rates for 350–2500 nm studies that used SDA feature selection, and the selection rate of all other feature selection methods combined.

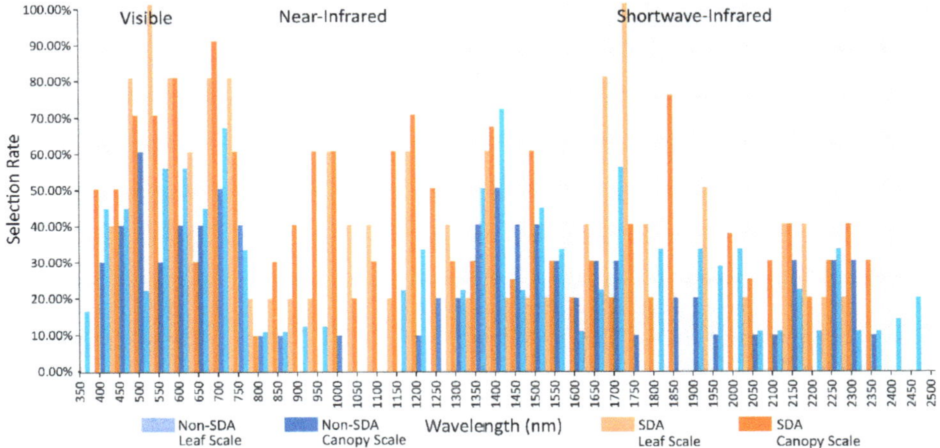

Figure 5. Feature selection rates for 350–2500 nm studies that used SDA feature selection subset by canopy and leaf scale spectra, and the selection rate of all other feature selection methods combined.

According to [91], "Stepwise analytic methods may be among the most popular research practices employed in both substantive and validity research". Despite this statement being made in the late 1980s, the use of SDA in approximately a third of the studies included in this review demonstrates its continued popularity, being by far the most used method encountered. However, the widespread use of stepwise methods has prompted strong arguments against its usage [90,92–94], particularly when utilised in a predictive discriminant analysis application such as feature selection for classification [95]. The studies that utilised SDA in this review made no mention of these criticisms and therefore no direct attempt to mitigate them. Despite this, [25] did validate their model with 20 repetitions of 1000 random samples, with the final feature subset being based on the selection rates of features across the repetitions, the consideration of important features identified in the literature from [6] and [47], as well as the results from principal component analysis (PCA). PCA is a mathematical transformation used to produce uncorrelated features from the spectral features, reducing dimensionality whilst retaining the most informative spectral data. Additionally, [47], and [5] included SDA as part of an ensemble of feature selection methods, again determining the final feature subset based on the selection rates of features across all methods within the ensemble. Although one of these ensemble methods (Lambda– Lambda plots) allows for the identification and removal of correlated features, in both cases, it was run in parallel to SDA with the removal of correlated features occurring after features had been selected. The remaining studies reported no efforts to mitigate the concerns of using SDA for feature selection [38,41–43,48,51].

It must be acknowledged that the sub-setting of reviewed studies into canopy and leaf scale, and then into SDA and non-SDA, meant each class was only represented by a small number of samples (~8 per class), though leaf-SDA was only represented by five studies extracted from two papers. As a result of this, a few outliers are evident, such as the 100% selection in bin 1700–1749 nm, and the 100% selection of the 500–549 nm bin, both associated with the low leaf-SDA sample size. Additionally, the comparison of SDA to non-SDA may disguise selection biases of the non-SDA methods as they are often only represented by one or two studies, with any bias they may exhibit being masked by the selection rates of the other methods.

4. Study Design Influence

All aspects of a study design influence waveband selection. However, many of these aspects may be outside the control or be heavily constrained for the researcher, such as target classes, number of samples and collection method, though the researcher often has control over data pre-processing, feature selection, and classification methods. Due to this, and the apparent influence of feature selectors previously described, we focus on how the choice of feature selection method effects waveband selection.

In order to ascertain any influence feature selection may have over waveband selection, some of the most common feature selection methods were applied to a synthesised dataset. A key requirement for these experiments is the need for a dataset with many species with a large number of samples, something generally lacking in vegetation hyperspectral data. To accomplish this, a hyperspectral synthesis method was created [20] to allow for the creation of any number of samples from 22 species of New Zealand plants. The synthesised dataset consisted of 500 samples per class with 540 wavebands from 350–2450 nm at 3-nm bandwidths, excluding regions of high noise.

Hyperparameters for the feature selectors were tuned via a holdout dataset, with the parameters that selected features resulting in the highest classification accuracy being used for all experiments (Table 3). This is crucial to ensure that the only variables that could affect waveband selection were constrained to either the feature selector (svm_*, sda_*,sffs_*,rf_*) or the dataset (*_0 ... *_9).

Table 3. Software packages and hyperparameters for each feature selection method.

Feature Selector	Software Package and Library	Hyperpaprametes
SVM	Python 3.6, scikit-learn v0.21.3	$C = \textbf{100}$, *class_weight* = **'balanced'**, *kernel* = **'linear'**
SDA	Python 3.6, milk v0.6.1	*tolerance* = **0.001**, *significance_in* = **0.01**, *significance_out* = **0.01**, *Metric* = **'Wilk's Lambda'**
SFFS	R 3.6.1, varSel v0.1	*Metric* = **"Jeffries-Matusita distance"**, *Strategy* = **"mean"**
RF	Python 3.6, scikit-learn v0.21.3	*n_estimators* = **100**, *criterion* = **'gini'**, *max_depth* = **None**, *min_samples_leaf* = **1**

Three experiments were devised. First, each feature selection method was performed on the same dataset, cross-validated 10 times (eg. rf_0, rf_1 ... rf_9), selecting the top 30 discriminative wavebands, thus revealing any possible biases in waveband selection resulting from the choice of feature selection method (Figures 6 and 7). Secondly, feature selection was performed on datasets consisting of different classes and samples to simulate many different studies, giving an idea if attributes of the samples affect the wavebands being selected, which will impact generalizability and transferability. Variants of this experiment were performed wherein the classes used remained the same as did the number of samples, though actual samples were randomly selected. Additionally an experiment with the same classes though with differing numbers of samples. Results for these variants did not significantly differ and therefore aren't shown here.

Each dataset produced significantly different waveband selections. This is especially evident in Figure 6b were the histogram is ordered by feature selector, placing each repetition with a new dataset next to each other. Here, it is clear the RF favours the red edge and NIR bands, essentially ignoring the SWIR. SFFS demonstrated higher selection in the VIS, especially at shorter wavelengths, minimal selection in the NIR, and medium selection in late SWIR. SDA and SVM are the most similar due to both selecting broadly and relatively evenly along the entire spectrum. Dimensionality reduction techniques offer a way to visualize the relationship between selection histograms (Figure 7). Due to their broad general selection, SDA and SVM are grouped close to each other with SFFS and RF adjacent though separate. Further, the histograms are clearly grouped by feature selection method rather than dataset, indicating that feature selection method is a dominant factor affecting the selection of wavebands.

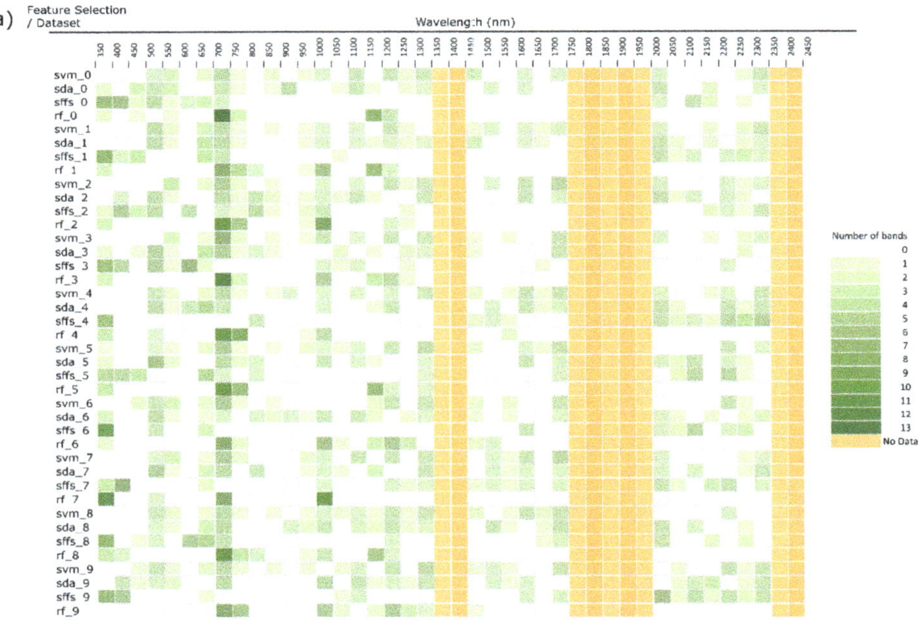

Figure 6. *Cont.*

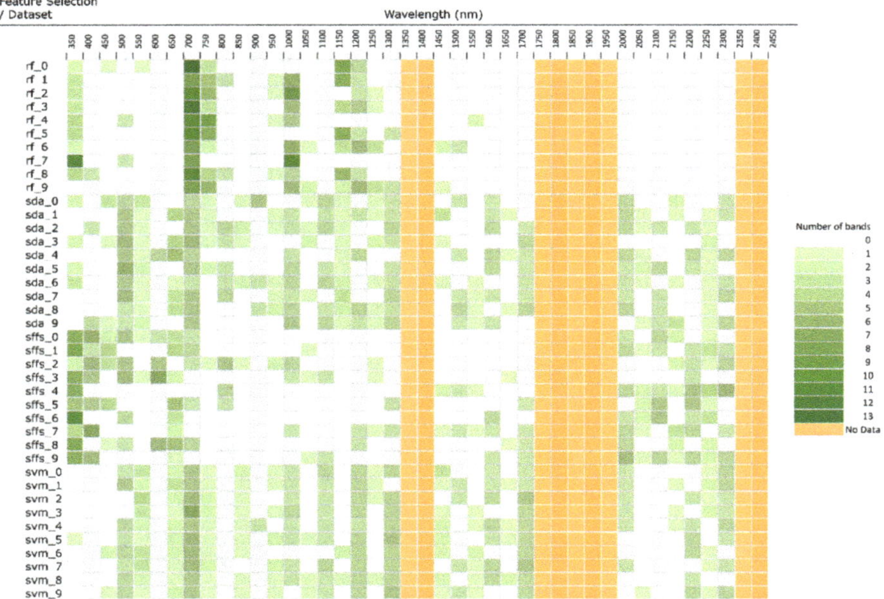

Figure 6. (**a**): Histogram of band feature selection binned at 50nm, ordered by dataset. Four feature selectors run on the same dataset 10 cross-validation (new dataset consisting of 10 classes and 200 samples for each cross-val.). (**b**): Results of Figure 5a ordered by feature selection method. (RF = random forest, SDA = stepwise discriminant analysis, SFFS = sequential floating feature selection, SVM = support vector machine).

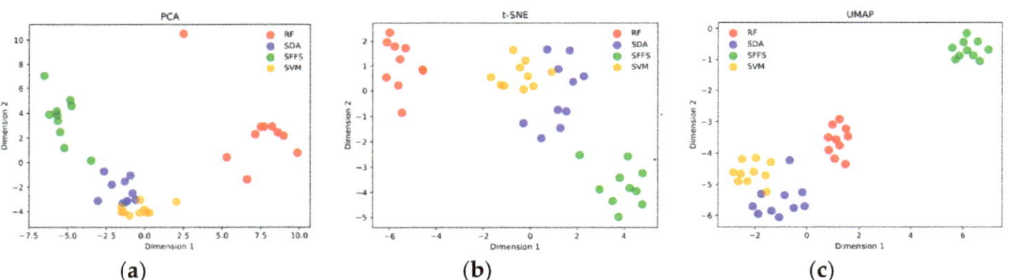

Figure 7. (**a**) PCA dimensional reduction of histogram waveband feature selection. (**b**) t-Distributed Stochastic Neighbor Embedding (T-SNE) dimensional reduction of histogram waveband feature selection. (**c**) Uniform Manifold Approximation and Projection (UMAP) dimensional reduction of histogram waveband feature selection.

5. Discussion

This review of hyperspectral vegetation classification literature has determined that every aspect of a study can greatly influence selected wavebands and classification performance. However, despite this, we have identified some important and consistent patterns that appear throughout the literature. Visible wavelengths and their associations with photosynthetic pigments have played important discriminatory roles in a wide range of studies, their high levels of selection clearly evident in this review (Figure 1). Selection rates in the VIS showed only minor variations between VIS/SWIR studies and the VIS/NIR (Figure 1), although the comparisons between canopy and leaf scale spectra demonstrated significant differences (Figure 3). The discriminatory value of the red edge has been well documented with its close relationship to chlorophyll concentration and structural features. This is reflected in the consistently high rates of selection of the red edge as well as the robustness of its selection with only minor variation in magnitude between the comparisons. The inclusion of structural features in canopy spectra can provide high levels of interspecific variation in the NIR, primarily in the form of differences in albedo, rather than spectral shape [68]. However, selection rates from the non-red edge NIR are low, with the selected bands generally being related to water absorption features and potentially high levels of within-class variability. Additionally, the NIR has demonstrated the greatest degree of variability between the canopy and leaf scale spectra studies. Wavebands selected in the SWIR are associated with water absorption and non-photosynthetic biochemicals, with selection rates heavily skewed towards the NSWIR over the FSWIR.

The reported importance of NIR bands [45,52] seems to be contentious, primarily being driven by the use of a single feature selection technique. Comparisons between selection rates for the NIR with and without the use of SDA as the feature selector are starkly contrasted, with the importance of NIR being significantly higher with the use of SDA. The criticisms of SDA and stepwise methods in general perhaps offer an answer to the selection biases presented in this review.

It is apparent that there is no single best feature selection method, with the same method performing very differently within and between studies. This suggests that either multiple methods should be applied to the data, or an ensemble of multiple methods may be the best practice, a conclusion recognized by this review, and previously suggested by some studies [36]. Additionally, multiple subsets of selected features have proven to discriminate species equally well [8], or alternatively, no feature selection, with the original data outperforming feature selected subsets [7,9,36]. Additionally, as computation power, dataset sizes, and machine learning techniques all increase, the need for feature selection as a data reduction technique becomes less necessary.

6. Conclusions

This review has established that the variability in waveband selection seen between studies, driven by study parameters beyond the characteristics of the target samples, prevents the determination of generalizable, high utility spectral regions for specific taxonomic discrimination. Broad trends such as the importance of VIS and red edge wavelengths are apparent, independent of plant groupings, though in and of themselves they are not sufficiently specific for taxonomic discrimination. The possibility of discriminatory spectral regions being associated with specific taxonomic, structural, or functional groupings of plants is inconclusive due to the large degree of variability between studies. This is further highlighted by the apparent dominance of feature selector choice over other parameters for waveband selection (Figures 6 and 7). Building on this review, future works could investigate variance in waveband selection caused by the hyperparameter choice of feature selectors, data preprocessing, as well as the inclusion of vegetation indices.

Author Contributions: Conceptualization, A.H.; methodology, A.H.; data curation, A.H.; formal analysis, A.H.; writing—original draft preparation, A.H., K.C. and M.L.; writing—review and editing, A.H., K.C., and M.L.; supervision, K.C., M.L. All authors have read and agreed to the published version of the manuscript.

Funding: This research received no external funding.

Acknowledgments: Financial support for this research was provided by the Australian Government Research Training Program Scholarship and the University of Adelaide School of Biological Sciences.

Conflicts of Interest: The authors declare no conflict of interest.

References

1. Fassnacht, F.E.; Latifi, H.; Stereńczak, K.; Modzelewska, A.; Lefsky, M.; Waser, L.T.; Straub, C.; Ghosh, A. Review of studies on tree species classification from remotely sensed data. *Remote Sens. Environ.* **2016**, *186*, 64–87. [CrossRef]
2. Fernandes, A.; Melo-Pinto, P.; Millan, B.; Tardaguila, J.; Diago, M. Automatic discrimination of grapevine (*Vitis vinifera* L.) clones using leaf hyperspectral imaging and partial least squares. *J. Agric. Sci.* **2015**, *153*, 455–465. [CrossRef]
3. Clark, M.L.; Roberts, D.A.; Clark, D.B. Hyperspectral discrimination of tropical rain forest tree species at leaf to crown scales. *Remote Sens. Environ.* **2005**, *96*, 375–398. [CrossRef]
4. Deng, W.; Huang, Y.; Zhao, C.; Chen, L.; Wang, X. Bayesian discriminant analysis of plant leaf hyperspectral reflectance for identification of weeds from cabbages. *Afr. J. Agric. Res.* **2016**, *11*, 551–562.
5. Mariotto, I.; Thenkabail, P.S.; Huete, A.; Slonecker, E.T.; Platonov, A. Hyperspectral versus multispectral crop-productivity modeling and type discrimination for the HyspIRI mission. *Remote Sens. Environ.* **2013**, *139*, 291–305. [CrossRef]
6. Lewis, M. Spectral characterization of Australian arid zone plants. *Can. J. Remote Sens.* **2002**, *28*, 219–230. [CrossRef]
7. Sommer, C.; Holzwarth, S.; Heiden, U.; Heurich, M.; Müller, J.; Mauser, W. Feature based tree species classification using hyperspectral and lidar data in the Bavarian Forest National Park. In Proceedings of the 9th EARSeL Imaging Spectroscopy Workshop, Luxembourg, France, 14–16 April 2015; Volume 14, pp. 49–70.
8. Prospere, K.; McLaren, K.; Wilson, B. Plant species discrimination in a tropical wetland using in situ hyperspectral data. *Remote Sens.* **2014**, *6*, 8494–8523. [CrossRef]
9. Alonzo, M.; Bookhagen, B.; Roberts, D.A. Urban tree species mapping using hyperspectral and lidar data fusion. *Remote Sens. Environ.* **2014**, *148*, 70–83. [CrossRef]
10. Cho, M.A.; Debba, P.; Mathieu, R.; Naidoo, L.; Van Aardt, J.; Asner, G.P. Improving discrimination of savanna tree species through a multiple-endmember spectral angle mapper approach: Canopy-level analysis. *IEEE Trans. Geosci. Remote Sens.* **2010**, *48*, 4133–4142. [CrossRef]
11. Dalponte, M.; Bruzzone, L.; Gianelle, D. Tree species classification in the Southern Alps based on the fusion of very high geometrical resolution multispectral/hyperspectral images and LiDAR data. *Remote Sens. Environ.* **2012**, *123*, 258–270. [CrossRef]
12. Alonso, M.C.; Malpica, J.A.; de Agirre, A.M. Consequences of the Hughes phenomenon on some classification techniques. In Proceedings of the ASPRS 2011 Annual Conference, Milwaukee, WI, USA, 1–5 May 2011; pp. 1–5.
13. Hughes, G. On the mean accuracy of statistical pattern recognizers. *IEEE Trans. Inf. Theory* **1968**, *14*, 55–63. [CrossRef]
14. Settles, B. *Active Learning Literature Survey*; University of Wisconsin-Madison Department of Computer Sciences: Madison, WI, USA, 2009.
15. Zhu, X.J. *Semi-Supervised Learning Literature Survey*; University of Wisconsin-Madison Department of Computer Sciences: Madison, WI, USA, 2005.
16. Van Dyk, D.A.; Meng, X.-L. The art of data augmentation. *J. Comput. Graph. Stat.* **2001**, *10*, 1–50. [CrossRef]
17. Fassnacht, F.E.; Neumann, C.; Förster, M.; Buddenbaum, H.; Ghosh, A.; Clasen, A.; Joshi, P.K.; Koch, B. Comparison of feature reduction algorithms for classifying tree species with hyperspectral data on three central European test sites. *IEEE J. Sel. Top. Appl. Earth Obs. Remote Sens.* **2014**, *7*, 2547–2561. [CrossRef]
18. Richter, R.; Reu, B.; Wirth, C.; Doktor, D.; Vohland, M. The use of airborne hyperspectral data for tree species classification in a species-rich central European forest area. *Int. J. Appl. Earth Obs. Geoinf.* **2016**, *52*, 464–474. [CrossRef]
19. Curran, P.J. Remote sensing of foliar chemistry. *Remote Sens. Environ.* **1989**, *30*, 271–278. [CrossRef]
20. Hennessy, A.; Lewis, M.; Clarke, K. Generative adversarial network synthesis of hyperspectral vegetation data. 2019; Unpublished manuscript, last modified 20 October 2019.
21. Hueni, A. Field Spectroradiometer Data: Acquisition, Organisation, Processing and Analysis on the Example of New Zealand Native Plants. Master's Thesis, Massey University, Palmerston North, New Zealand, 2006.
22. Aneece, I.; Epstein, H. Distinguishing early successional plant communities using ground-level hyperspectral data. *Remote Sens.* **2015**, *7*, 16588–16606. [CrossRef]

23. Cao, J.; Liu, K.; Liu, L.; Zhu, Y.; Li, J.; He, Z. Identifying mangrove species using field close-range snapshot hyperspectral imaging and machine-learning techniques. *Remote Sens.* **2018**, *10*, 2047. [CrossRef]

24. Dian, Y.; Fang, S.; Le, Y.; Xu, Y.; Yao, C. Comparison of the different classifiers in vegetation species discrimination using hyperspectral reflectance data. *J. Indian Soc. Remote Sens.* **2014**, *42*, 61–72. [CrossRef]

25. Eddy, P.; Smith, A.; Hill, B.; Peddle, D.; Coburn, C.; Blackshaw, R. Weed and crop discrimination using hyperspectral image data and reduced bandsets. *Can. J. Remote Sens.* **2014**, *39*, 481–490. [CrossRef]

26. Fung, T.; Yan Ma, H.F.; Siu, W.L. Band selection using hyperspectral data of subtropical tree species. *Geocarto Int.* **2003**, *18*, 3–11. [CrossRef]

27. Gross, J.W.; Heumann, B.W. Can flowers provide better spectral discrimination between herbaceous wetland species than leaves? *Remote Sens. Lett.* **2014**, *5*, 892–901. [CrossRef]

28. Hoa, P.; Giang, N.; Binh, N.; Hieu, N.; Trang, N.; Toan, L.; Long, V.; Ai, T.; Hong, P.; Hai, L. Mangrove species discrimination in southern Vietnam based on in-situ measured hyperspectral reflectance. *Int. J. Geoinform.* **2017**, *13*, 25–35.

29. Naidoo, L.; Cho, M.A.; Mathieu, R.; Asner, G. Classification of savanna tree species, in the greater Kruger national park region, by integrating hyperspectral and LiDAR data in a random forest data mining environment. *ISPRS J. Photogramm. Remote Sens.* **2012**, *69*, 167–179. [CrossRef]

30. Peerbhay, K.Y.; Mutanga, O.; Ismail, R. Commercial tree species discrimination using airborne AISA Eagle hyperspectral imagery and partial least squares discriminant analysis (PLS-DA) in KwaZulu–Natal, South Africa. *ISPRS J. Photogramm. Remote Sens.* **2013**, *79*, 19–28. [CrossRef]

31. Pu, R.; Bell, S.; Baggett, L.; Meyer, C.; Zhao, Y. Discrimination of seagrass species and cover classes with in situ hyperspectral data. *J. Coast. Res.* **2012**, *28*, 1330–1344. [CrossRef]

32. Adam, E.; Mutanga, O. Spectral discrimination of papyrus vegetation (Cyperus papyrus L.) in swamp wetlands using field spectrometry. *ISPRS J. Photogramm. Remote Sens.* **2009**, *64*, 612–620. [CrossRef]

33. Adam, E.; Mutanga, O.; Rugege, D.; Ismail, R. Discriminating the papyrus vegetation (Cyperus papyrus L.) and its co-existent species using random forest and hyperspectral data resampled to HYMAP. *Int. J. Remote Sens.* **2012**, *33*, 552–569. [CrossRef]

34. Aneece, I.; Thenkabail, P. Accuracies achieved in classifying five leading world crop types and their growth stages using optimal earth observing-1 hyperion hyperspectral narrowbands on google earth engine. *Remote Sens.* **2018**, *10*, 2027. [CrossRef]

35. Beh, B.C.; Tan, K.C.; Jafri, M.Z.M.; San Lim, H. Comparison of different discriminant functions for mangrove species analysis in Matang Mangrove Forest Reserve (MMFR), Perak based on statistical approach. In Proceedings of the Remote Sensing for Agriculture, Ecosystems, and Hydrology XIX, Maspalomas, Spain, 14–16 September 2004; p. 104211U.

36. Chan, J.C.-W.; Paelinckx, D. Evaluation of random forest and adaboost tree-based ensemble classification and spectral band selection for ecotope mapping using airborne hyperspectral imagery. *Remote Sens. Environ.* **2008**, *112*, 2999–3011. [CrossRef]

37. Das, B.; Sahoo, R.N.; Biswas, A.; Pargal, S.; Krishna, G.; Verma, R.; Chinnusamy, V.; Sehgal, V.K.; Gupta, V.K. Discrimination of rice genotypes using field spectroradiometry. *Geocarto Int.* **2018**, *35*, 64–77. [CrossRef]

38. Datt, B. Recognition of eucalyptus forest species using hyperspectral reflectance data. In Proceedings of the IGARSS 2000. IEEE 2000 International Geoscience and Remote Sensing Symposium, Honolulu, HI, USA, 24–28 July 2000; pp. 1405–1407.

39. Fernandes, M.R.; Aguiar, F.C.; Silva, J.M.; Ferreira, M.T.; Pereira, J.M. Spectral discrimination of giant reed (Arundo donax L.): A seasonal study in riparian areas. *ISPRS J. Photogramm. Remote Sens.* **2013**, *80*, 80–90. [CrossRef]

40. Ferreira, M.P.; Zortea, M.; Zanotta, D.C.; Shimabukuro, Y.E.; de Souza Filho, C.R. Mapping tree species in tropical seasonal semi-deciduous forests with hyperspectral and multispectral data. *Remote Sens. Environ.* **2016**, *179*, 66–78. [CrossRef]

41. George, R.; Padalia, H.; Kushwaha, S. Forest tree species discrimination in western Himalaya using EO-1 Hyperion. *Int. J. Appl. Earth Obs. Geoinf.* **2014**, *28*, 140–149. [CrossRef]

42. Jones, T.G.; Coops, N.C.; Sharma, T. Employing ground-based spectroscopy for tree-species differentiation in the Gulf Islands National Park Reserve. *Int. J. Remote Sens.* **2010**, *31*, 1121–1127. [CrossRef]

43. Papeş, M.; Tupayachi, R.; Martinez, P.; Peterson, A.; Powell, G. Using hyperspectral satellite imagery for regional inventories: A test with tropical emergent trees in the Amazon basin. *J. Veg. Sci.* **2010**, *21*, 342–354. [CrossRef]

44. Raczko, E.; Zagajewski, B. Tree species classification of the UNESCO man and the biosphere karkonoski national park (poland) using artificial neural networks and APEX hyperspectral images. *Remote Sens.* **2018**, *10*, 1111. [CrossRef]

45. Schmidt, K.; Skidmore, A. Spectral discrimination of vegetation types in a coastal wetland. *Remote Sens. Environ.* **2003**, *85*, 92–108. [CrossRef]

46. Shang, X.; Chisholm, L.A. Classification of Australian native forest species using hyperspectral remote sensing and machine-learning classification algorithms. *IEEE J. Sel. Top. Appl. Earth Obs. Remote Sens.* **2014**, *7*, 2481–2489. [CrossRef]

47. Thenkabail, P.S.; Enclona, E.A.; Ashton, M.S.; Van Der Meer, B. Accuracy assessments of hyperspectral waveband performance for vegetation analysis applications. *Remote Sens. Environ.* **2004**, *91*, 354–376. [CrossRef]

48. Thenkabail, P.S.; Mariotto, I.; Gumma, M.K.; Middleton, E.M.; Landis, D.R.; Huemmrich, K.F. Selection of hyperspectral narrowbands (HNBs) and composition of hyperspectral twoband vegetation indices (HVIs) for biophysical characterization and discrimination of crop types using field reflectance and Hyperion/EO-1 data. *IEEE J. Sel. Top. Appl. Earth Obs. Remote Sens.* **2013**, *6*, 427–439. [CrossRef]

49. Vaiphasa, C.; Ongsomwang, S.; Vaiphasa, T.; Skidmore, A.K. Tropical mangrove species discrimination using hyperspectral data: A laboratory study. *Estuar. Coast. Shelf Sci.* **2005**, *65*, 371–379. [CrossRef]

50. Vaiphasa, C.; Skidmore, A.K.; de Boer, W.F.; Vaiphasa, T. A hyperspectral band selector for plant species discrimination. *ISPRS J. Photogramm. Remote Sens.* **2007**, *62*, 225–235. [CrossRef]

51. Van Aardt, J.; Wynne, R. Examining pine spectral separability using hyperspectral data from an airborne sensor: An extension of field-based results. *Int. J. Remote Sens.* **2007**, *28*, 431–436. [CrossRef]

52. Wang, J.; Xu, R.; Yang, S. Estimation of plant water content by spectral absorption features centered at 1450 nm and 1940 nm regions. *Environ. Monit. Assess.* **2009**, *157*, 459–469. [CrossRef] [PubMed]

53. Asner, G.P. Biophysical and biochemical sources of variability in canopy reflectance. *Remote Sens. Environ.* **1998**, *64*, 234–253. [CrossRef]

54. Rivard, B.; Sanchez-Azofeifa, G.; Foley, S.; Calvo-Alvarado, J. Species classification of tropical tree leaf reflectance and dependence on selection of spectral bands. *Hyperspect. Remote Sens. Trop. Sub-Trop. For.* **2008**, *6*, 141–159.

55. Ollinger, S.V. Sources of variability in canopy reflectance and the convergent properties of plants. *New Phytol.* **2011**, *189*, 375–394. [CrossRef]

56. Thenkabail, P.; Smith, R.; De Pauw, E. *Hyperspectral Vegetation Indices for Determining Agricultural Crop Characteristics, CEO Research Publication Series No. 1*; Center for Earth Observation, Yale University Press: New Haven, CT, USA, 1999.

57. Thenkabail, P.S.; Smith, R.B.; De Pauw, E. Evaluation of narrowband and broadband vegetation indices for determining optimal hyperspectral wavebands for agricultural crop characterization. *Photogramm. Eng. Remote Sens.* **2002**, *68*, 607–622.

58. Ferreira, M.P.; Grondona, A.E.B.; Rolim, S.B.A.; Shimabukuro, Y.E. Analyzing the spectral variability of tropical tree species using hyperspectral feature selection and leaf optical modeling. *J. Appl. Remote Sens.* **2013**, *7*, 073502. [CrossRef]

59. Galvão, L.S.; Roberts, D.A.; Formaggio, A.R.; Numata, I.; Breunig, F.M. View angle effects on the discrimination of soybean varieties and on the relationships between vegetation indices and yield using off-nadir Hyperion data. *Remote Sens. Environ.* **2009**, *113*, 846–856. [CrossRef]

60. Blackburn, G.A. Hyperspectral remote sensing of plant pigments. *J. Exp. Bot.* **2006**, *58*, 855–867. [CrossRef] [PubMed]

61. Thomas, J.; Gausman, H. Leaf reflectance vs. leaf chlorophyll and carotenoid concentrations for eight crops. *Agron. J.* **1977**, *69*, 799–802. [CrossRef]

62. Castro-Esau, K.L.; Sánchez-Azofeifa, G.A.; Rivard, B.; Wright, S.J.; Quesada, M. Variability in leaf optical properties of Mesoamerican trees and the potential for species classification. *Am. J. Bot.* **2006**, *93*, 517–530. [CrossRef]

63. Pu, R. Broadleaf species recognition with in situ hyperspectral data. *Int. J. Remote Sens.* **2009**, *30*, 2759–2779. [CrossRef]

64. Demmig-Adams, B.; Adams, W.W. The role of xanthophyll cycle carotenoids in the protection of photosynthesis. *Trends Plant Sci.* **1996**, *1*, 21–26. [CrossRef]

65. Gamon, J.; Penuelas, J.; Field, C. A narrow-waveband spectral index that tracks diurnal changes in photosynthetic efficiency. *Remote Sens. Environ.* **1992**, *41*, 35–44. [CrossRef]

66. Gitelson, A.A.; Merzlyak, M.N.; Chivkunova, O.B. Optical properties and nondestructive estimation of anthocyanin content in plant leaves. *Photochem. Photobiol.* **2001**, *74*, 38–45. [CrossRef]

67. Gong, P.; Pu, R.; Yu, B. Conifer species recognition: An exploratory analysis of in situ hyperspectral data. *Remote Sens. Environ.* **1997**, *62*, 189–200. [CrossRef]

68. Van Aardt, J.A. Spectral separability among six southern tree species. *Photogramm. Eng. Remote Sens.* **2001**, *67*, 1367–1375.

69. Karlovska, A.; Grīnfelde, I.; Alsiņa, I.; Priedītis, G.; Roze, D. Plant reflected spectra depending on biological characteristics and growth conditions. In Proceedings of the International Scientific Conference Rural Development, Akademija, Lithuania, 23–24 November 2017.

70. Clevers, J.; De Jong, S.; Epema, G.; Van Der Meer, F.; Bakker, W.; Skidmore, A.; Scholte, K. Derivation of the red edge index using the MERIS standard band setting. *Int. J. Remote Sens.* **2002**, *23*, 3169–3184. [CrossRef]

71. Dawson, T.; Curran, P. Technical note A new technique for interpolating the reflectance red edge position. *Int. J. Remote Sens.* **1998**, *19*, 2133–2139. [CrossRef]

72. Gholizadeh, A.; Mišurec, J.; Kopačková, V.; Mielke, C.; Rogass, C. Assessment of red-edge position extraction techniques: A Case study for norway spruce forests using hymap and simulated sentinel-2 data. *Forests* **2016**, *7*, 226. [CrossRef]

73. Dalponte, M.; Bruzzone, L.; Vescovo, L.; Gianelle, D. The role of spectral resolution and classifier complexity in the analysis of hyperspectral images of forest areas. *Remote Sens. Environ.* **2009**, *113*, 2345–2355. [CrossRef]

74. Cochrane, M. Using vegetation reflectance variability for species level classification of hyperspectral data. *Int. J. Remote Sens.* **2000**, *21*, 2075–2087. [CrossRef]

75. Rock, B.; Hoshizaki, T.; Miller, J. Comparison of in situ and airborne spectral measurements of the blue shift associated with forest decline. *Remote Sens. Environ.* **1988**, *24*, 109–127. [CrossRef]

76. Knipling, E.B. Physical and physiological basis for the reflectance of visible and near-infrared radiation from vegetation. *Remote Sens. Environ.* **1970**, *1*, 155–159. [CrossRef]

77. Kumar, L. A comparison of reflectance characteristics of some Australian eucalyptus species based on high spectral resolution data—Discriminating using the visible and NIR regions. *J. Spat. Sci.* **2007**, *52*, 51–64. [CrossRef]

78. Kumar, L.; Skidmore, A.K.; Mutanga, O. Leaf level experiments to discriminate between eucalyptus species using high spectral resolution reflectance data: Use of derivatives, ratios and vegetation indices. *Geocarto Int.* **2010**, *25*, 327–344. [CrossRef]

79. Elvidge, C.D. Visible and near infrared reflectance characteristics of dry plant materials. *Remote Sens.* **1990**, *11*, 1775–1795. [CrossRef]

80. Lehmann, J.R.K.; Große-Stoltenberg, A.; Römer, M.; Oldeland, J. Field spectroscopy in the VNIR-SWIR region to discriminate between Mediterranean native plants and exotic-invasive shrubs based on leaf tannin content. *Remote Sens.* **2015**, *7*, 1225–1241. [CrossRef]

81. Kokaly, R.F.; Asner, G.P.; Ollinger, S.V.; Martin, M.E.; Wessman, C.A. Characterizing canopy biochemistry from imaging spectroscopy and its application to ecosystem studies. *Remote Sens. Environ.* **2009**, *113*, S78–S91. [CrossRef]

82. Alonso-Atienza, F.; Rojo-Álvarez, J.L.; Rosado-Muñoz, A.; Vinagre, J.J.; García-Alberola, A.; Camps-Valls, G. Feature selection using support vector machines and bootstrap methods for ventricular fibrillation detection. *Expert Syst. Appl.* **2012**, *39*, 1956–1967. [CrossRef]

83. Pudil, P.; Novovičová, J.; Kittler, J. Floating search methods in feature selection. *Pattern Recognit. Lett.* **1994**, *15*, 1119–1125. [CrossRef]

84. Guyon, I.; Weston, J.; Barnhill, S.; Vapnik, V. Gene selection for cancer classification using support vector machines. *Mach. Learn.* **2002**, *46*, 389–422. [CrossRef]

85. Deng, H.; Runger, G. Feature selection via regularized trees. In Proceedings of the 2012 International Joint Conference on Neural Networks (IJCNN), Dallas, TX, USA, 4–9 August 2013; pp. 1–8.

86. Melgani, F.; Bruzzone, L. Classification of hyperspectral remote sensing images with support vector machines. *IEEE Trans. Geosci. Remote Sens.* **2004**, *42*, 1778–1790. [CrossRef]

87. Pal, M.; Mather, P.M. Assessment of the effectiveness of support vector machines for hyperspectral data. *Future Gener. Comput. Syst.* **2004**, *20*, 1215–1225. [CrossRef]

88. Pal, M.; Foody, G.M. Feature selection for classification of hyperspectral data by SVM. *IEEE Trans. Geosci. Remote Sens.* **2010**, *48*, 2297–2307. [CrossRef]

89. Breiman, L. Random forests. *Mach. Learn.* **2001**, *45*, 5–32. [CrossRef]

90. Huberty, C.J. *Applied Discriminant Analysis*; Wiley-Interscience: New York, NY, USA, 1994.

91. Thompson, B. *Why Wont Stepwise Methods Die*; American Counseling Association: Alexandria, VA, USA, 1989.

92. Thompson, B. *Stepwise Regression and Stepwise Discriminant Analysis Need Not Apply Here: A Guidelines Editorial*; SAGE Publications: Thousand Oaks, CA, USA, 1995.

93. Flom, P.L.; Cassell, D.L. Stopping stepwise: Why stepwise and similar selection methods are bad, and what you should use. In Proceedings of the NorthEast SAS Users Group Inc 20th Annual Conference, Baltimore, MD, USA, 11–14 November 2007.

94. Whitaker, J.S. *Use of Stepwise Methodology in Discriminant Analysis*; The Annual Meeting of the Southwest Educational Research Association; Texas A & M University: College Station, TX, USA, 1997; p. 18.

95. Huberty, C.J.; Barton, R.M. An Introduction to Discriminant Analysis. *Meas. Eval. Couns. Dev.* **1989**, *22*, 158–168. [CrossRef]

 *remote sensing*

Article

Comparison of Support Vector Machine and Random Forest Algorithms for Invasive and Expansive Species Classification Using Airborne Hyperspectral Data

Anita Sabat-Tomala *, Edwin Raczko and Bogdan Zagajewski

Department of Geoinformatics, Faculty of Geography and Regional Studies, University of Warsaw, Cartography and Remote Sensing, ul. Krakowskie Przedmieście 30, 00-927 Warsaw, Poland; edwin.raczko@uw.edu.pl (E.R.); bogdan@uw.edu.pl (B.Z.)
* Correspondence: anita.sabat@uw.edu.pl; Tel.: +48-225-520-654

Received: 17 December 2019; Accepted: 3 February 2020; Published: 5 February 2020

Abstract: Invasive and expansive plant species are considered a threat to natural biodiversity because of their high adaptability and low habitat requirements. Species investigated in this research, including *Solidago* spp., *Calamagrostis epigejos*, and *Rubus* spp., are successfully displacing native vegetation and claiming new areas, which in turn severely decreases natural ecosystem richness, as they rapidly encroach on protected areas (e.g., Natura 2000 habitats). Because of the damage caused, the European Union (EU) has committed all its member countries to monitor biodiversity. In this paper we compared two machine learning algorithms, Support Vector Machine (SVM) and Random Forest (RF), to identify *Solidago* spp., *Calamagrostis epigejos*, and *Rubus* spp. on HySpex hyperspectral aerial images. SVM and RF are reliable and well-known classifiers that achieve satisfactory results in the literature. Data sets containing 30, 50, 100, 200, and 300 pixels per class in the training data set were used to train SVM and RF classifiers. The classifications were performed on 430-spectral bands and on the most informative 30 bands extracted using the Minimum Noise Fraction (MNF) transformation. As a result, maps of the spatial distribution of analyzed species were achieved; high accuracies were observed for all data sets and classifiers (an average F1 score above 0.78). The highest accuracies were obtained using 30 MNF bands and 300 sample pixels per class in the training data set (average F1 score > 0.9). Lower training data set sample sizes resulted in decreased average F1 scores, up to 13 percentage points in the case of 30-pixel samples per class.

Keywords: Natura 2000; invasive species; expansive species; support vector machine; random forest; biodiversity

---

1. Introduction

The spread of invasive and expansive species is one of the main threats to biodiversity and functioning of ecosystems [1]. This results in transformation of natural habitats, displacement of native species, and degrading environmental conditions (e.g., number of existing micro- and macrophytes). It also generates economic losses by degrading the quality of soil and destroying road and railway infrastructure [2]. In the European Union (EU), it is estimated that the cost of controlling and combating invasive species amounts to approximately 12 billion EUR per year [3]. Implementation of appropriate remedial strategies and effective limitation of the invasion's effects require constant monitoring, which is emphasized in the EU Regulation No. 1143/2014.

The species that pose a threat to natural habitats protected under the Natura 2000 program in Poland include, for example, native expansive plants such as blackberry shrubs (*Rubus* spp. L.), perennial wood small-reed (*Calamagrostis epigejos* (L.) Roth), and foreign invasive goldenrod species (*Solidago* spp. L). These species do not have high requirements concerning their habitat; they also reproduce quickly, both in terms of vegetative and generative reproduction, and they stifle other plants [4]. They negatively impact valuable natural habitats, such as inland sand calcareous grasslands, mountain and lowland *Nardus* grasslands, *Molinia* meadows, and alluvial meadows. They are extensively used in fresh low pastures in mountain hay and bent-grass meadows [5–7]. In order to prevent further changes in the vegetation, these harmful species should be identified and removed preferably at the early stages of invasion.

The current monitoring of plant species changes is based on fixed target areas. Individual specimens of the species found in target areas are counted, and the observed regularities are extrapolated to the whole area, which can differentiate due to, for example, environmental components or land use. In comparison to traditional fields, remote sensing allows for objective and repetitive monitoring that can be conducted both on local and global scales [8,9]. Considering the complexity of class distinctions, both intra-class similarities and differences between classes, the data which can be used for this purpose are multispectral, such as Landsat [10], WorldView-2 [11], or hyperspectral data (e.g., HyMap) [12]. As hyperspectral data constitute a source of ongoing information about spectral reflection, they provide a lot of information about the biophysical and chemical characteristics of the analyzed vegetation [13–15]. Either hyperspectral satellite data (e.g., Hyperion [16] and CHRIS [15,17]) or aerial data (e.g., APEX [18] and AISA [19,20]) are used, depending on the size of the research area and the canopy characteristics of the identified vegetation. Airborne data are more useful for the detection of small, less compact patches of plant species because of their high spatial resolution [16]. The study of Mediterranean plants in southern France confirms that spectral and spatial resolution influence the accuracy of vegetation mapping [21]. The highest accuracy of classification of five vegetation types was obtained using the airborne hyperspectral imaging sensor, HyMap. Depending on the classification method used, the overall accuracies (OAs) ranged from 62.3% for k-nearest neighbor (k-nn), 67.7% for Random Forest (RF), and 70.2% for Support Vector Machine (SVM), up to 72.5% for Artificial Neural Networks (ANNs), while the use of ASTER satellite data resulted in slightly lower accuracy levels (from 60.3%), and the worst results were obtained using multispectral data Landsat 7 ETM + (59.3%).

Multi-dimensional, large-scale image data can be used effectively when their use is based on modern classification methods, i.e., Support Vector Machine (SVM) [22] or Random Forest (RF) [23]. Both are considered to be among the most effective classification methods [21]. The SVM algorithm transforms the original space and then constructs an optimal hyperplane in the multi-dimensional feature space, which divides the data into different classes with the largest possible margin of separation. The algorithm works well on noisy data and small numbers of training pixels; it is sufficient to develop support vectors and usually has a higher level of accuracy than other classification algorithms [21,24]. The SVM method was compared with different types of neural networks (MLP, multilayer perceptrons; CANFIS, co-active neurofuzzy inference systems) used for classifying five types of cultivated plants in Spain using HyMap data [25]. Results have shown that, despite small differences in the classification accuracy (OA_{SVM} = 96,4% 29, OA_{MLP} = 94,5%, OA_{RBF} = 94,1%, OA_{CANFIS} = 94,2%), the SVM algorithm is more efficient than neural networks in terms of stability, reliability, simplicity, as well as the speed of the classification process. Moreover, SVM achieved very high accuracies (OA = 93%) during the detection of invasive *Solanum mauritianum* shrubs on *Pinus patula* plantations in southern Africa on the basis of AISA Eagle images [20].

On the other hand, the RF algorithm works by creating many decision trees based on a random subset of training data, and the final decision is made by combining individual tree votes [23]. The advantage of this method is its resistance to overfitting of the training set and its short classification time. Good results were achieved by using the RF method to study the invasion of *Euphorbia escula* and *Centaurea maculosa* in Montana [15]. The accuracy levels of classification based on the airborne hyperspectral HySpex images for the mentioned plant species were 86% and 84%, respectively. Additionally, the Random Forest algorithm has proved its worth in identifying two expansive grassland species, *Molinia caerulea* and *Calamagrostis epigejos*, in the Silesia Upland in Poland. HySpex and LiDAR (light detection and ranging) products from the Riegl LMS-Q680i scanner were used in the study, obtaining the highest median Kappa of 0.85 (F1 = 0.89, which is a mathematical product of the user (UA) and producer accuracies (PA)) for *M. caerulea* identification and 0.65 (F1 = 0.73) for *C. epigejos* [26].

The use of SVM and RF methods yielded good results during the classification of 20 types of grassy vegetation in the Hortobágy National Park in eastern Hungary on the basis of AISA Eagle II data [27]. The highest accuracy of classification was obtained on the first nine Minimum Noise Fraction (MNF) transformation bands of the hyperspectral image and by using 30 random training pixels (OA_{SVM} = 82.06%, OA_{RF} = 79.14%, OA_{ML} = 80.78%). However, when the training set was reduced to 10 pixels, SVM and RF methods still maintained high levels of accuracy (79.57% and 76.55%, respectively), while the ML accuracy dropped significantly to 52.56%. The low level of sensitivity to the training sample size is a big advantage of these algorithms, especially SVM. On the other hand, the RF algorithm had a short image classification time (3 minutes) compared to the other methods used on the same data set (SVM = 16 min, ML = 8 min). Studies of Mediterranean vegetation (mainly shrubs varying in height from about 0.5 m to almost 5 m) that were carried out in Languedoc in southern France demonstrated that RF and SVM methods obtained better information from hyperspectral data than any traditional classifiers (e.g., classification tree (CT), linear discriminant analysis (LDA), quadratic discriminant analysis (QDA), and k-nearest neighbor (k-nn)), especially when the spectral differences between classes were small [21]. When distinguishing 15 species of plants, the overall accuracies of the classification for modern methods, i.e., SVM and RF (OA_{SVM} = 39.2–47.9%, OA_{RF} = 39.3–49.5%), were higher than those recorded for traditional methods (OA_{CT} = 28.6–44.4%, OA_{LDA} = 37–45.1%, OA_{QDA} = 37.5–39.3%, $OA_{k\text{-}nn}$ = 18–28.8%), depending on the set of input data. The artificial neural network (ANN) method was also used to identify plant species; however, this experiment did not lead to satisfactory results.

The aim of the current analysis was to verify whether the expansive/invasive *Rubus* spp., *Calamagrostis epigejos*, and *Solidago* spp. were characterized by a specific set of spectral characteristics that allowed them to be distinguished from the surrounding species, which altogether create a mix of fuzzy, covered patterns. Moreover, an analysis of the impact of the number of pixels in training data set on the classification accuracy was performed. Well-known reference classification algorithms were applied, SVM and RF methods, which are commonly used because of their effectiveness.

The proposed method could be applied in extensively used agricultural areas (considering traditionally used farming methods), and not limited to only selected test areas.

2. Materials and Methods

2.1. Study Area and Objects of the Study

The research area was located in southern Poland near the town of Malinowice (Silesian Province) and covered an area of approximately 10.6 km^2 of the Natura 2000 habitat (Figure 1). This is an upland area covering the Tarnogóra Hummock and the Katowice Upland and is in a transitional temperate climate. This area is dominated by grasslands, meadows, and forests. Blackberry (mainly *Rubus caesius* L., European dewberry), various species of goldenrod (*Solidago* spp.), and wood small-reed grass (*Calamagrostis epigejos*) occur very frequently in this area.

Rubus spp. L., a genus of plant in the *Rosacee* family commonly called bramble, is one of the most important expansive species [28]. Blackberries are native to Asia, Europe, and North and South America [29], and they often pose a threat to young forest crops and habitats protected under the Natura 2000 program. They are typically shrubs (can be up to 3 meters high) with perennial roots, biennial prickly stems, and edible fruits which are aggregates of drupelets [29]. Blackberries can be found in all kinds of environments, including forests, shrubs, meadows, wastelands, and roadsides. Vegetative reproduction and production of a large number of seeds that are spread by birds and other animals allows them to quickly colonize new areas [30]. They bloom from May to September. According to the latest data, there are 105 *Rubus* species in Poland alone [31]. *Rubus* spp. L. is linked to negative economic and environmental consequences (e.g., changes in the dominant type of vegetation, soil depletion, or increased susceptibility to fires) [32]. The spectral characteristics of *Rubus* spp. are very similar, which is why they were identified collectively in the paper without division into individual species.

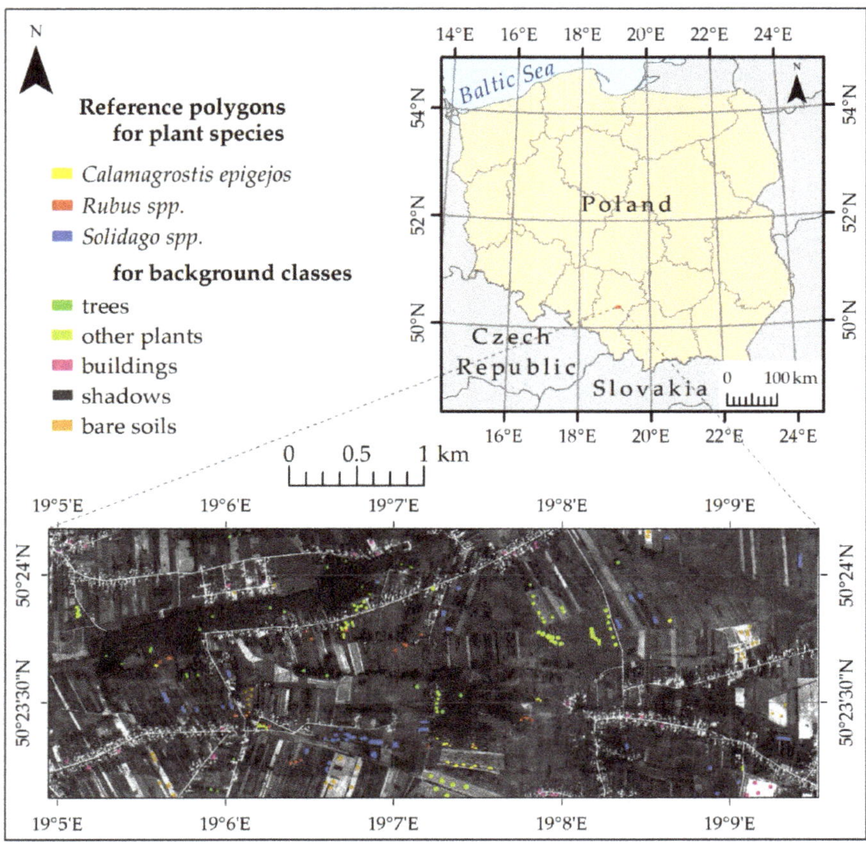

Figure 1. Field research polygons on the Malinowice area.

Another widespread, expansive species that degrades grassland and meadow communities is *Calamagrostis epigejos* (L.) Roth, commonly referred to as wood small-reed [33]. It is a perennial grass in the *Poaceae* family, which is native to the Eurasian area [5], and has spread to North America [34]. The plant has thick and rigid blades that can be up to 2 meters high and has complex inflorescences in the form of a panicle. Wood small-reed propagates vegetatively, through numerous stolons, as well as generatively, through seeds (i.e., kernels) [35]. It blooms from July to September, often forming extensive single-species fields whose colors vary from green to brown to purple. Wood small-reed grows in meadows, forests, urban areas, along railways, and on the roadsides. A large amount of reed biomass is deposited in non-hay areas, and its lengthy decomposition time causes acidification of the substrate and hinders development of other plants [36].

Some of the most invasive plants that pose a huge threat to native species and biodiversity of entire ecosystems are representatives of the goldenrod genus (*Solidago* spp. L.). They are perennials from the *Asteraceae* family, imported from North America to Europe as decorative plants [37]. Goldenrod occurs in the form of three invasive species: *Solidago canadensis* (Canadian goldenrod), *Solidago gigantea* (tall goldenrod), and *Solidago graminifolia* (grass-leaved goldenrod) [38,39]. These plants have stiff sprouts that can be up to 2 meters tall, ending in pyramidal panicle clusters, which are formed by flowers clustered in heads [40]. They propagate vegetatively, thanks to underground rhizomes, and generatively with the help of light seeds (achenes with pappus) that can be spread over considerable distances [41]. They quickly begin to dominate and often form dense single-species patches. They bloom from July to October, forming characteristic yellow inflorescences. Goldenrods have a high tolerance for various soil types, but they require exposure to full sun [42]. They grow in open habitats such as meadows, wastelands, anthropogenic areas, and along roads and river banks [2].

2.2. Field Measurements

The field studies were conducted in the summer of 2017. Compact polygons (in the shape of circles with a radius of 3 meters) of *Rubus* spp., *Solidago* spp., *Calamagrostis epigejos*, and other background plants were located within the research area with the help of the Leica CS20 GNSS device (Figure 2). The number of polygons was proportional to the prevalence of species in the research area and amounted to 50 polygons for blackberry and wood small-reed, 60 polygons for goldenrod, and 100 polygons for background plants.

Figure 2. Collection of reference plot locations—plot for *Calamagrostis epigejos* shown in above picture.

Then, the polygons were transferred to ArcMap 10.3, where photo interpretation techniques were used to create an additional 30 reference polygons for other types of land cover (i.e., trees, buildings, bare soil, and shaded areas). These additional classes were meant to indicate for the algorithm the spectral properties of objects that occurred in the research area and constituted non-forest vegetation. Finally, reference polygons were created for eight classes: *Calamagrostis epigejos*, *Solidago* spp., *Rubus* spp., other plants, trees, bare soils, buildings, and shadows (Figure 1).

2.3. Airborne Hyperspectral HySpex Data

Aerial hyperspectral data were obtained by MGGP Aero Sp. z o.o. on 29 August 2017 using sensors located on a Cessna 402B aircraft. A hyperspectral image with a 16-bit radiometric resolution was recorded with two HySpex Visible and Near Infrared (VNIR-1800) scanners and a Shortwave Infrared (SWIR-384) scanner. The specification of both sensors is provided in the table below (Table 1).

Table 1. Main technical characteristics of HySpex scanners used in this work.

Parameters	HySpex VNIR-1800	HySpex SWIR-384
Spectral range	416–995 nm	954–2510 nm
Number of spectral bands	182 (163 [1])	288
Spectral sampling	3.26 nm	5.45 nm
Spatial resolution	0.5 m	1 m
Spatial pixels [2]	1800	384
Field of View (FOV) across track	17–34o	16–32o
Instantaneous Field of View (IFOV)	0.01–0.04o	0.04–0.08o

[1] A number of spectral bands were deleted due to overlapping spectral ranges between VNIR-1800 and SWIR-384 sensors.
[2] a number of pixels per scan line.

The aerial hyperspectral images were prepared for further processing in accordance with the diagram presented in the schema (Figure 3). The data obtained by hyperspectral sensors were converted to radiance units with HySpex RAD software. Then, the hyperspectral image was subjected to geometric correction, which employed the digital surface model in PARGE (PARametric GEocoding) software (ReSe Applications LLC, Wil, Switzerland), and atmospheric correction was performed with the MODTRAN5 model in ATCOR4 (ATmospheric CORrection) software (ReSe Applications LLC, Wil, Switzerland). Nine flight lines were mosaicked and re-sampled to achieve a uniform spatial resolution of 1 m. Next, the last 21 bands in the SWIR range were removed because of the high level of noise caused by the sensor's lower SNR (Signal to Noise Ratio) at the extreme ranges of the imaged spectrum, which ultimately resulted in a 430-band image in the spectral range of 416.18–2396.44 nm.

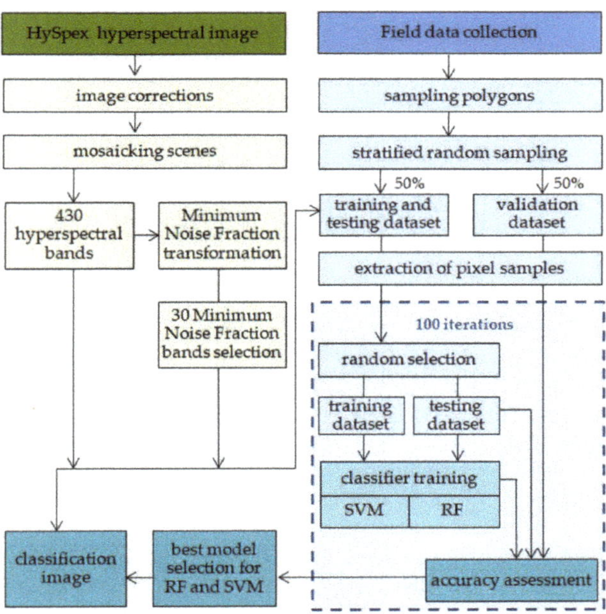

Figure 3. Research algorithm.

In order to reduce HySpex data dimensionality, Minimum Noise Fraction (MNF) transformation was applied. Based on MNF bands, eigenvalues, and visual assessment of transformed bands, the first 30 MNF bands were selected for further processing. Finally, two data sets were prepared for species classification: the first one contained 430 HySpex bands while the second one contained 30 MNF bands.

2.4. Classification Process and Accuracy Assessment

One of our goals was to analyze the impact of pixel number in the training data sets on classification accuracy; hence, we created training data sets with a set number of pixels per class. Using stratified random sampling, 50% of all reference polygons were selected to create a training test data set, while remaining polygons were used to create a validation data set.

The training test data set was used to create subsets (training data sets with a set number of pixels per class), and all remaining samples ended up in the test data set that was used for preliminary accuracy assessment. The validation data set was created to eliminate spatial autocorrelation with the training test data set (randomly selected pixels were used in the training and validation sets). This allowed to create a spatially independent and stable validation data set, which was used to assess final results.

To investigate the influence of training data set size on achieved classification results, the training test data set was sub-sampled to create training data sets that contained exactly 30, 50, 100, 200, and 300 pixels per class. These sub-sampled data sets will be used for classifier training. If a given class had fewer total available samples than required, random sampling with replacement was used, otherwise random sampling without replacement was employed. If all available pixels for given class were selected for training purposes, a copy of training data for this class was used instead.

An iterative accuracy assessment was used in order to objectively compare achieved results. This was a procedure consisting of the following steps repeated 100 times:

1. Sub-sample the training test data set in order to create a training data set with a set number of samples per class;
2. Train SVM and RF classifiers;
3. Assess accuracy using test and validation data sets; and
4. Save trained classifier models and accuracy measures for further analysis.

Pixel classification was carried out on the basis of the Support Vector Machine and Random Forest algorithms in R software. The first stage of the training process was to optimize the learning parameters of these algorithms in order to obtain the best possible settings. This task was completed on the training and test sets before the division. A radial basis function was chosen for the SVM algorithm because of its proven efficiency [43] and smaller number of computational difficulties [44]. The learning parameters of the compared classification algorithms were subjected to a tuning process. A gamma value of 0.1 and cost of 1000 was obtained for the SVM algorithm. In the case of the Random Forest algorithm, on the basis of the out-of-bag (OOB) error analysis, the mtry parameter (the number of features randomly sampled at each split) was set at 140 for classification on 430 hyperspectral bands and at 13 for classification on the set of the first 30 MNF transformation bands. In both cases, the number of random trees (ntree) amounted to 500.

In this work, we compared two classification algorithms (SVM and RF), two different data sets (430 original hyperspectral bands, 430 HS, and 30 Minimum Noise Fraction bands, 30 MNF), and five different sample sizes per class in the training data set (30, 50, 100, 200, and 300 pixels). Due to the unavailability of the larger continuous areas of invasive plants on our study area, we have limited the analysis to 300 pixels. All combinations of the above parameters were tested, resulting in 20 different classification scenarios.

Accuracy of the performed classifier training was assessed with the set of test data and the data spatially separated from the training and test set (i.e., on pixels of the validation data set), which was constant for all scenarios. The algorithms were compared, and the best combination of image data set and classifier was determined based on validation performance. The following accuracy parameters were calculated on the basis of the error matrix:

- Overall accuracy—the ratio of the total number of correctly classified pixels to the total number of reference pixels [45];
- Cohen's Kappa—the similarity of the analyzed classification compared to the random classification (a Kappa value of 0 means full similarity while 1 means no similarity) [46];
- Producer's Accuracy (PA)—the ratio of correctly classified pixels of a given class to all pixels in the validation data set for this class [45];
- User's Accuracy (UA)—the ratio of pixels correctly classified in a given class to all pixels classified in this category [45]; and
- F1 score sensitivity, measured using harmonic mean of precision (P), positive predictive value, and recall (R) as in Equation (1) [47,48]:

$$F = 2PR/(P+R) \tag{1}$$

Afterwards, the best models for each classifier and data set were selected on the basis of the mean F1 scores for all classes (based on the validation data), and the images were classified. The significance of statistical differences between the accuracy of the models was checked using the Mann–Whitney–Wilcoxon test [49] (significance level = 0.05). The Mann–Whitney–Wilcoxon test is well suited for testing differences between non-normally distributed populations [26,50]. Distributions of achieved accuracy measures for all classification scenarios were visualized using box plots. A detailed explanation of boxes used in box plots is shown in Figure 4.

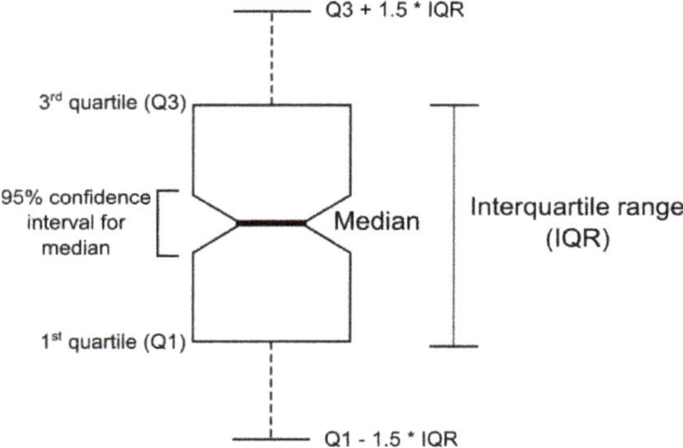

Figure 4. Explanation of structural elements of boxes used in box plot.

Moreover, classifier training was performed on nine classes, each with an identical number of training samples to reduce any effect of unbalanced training data. After classifier training, background classes were considered as one class with relation to plant classes. Such steps allowed us to properly assess classification quality (which classes are confused with which) and helped us achieve the most accurate results. In our work we assumed that confusion between background classes was acceptable, while confusion between plant species and background classes or other plant species would be a concern that would need to be addressed and reported. When classifying plant species, it is important to deliver a suitable and representative sample of pixels that characterize objects other than object of the study. Such classes can be oftentimes referred to as background classes. Since our study aimed to investigate the influence of training data set size, it would be insufficient to perform classification of four classes, that is, three plant species and one class with background objects. This mainly is due to difficulties in randomly sampling background classes in such a way that, for example, 30 pixels will represent them all. In fact, such an approach would almost guarantee that pixels for background classes covering a relatively small area would not be included in the training data set with a sufficient number of samples, which in turn would destroy any credibility of such work. To address this issue when creating the training data set, each background class (shadows, trees, other plants, soils, and buildings) had the same number of training samples, equal to the number of samples used for each plant species class. This is to ensure our background classes had similar representation to the plant species classes during classifier training.

3. Results

3.1. Statistical Analysis of Investigated Classification Scenarios

The mean F1 score was calculated for all classes on two sets: the test set and the validation set. The test set was dependent on the training set—the pixels in these sets were drawn from the same polygons, so the number of pixels in the test set decreased with an increasing number of pixels in the training set (Table 2). The high accuracy level obtained for this set is, therefore, not surprising, nor can it be used to compare the classifiers.

In contrast, the validation set had a fixed number of observations (4835 pixels) and was spatially independent of the other data sets. Regardless of the classifier used, higher mean F1 scores for all classes based on the validation set were obtained for classifications performed on 30 MNF transformation bands (0.854–0.918) compared to that of the 430 hyperspectral data bands (0.760–0.853).

Table 2. Classifier training parameters and their average F1 scores.

| Data Used | Algorithm (Parameters) | Number of Pixels | | Testing Set | Mean for 100 Iterations | | | |
| | | Training Set per Class/All | Testing Set | F1 for All Classes | Validation Set (4835 pixels) | | | |
					F1 for all Classes	F1 for 3 Plant Species	F1 for Background Classes	Kappa
430 hyperspectral bands	RF (mtry =140; ntree = 500)	30/240	4612	0.847	0.784	0.789	0.856	0.759
		50/400	4452	0.876	0.803	0.811	0.871	0.785
		100/800	4052	0.905	0.823	0.831	0.866	0.808
		200/1600	3252	0.931	0.843	0.851	0.869	0.832
		300/2400	2452	0.935	0.853	0.859	0.869	0.842
	SVM (kernel = radial; cost = 1000; gamma = 0.1)	30/240	4612	0.843	0.760	0.803	0.807	0.737
		50/400	4452	0.888	0.787	0.833	0.853	0.771
		100/800	4052	0.935	0.817	0.867	0.820	0.808
		200/1600	3252	0.964	0.840	0.893	0.827	0.833
		300/2400	2452	0.972	0.852	0.907	0.826	0.847
30 MNF bands	RF (mtry = 13; tree = 500)	30/240	4612	0.952	0.869	0.868	0.965	0.853
		50/400	4452	0.974	0.893	0.890	0.952	0.888
		100/800	4052	0.988	0.910	0.908	0.942	0.906
		200/1600	3252	0.994	0.917	0.920	0.940	0.917
		300/2400	2452	0.995	0.918	0.926	0.932	0.92
	SVM (kernel = radial; cost = 1000; gamma = 0.1)	30/240	4612	0.961	0.854	0.918	0.860	0.850
		50/400	4452	0.980	0.871	0.933	0.850	0.874
		100/800	4052	0.993	0.881	0.943	0.847	0.885
		200/1600	3252	0.998	0.883	0.949	0.846	0.889
		300/2400	2452	0.999	0.883	0.951	0.838	0.891

The accuracy level for both classifiers increased with the number of training pixels used for classification (Figure 5). The distributions of the mean F1 score for all classes revealed that when the number of training pixels increased, the interquartile range of the obtained accuracies decreased, so the results obtained in 100 iterations were more stable What is more, the use of a smaller number of training pixels caused a greater decrease in the accuracy of classifications performed on the original hyperspectral bands than in the case of classifications performed on the MNF transformation bands. The most stable distributions and the highest F1 scores for all classes were obtained by the classifications performed on a set of 30 MNF transformation bands and 300 training pixels (the median F1 for RF was about 0.92, while the median F1 for SVM was about 0.88).

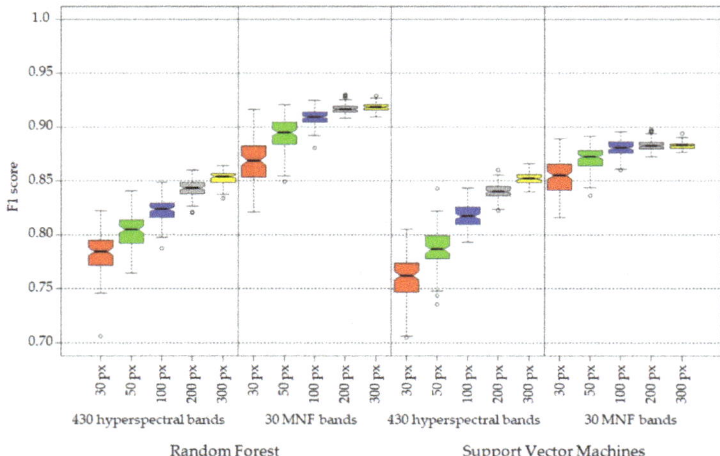

Figure 5. Distributions of mean F1 scores for all classes calculated on the validation data set for SVM and RF classifiers; both analyzed raster data sets and a different number of training pixels. Explanations are presented in Figure 4.

In order to check if there are statistically significant differences in the F1 scores of all the tested scenarios, the Mann–Whitney–Wilcoxon test was carried out at the significance level of 0.05 (Figure 6). There were statistically significant differences between most of the considered scenarios. The SVM classifications on MNF bands using 200 and 300 pixels for classifier training were the only exception. There were no statistically significant differences found for the RF classification performed on 430 hyperspectral bands using 300 training pixels and the SVM classification on a very limited data set consisting of 30 MNF bands and 30 training pixels.

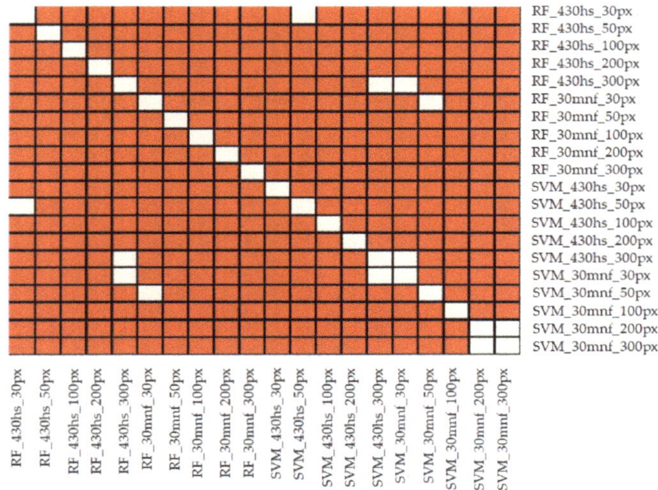

Figure 6. Matrix of statistical significance between scenarios calculated on the basis of F1 accuracy for all classes using the U Mann–Whitney–Wilcoxon test (red fields indicate significant differences between populations at the 0.05 significance level). Names of scenarios contain an acronym of the classification algorithm (RF or SVM), information about raster data (430 HS or 30 MNF), and size of the training data set in pixels.

An analysis of the distribution of F1 scores for individual classes of identified species (Figure 7) makes it possible to draw conclusions about the best data sets and algorithms for classifying each class.

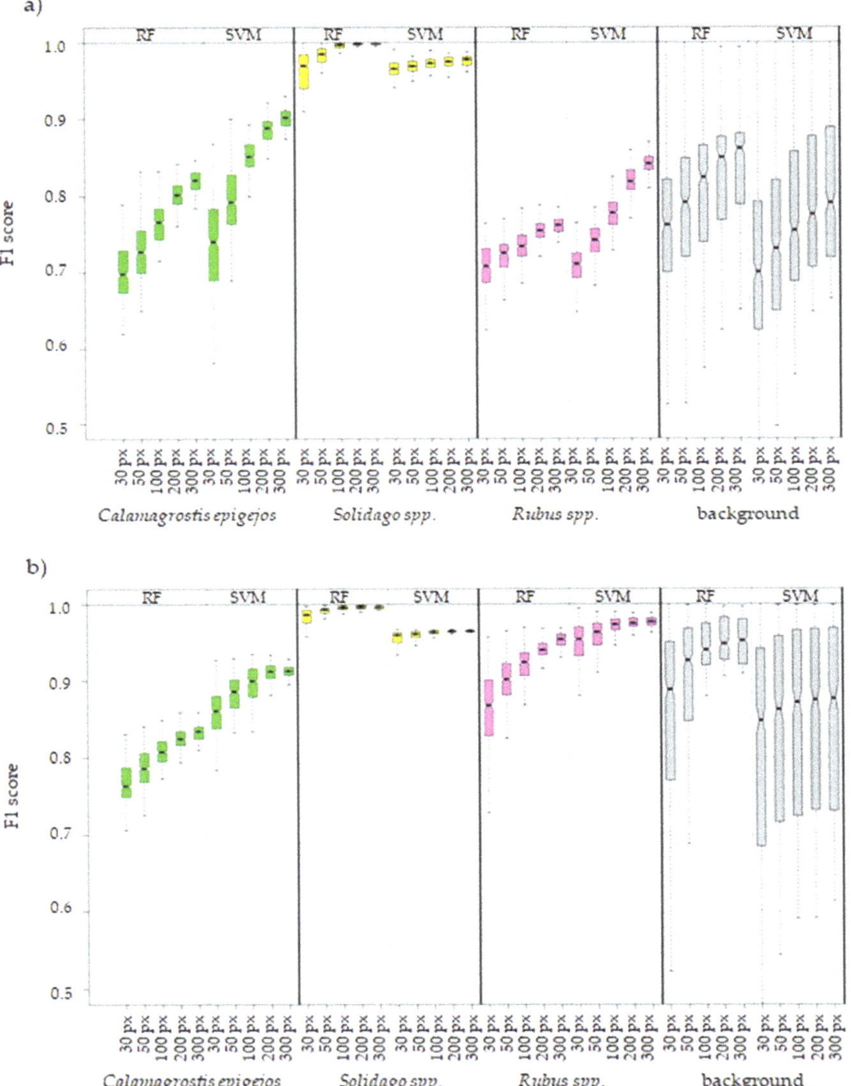

Figure 7. F1 score distribution for validation data set (**a**) 430 bands and (**b**) 30 MNF bands. The horizontal axis of the charts indicates the number of pixels in the training set used to classify the given species using RF or SVM classifiers. The vertical axis shows the accuracy of the scenarios.

The *Solidago* spp. class identified well with all classifiers and raster data sets (the F1 score was above 0.95). The accuracy levels increased with an increasing number of training pixels, whereas the differences in accuracy levels resulting from the change in the size of the training sets were small. However, slightly higher mean F1 scores were recorded for the Random Forest classifier. *Solidago* are marked by their very characteristic yellow color and spectral properties, which distinguished them from other classes in the imaging, and additionally tend to form large, uniform fields, so the almost perfect identification of this species was not surprising.

In the case of the *Rubus* spp. class, the best identification results were obtained for the SVM classification on 30 MNF bands using 300 training pixels (F1 = 0.97), but application of the same classifier with the number of training pixels reduced to 100 resulted in a similar accuracy level. Good results were also obtained for the RF classification on the same raster data set and 300 training pixels (F1 = 0.95). The F1 scores obtained on 430 hyperspectral data bands were lower (F1 RF from 0.7 to 0.76, and F1 SVM from 0.71 to 0.84).

Calamagrostis epigejos was a more difficult plant species to identify. However, high F1 scores of around 0.91 were obtained using the SVM algorithm, 30 MNF transformation bands, and sets of 200 and 300 training pixels. A similar accuracy level was also obtained for the SVM classification and 300 training pixels on 430 hyperspectral bands (F1 = 0.9). The Random Forest classification resulted in lower accuracy levels for this species, with F1 scores between 0.7 and 0.82 on the hyperspectral data set, and between 0.76 and 0.83 on the MNF transformation bands. The accuracy increased with the growth of the number of training pixels.

Considering the mean accuracy level for three species identified in the research area, it can be concluded that the best spatial distribution was obtained using the SVM algorithm and 200 or 300 training pixels (F1 = 0.95). For the other classes distinguished in the image (i.e., plant background, forests, buildings, bare soil, and shadows), the best F1 scores (from 0.93 to 0.96) were obtained with the RF algorithm. However, in terms of accuracy for all the classes together, the best accuracy (Kappa = 0.92, F1 for all classes = 0.92) was obtained for the RF classifier, 30 MNF bands, and sets of 200 and 300 training pixels.

To sum up, the SVM algorithm and the data set consisting of 30 MNF bands and 300 training pixels proved to be the best for identifying the *Calamagrostis* and *Rubus* classes. In the case of *Solidago* and background classes, better results were obtained with the Random Forest classifier. However, goldenrod classified well (mean F1> 0.95) on both sets of raster data and with a different numbers of training pixels. On the other hand, in the case of background classes, the best results were obtained for 30 MNF bands and 200 training pixels. This may indicate that the Random Forest method works better for the classification of spectrally uniform, large forms of land use, which differ significantly from their surroundings, while the SVM method is better for identifying plant species that are more spectrally different and similar to the background classes.

3.2. Best Model Plant Species Identification Accuracy

A set of data consisting of 30 MNF bands and 300 training pixels was selected on the basis of the analysis of statistical accuracy to develop images showing spatial distributions of the analyzed species in the research area. Figure 8 presents distributions of the producer and user accuracies for 100 iterations of classifications performed on a selected set of data using both classifiers.

For the *Rubus* spp. class, the RF classifier yielded a lower median user's accuracy than that of SVM by three percentage points, while the differences in the producer's accuracy levels between the classifiers were small. Both the producer's and user's accuracies for *Solidago* spp. were very high (close to 100%), a slight underestimation was detected only in the case of the SVM classification (producer's accuracy about 93%). In contrast, the *Calamagrostis epigejos* class achieved the lowest median producer and user accuracies of all classes. The SVM classifier achieved higher producer and user accuracy levels for *C. epigejos* (PA = 96%, UA = 87%) than the RF classifier (PA = 88%, UA = 78%).

The resulting images for both classification methods prepared for the best mean F1 scores for all iteration classes are presented and compared below (Figure 9). The correctness of species identification was also assessed on the basis of the confusion matrix (Tables 3 and 4).

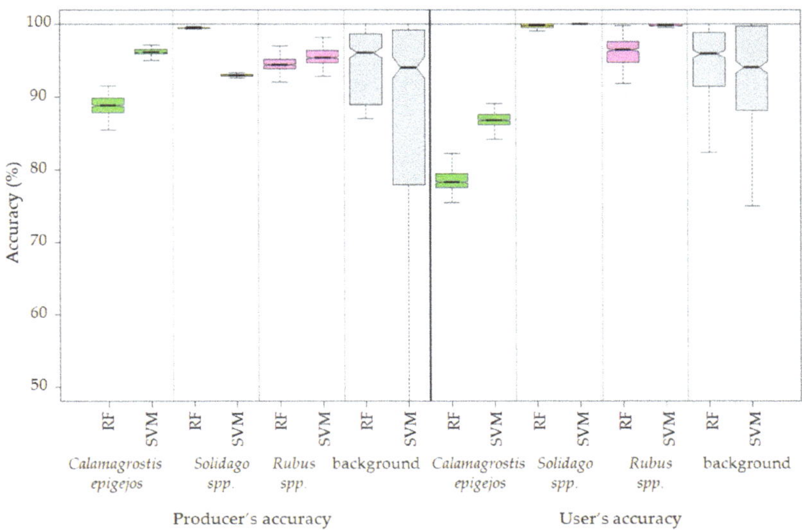

Figure 8. User and producer accuracies of the 300 pixel training set and 30 MNF bands classification.

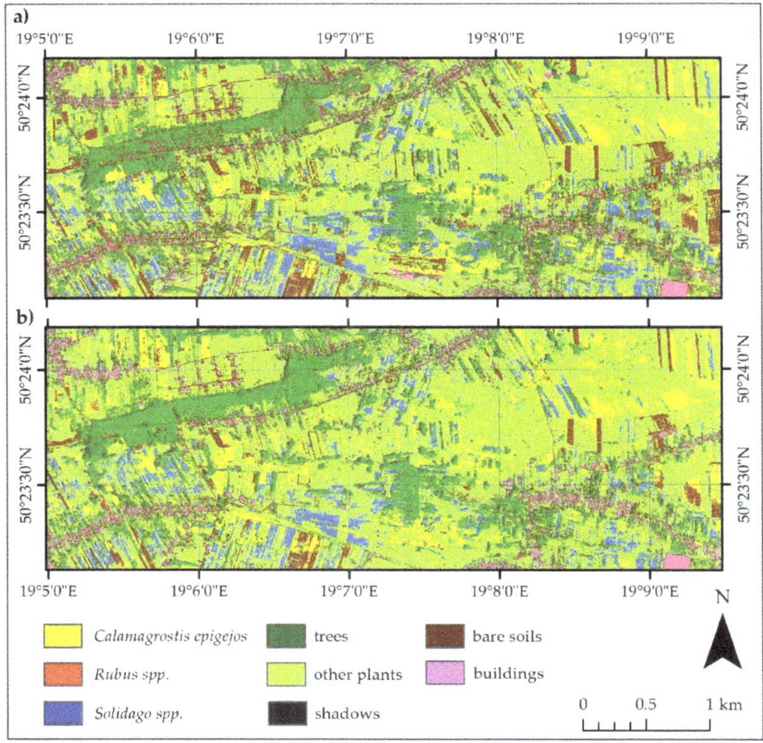

Figure 9. Classification results of the (**a**) SVM and (**b**) RF based on 30 MNF bands and 300-pixel training sets; SVM: Kappa coefficient = 0.89, OA = 91.21; RF: Kappa coefficient = 0.92, OA = 93.23%.

Table 3. Confusion matrix of the SVM classification with 30-MNF bands and the 300 pixel training set (Kappa coefficient = 0.89, OA= 91.21%).

Class	*C. epigejos*	*Rubus* spp.	*Solidago* spp.	Shadows	Trees	Other Plants	Soils	Buildings	Total	UA (%)
C. epigejos	454	0	0	0	0	21	66	0	541	83.92
Rubus spp.	0	406	0	0	0	0	0	0	406	100.00
Solidago spp.	0	0	781	0	0	0	0	0	781	100.00
Shadows	0	0	0	415	0	0	0	5	420	98.81
Trees	0	6	0	0	344	2	0	0	352	97.73
Other plants	18	14	0	0	11	1375	44	0	1462	94.05
Soils	3	3	56	0	12	0	309	87	470	65.74
Buildings	9	2	3	3	60	0	0	326	403	80.89
Total	484	431	840	418	427	1398	419	418	4835	
PA (%)	93.80	94.20	92.98	99.28	80.56	98.35	73.75	77.99		

Table 4. Confusion matrix of the RF classification with 30 MNF bands and the 300 pixel training set (Kappa coefficient = 0.92, OA = 93.23%.).

Class	*C. epigejos*	*Rubus* spp.	*Solidago* spp.	Shadows	Trees	Other Plants	Soils	Buildings	Total	UA (%)
C. epigejos	437	0	0	0	0	33	84	0	554	78.88
Rubus spp.	0	394	0	0	2	0	23	0	419	94.03
Solidago spp.	0	0	833	0	0	0	0	0	833	100.00
Shadows	0	0	0	418	0	0	0	10	428	97.66
Trees	0	2	1	0	414	5	0	0	422	98.10
Other plants	23	35	6	0	11	1360	31	27	1493	91.09
Soils	24	0	0	0	0	0	278	5	307	90.55
Buildings	0	0	0	0	0	0	3	376	379	99.21
Total	484	431	840	418	427	1398	419	418	4835	
PA (%)	90.29	91.42	99.17	100.00	96.96	97.28	66.35	89.95		

Rubus spp. was identified near forest borders and buildings, and its spatial distribution for the SVM method reflected reality more accurately than the result of using the RF method (Figure 9). There was a slight overestimation of this species in the case of the RF method, especially in places with trees and bushes near buildings (Table 4). The *Calamagrostis epigejos* and *Solidago* spp. classes can be found in the open spaces of non-agricultural meadows. The spatial distribution of *Solidago* in the image resulting from the use of the RF method reflected reality almost perfectly, and in the case of the SVM method, the underestimation of this species applied mainly to uncut meadows in the south of the area. On the other hand, the *Calamagrostis epigejos* class was slightly overestimated in the results of both classifications, especially in places with dry or mowed meadows. The SVM classification image presents the spatial distribution of this species in the research area with greater precision (Table 3), and its estimations were more accurate, especially in places with bare soils, which have a similar spectral response.

4. Discussion

The effects of the raster data set and the number of training pixels on the classification accuracy of three invasive or expansive plant species were tested in this paper using the Random Forest and Support Vector Machine methods. The method we used to divide the patterns into three sets—the training set, the test set, and the spatially independent validation set—allows for reliable assessment of the classification accuracy. Balanced training sets of 30, 50, 100, 200, and 300 pixels per class were tested in this paper. The test set was strongly spatially correlated with the training set, which led to inflated accuracy results; therefore, it was used only for the initial accuracy assessment. However, surprisingly accurate measures (PA, UA, F1) calculated on the test data set increased, despite the decrease in the number of patterns in the test set. This highlights the importance of using spatially separate data set for proper accuracy assessments. A constant set of validation pixels that remained both unchanged between iterations and was spatially separate from training data allowed us to reliably assess the accuracy of classification. Spatial separation of the data sets used to assess classification results and train classifiers allowed us to avoid artificial inflation due to spatial correlation between pixels belonging to the same reference polygon. Such a method allows for more objective comparisons of classification algorithms and data sets, while delivering more trustworthy accuracy metrics. The very act of creating training and test or validation data sets introduces human or random bias into any comparison. In order to decrease such bias of our method, training and validation data sets were created multiple times. Such approaches were already used multiple times in the past [24,51,52] and are proven to be more reliable when it comes to classifier comparison. The accuracy of any machine learning procedure is directly related to the quality of samples used for training and validation of a given classifier. In order to decrease the impact of human or random bias in creating the data sets, training and validation data sets were created multiple times. Repeated sampling of pixels for the reference sets and assessing classification accuracy minimized the impact of pixel selection for training on the classification accuracy and allowed an objective assessment of the impact of the tested data sets on the effectiveness of species identification [26,52,53].

The analyses showed that, regardless of the selected classifier, a higher F1 score for all classes was obtained for classifications performed on 30 MNF transformation bands (0.854–0.918) than those on 430 hyperspectral data bands (0.760–0.853). The reduction in the number of input layers to several dozen of the most informative bands is recommended for the Random Forest and Support Vector Machine algorithms, as it allows one to obtain higher accuracy levels and significantly shortens the classification time [51,54–56]. During the classification of herbaceous vegetation in the Hortobágy National Park (Eastern Hungary), a higher overall accuracy level was obtained for nine MNF transformation bands (SVM = 82.06%, RF = 79.14%) than for 128 original bands of AISA Eagle (SVM = 72.85%, RF = 72.89%) [27]. Similarly, when identifying tree species based on AISA Eagle data using the SVM algorithm, classification of the MNF-transformed data resulted in an increase of about 30% in the classification agreement compared to the classification performed on the original bands [57]. The first 30 MNF transformation bands were used, for example, to identify four invasive or expansive species in central Poland, obtaining high F1 scores of identification: about 0.80 for *Filipendula ulmaria* and *Molinia caerulea*, about 0.79 for *Phragmites australis*, and about 0.73 for *Solidago gigantean* [58].

The increase in the number of pixels used to train the F1 score classification for the three species analyzed in this article resulted in an increase of these values, but also a simultaneous decrease in their distribution width, which indicates stabilization of the results. Our observations indicate that the preferred number of training patterns is at least 300 pixels per class, regardless of the classifier used. In the case of 30 MNF and the SVM algorithm, 300 was the optimal value because there were no statistically significant differences between training data sets containing 200 and 300 pixels per class (Figure 6). Due to the unavailability of the larger continuous areas of invasive plants on our research area, we have limited the analysis to 300 pixels, and therefore we were unable to assess impact of larger number of pixels per class in training data set on achieved classification results. A similar trend was noticed by testing different sets of training pixels (from 10 to 30 pixels) and raster data for the classification of 20 herbaceous species in Eastern Hungary by means of the SVM and RF algorithms [27]. Moreover, the highest overall accuracy (SVM: 82.06%; RF: 79.14%) was obtained using the largest of the tested sets of patterns (30 training pixels). The overall classification accuracy decreased with a decreasing number of training pixels (lower by about 2 percentage points for the set of 10 training pixels).

After a detailed analysis, it can be concluded that the Support Vector Machine algorithm was more resistant to smaller numbers of training patterns and allowed to obtain a higher mean F1 score for three plant species (F1 SVM = 0.95) compared to the Random Forest algorithm (F1 RF = 0.92) on the best data set (30 MNF, 300 training pixels). Lower mean F1 scores for background classes (F1 SVM = 0.82, F1 RF = 0.91) were noted in the SVM result image, but classification errors occurred mainly between different background classes and not between the background and plant species.

Visual interpretation of the result images and statistical accuracy analyses indicated that both classifiers detected the plant species of this study in the research area with a very high level of accuracy. Correct identification of species was also confirmed by additional field verifications carried out after the analyses. High classification accuracy levels obtained for the analyzed scenarios may also be due to the optimal time in which the imaging was obtained [26,59]. The analyzed species are in their flowering and fruiting phases at the turn of August and September, which makes them more distinctive thanks to their characteristic colors of inflorescences, fruits, and leaves (Table 5).

The classification accuracy of the *Solidago* spp. species was very high (F1> 0.95) for both classifiers and the raster data. This is not surprising because this plant's yellow inflorescences form homogeneous fields, which are easy to distinguish from other objects in images, and it would probably be even possible to use photointerpretation for this task. The *Solidago gigantea* species was identified in central Poland using 30 MNF transformation bands (a mosaic of hyperspectral data from the same HySpex sensors) and the Random Forest method; a lower F1 score for the species, about 0.73, and a slightly higher F1 score for the background, about 0.94, were obtained [58]. *Solidago* spp. has also been classified with high accuracy (F1 about 0.83, UA = 0.71, PA = 1.0) on the Hungarian–Slovak cross-border site using 15 MNF bands (a mosaic of hyperspectral data from AISA Eagle II) and the maximum likelihood method [61]. High identification accuracy of one of the goldenrod species, *Solidago altissima* (F1 score of about 0.86, UA = 0.94, PA = 0.80), was also obtained during the research conducted in Watarase wetlands in Japan with the help of only 3 MNF transformation bands (a mosaic of hyperspectral data from AISA Eagle) and generalized linear models [19].

Table 5. Comparison of acquired results with references.

Plant Species	Sensor	Raster Data	Algorithm	UA (%)	PA (%)	F1 (%)	Reference
Calamagrostis epigejos	HySpex	430 HS	RF	77	87	82	Present paper
			SVM	89	92	90	
	HySpex	30 MNF	RF	79	90	83	
			SVM	84	94	91	
Calamagrostis epigejos	HySpex	30 MNF + 42 discrete LiDAR data	RF	88	63	73	[26]
Calamagrostis villosa	APEX	30 MNF	SVM	51	68		[60]
Solidago **spp.**	HySpex	430 HS	RF	99	99	99	Present paper
			SVM	97	98	97	
	HySpex	30 MNF	RF	100	99	99	
			SVM	100	93	96	
Solidago gigantea	HySpex	30 MNF	RF			73	[58]
Solidago **spp.**	AISA Eagle	15 MNF	Maximum Likelihood	71	100		[61]
		129 HS	Spectral Angle Mapper	61	69		
Solidago altissima	AISA Eagle	3 MNF	Generalized Linear Models	94	80		[19]
Rubus **spp.**	HySpex	430 HS	RF	70	83	76	Present paper
			SVM	79	90	84	
		30 MNF	RF	94	92	95	
			SVM	100	94	97	
Rubus fruticosus sp. agg.	HyMap	20 HS	MTMF	81	92		[32]
			MF	61	53		
			SAM	71	58		
			MTMF	90	77		
		128 HS	MF	49	35		
			SAM	75	45		

Rubus spp. was classified in the research area with F1 scores ranging from 0.70 to 0.97, with the highest accuracy obtained for the Support Vector Machine method and 30 MNF transformation bands. High accuracy (OA = 87.8% and Kappa = 0.75) was also obtained during the detection of *Rubus armeniacus* in open areas in Surrey, BC, Canada, by means of a combination of CASI hyperspectral imagery with LiDAR data and the Random Forest algorithm [62]. Similarly, when identifying *Rubus fruticosus sp. agg.* in the Kosciuszko National Park in Australia, a F1 score of about 0.83 was obtained for blackberry using 23 bands of a mosaic of hyperspectral data from HyMap after MNF transformation and the Mixture-Tuned Matched Filter (MTMF) algorithm [32]. On the other hand, research on the identification of *Rubus cuneifolius* species in the eastern parts of South Africa using the SVM algorithm and multispectral data led to results that were much lower in accuracy: the F1 scores for the Landsat data varied from 0.33 to 0.48, while the scores for the Sentinel-2 data were between 0.34 and 0.58, which confirms that hyperspectral data allow for much more accurate detection of blackberries [63].

Identification of *Calamagrostis epigejos* resulted in F1 scores between 0.70 and 0.91, depending on the algorithm and data set used. As before, the best data set for wood small-reed classification turned out to be the SVM algorithm and MNF-transformed bands (F1 scores from 0.86 to 0.91), while the RF method resulted in F1 scores between 0.76 and 0.83, depending on the number of pixels used for training. By carrying out *C. epigejos* classifications at various growth stages, it was confirmed that flowering time (around September) facilitated correct identification of wood small-reed [26]. In addition, the use of the Random Forest method and MNF transformation bands on the HySpex hyperspectral data led to an F1 score of 0.72, which is an accuracy level close to the one obtained for wood small-reed in our research. Lower accuracy was obtained (producer accuracy 68%, and user accuracy 51%) in the classification of plant communities representing the *Calamagrostis villosa* species when the APEX data and the SVM method were used [60]. However, an average PA of about 82% and UA of about 75% were obtained for wood small-reed grasses during the classification of high-mountain vegetation communities using 40 MNF transformation bands of the DAIS 7915 data and neural networks [64]. This was similar to the results obtained in our work on 30 MNF bands with the RF algorithm (PA of about 88%, UA of about 78%) and was lower than the results for SVM (PA of about 96%, UA of about 87%).

5. Conclusions

The above-presented research concerning identification of three species of invasive or expansive plants using the Random Forest and Support Vector Machine classification methods, as well as various sets of input data, has led to the following conclusions:

- The accuracy assessment method presented in the paper allows us to confirm that all analyzed species can be identified in heterogenous habitats (achieved classification results F1 oscillated around 0.90). The species created a unique set of spectral properties, which are recognizable by the SVM and RF classifiers, and separating training and validation sets at the level of the reference polygons, and not at the level of individual pixels, is justified. This allows one to avoid overestimating the accuracy of the results due to spatial correlation of pixels from the same reference polygon. We have shown a clear need to divide classes into training and validation at the polygon level in order to minimize spatial correlation between samples and in order to achieve unbiased and accurate classification metrics. A spatially separate and unchanging validation set can be used to improve the quality of the obtained accuracy scores and compare the results more objectively. Unfortunately, this type of approach makes it more difficult to use iterative methods of assessing accuracy or significantly reduces the number of observations in a data set that can be used for classifier training. What is more, the principles of a constant and unchanging validation set are not optimal, which may negatively affect the quality of the resulting post-classification images. A set of 30 MNF bands allows for more accurate identification of the analyzed invasive and expansive plant species than that of the 430 original spectral bands of the HySpex image.
- Increase of number of pixels per class in training data set has a greater effect on achieved accuracy measures in the case of 430 spectral bands data set (difference in medians between 30 and 300 pixel data sets around 8 percentage points (p.p.) in the case of RF and 9 p.p. in the case of SVM algorithm) then in the case of 30 MNF bands data set (median difference between 30 and 300 pixel data set: 5 p. p. for RF and 3 p.p for SVM, Figure 7).

- Three hundred pixels per class is the preferred number of samples in the training set for classification of the analyzed plant species with the help of the SVM and RF methods. Fewer pixels result in a significant decrease in classification accuracy and less stable results. In our case, we managed to find the optimal number of pixels in training data sets per class only in the case of the SVM classifier applied to MNF data. Figure 6 shows that there was no significant statistical difference between tests performed on MNF bands with 200 and 300 pixel samples per class. Hence, 200 pixels per class in the training data set for 30 MNF bands and the SVM classifier is optimal.

- Both the Support Vector Machine method and the Random Forest method allowed us to obtain very accurate images of the distribution of analyzed species in the research area. However, the SVM classifier worked better for the classification of blackberry and wood small-reed (i.e., for classes that are not uniform and do not differ spectrally from their surroundings). On the other hand, the Random Forest algorithm allows one to obtain a higher accuracy for homogeneous classes that stand out spectrally (i.e., goldenrod and background classes). Still, the SVM image was found to be more reliable, despite its relatively lower accuracy for the background classes. Most classification errors occurred between background classes rather than individual species.

Author Contributions: Conceptualization, A.S.-T. and B.Z.; methodology, A.S.-T. and E.R.; software, E.R.; validation, E.R. and A.S.-T.; formal analysis, A.S.-T. and E.R.; investigation, A.S.-T. and E.R.; data curation, A.S.-T.; writing—original draft preparation, A.S.-T.; writing—review and editing, B.Z. and E.R.; visualization, A.S.-T. and E.R.; supervision, B.Z. All authors have read and agreed to the published version of the manuscript.

Funding: This research has been carried out under the Biostrateg Programme of the Polish National Centre for Research and Development (NCBiR), "Natural Environment, Agriculture and Forestry" project DZP/BIOSTRATEG-II/390/2015: The innovative approach supporting monitoring of non-forest Natura 2000 habitats, using remote sensing methods (HabitARS, Consortium Leader: MGGP Aero; the project partners include the University of Lodz, the University of Warsaw, Warsaw University of Life Sciences, the Institute of Technology and Life Sciences, the University of Silesia in Katowice, Warsaw University of Technology). The publishing costs were covered by the Polish Ministry of Science and Higher Education, project the theme No. 500-D119-12-1190000.

Acknowledgments: The authors would like to thank the HabitARS Consortium, especially MGGP Aero for acquiring, pre-processing, and sharing the HySpex data. Data fusion procedures were conducted in the frame of the H2020-MSCA-RISE-2016: *innoVation in geOspatiaL and 3D daTA* (VOLTA), Reference GA No. 734687. H2020 VOLTA activities are supported by the Polish Ministry of Science and Higher Education in the frame of H2020 co-financed projects No. 3934/H2020/2018/2 and 379067/PnH/2017. We are also grateful to the editors and anonymous reviewers for their constructive comments and suggestions that helped to improve this manuscript.

Conflicts of Interest: The authors declare no conflicts of interest.

References

1. Mooney, H.A.; Cleland, E.E. The evolutionary impact of invasive species. *Proc. Natl. Acad. Sci. USA* **2001**, *98*, 5446–5451. [CrossRef] [PubMed]

2. Tokarska-Guzik, B.; Dajdok, Z.; Zając, M.; Zając, A.; Urbisz, A.; Danielewicz, W.; Hołdyński, C. *Rośliny Obcego Pochodzenia w Polsce ze Szczególnym Uwzględnieniem Gatunków Inwazyjnych*; Generalna Dyrekcja Ochrony Środowiska: Warszawa, Poland, 2012; ISBN 978-83-62940-34-9.

3. Hulme, P.E.; Pyšek, P.; Nentwig, W.; Vilà, M. Will Threat of Biological Invasions Unite the European Union? *Science* **2009**, *324*, 40–41. [CrossRef] [PubMed]

4. Holub, P.; Tůma, I.; Fiala, K. The effect of nitrogen addition on biomass production and competition in three expansive tall grasses. *Environ. Pollut.* **2012**, *170*, 211–216. [CrossRef] [PubMed]

5. Pruchniewicz, D.; Żołnierz, L. The influence of environmental factors and management methods on the vegetation of mesic grasslands in a central European mountain range. *Flora-Morphol. Distrib. Funct. Ecol. Plants* **2014**, *209*, 687–692. [CrossRef]

6. Babai, D.; Molnár, Z. Small-scale traditional management of highly species-rich grasslands in the Carpathians. *Agric. Ecosyst. Environ.* **2014**, *182*, 123–130. [CrossRef]

7. Sedláková, I.; Fiala, K. Ecological problems of degradation of alluvial meadows due to expanding Calamagrostis epigejos. *Ekol. Bratisl.* **2001**, *20*, 226–233.

8.   Huang, C.-Y.; Asner, G. Applications of Remote Sensing to Alien Invasive Plant Studies. *Sensors* **2009**, *9*, 4869–4889. [CrossRef]
9.   Sabat-Tomala, A.; Jarocińska, A.M.; Zagajewski, B.; Magnuszewski, A.S.; Sławik, Ł.M.; Ochtyra, A.; Raczko, E.; Lechnio, J.R. Application of HySpex hyperspectral images for verification of a two-dimensional hydrodynamic model. *Eur. J. Remote Sens.* **2018**, *51*, 637–649. [CrossRef]
10.   Curatola Fernández, G.; Obermeier, W.; Gerique, A.; López Sandoval, M.; Lehnert, L.; Thies, B.; Bendix, J. Land Cover Change in the Andes of Southern Ecuador—Patterns and Drivers. *Remote Sens.* **2015**, *7*, 2509–2542. [CrossRef]
11.   Rapinel, S.; Clément, B.; Magnanon, S.; Sellin, V.; Hubert-Moy, L. Identification and mapping of natural vegetation on a coastal site using a Worldview-2 satellite image. *J. Environ. Manag.* **2014**, *144*, 236–246. [CrossRef]
12.   Hestir, E.L.; Khanna, S.; Andrew, M.E.; Santos, M.J.; Viers, J.H.; Greenberg, J.A.; Rajapakse, S.S.; Ustin, S.L. Identification of invasive vegetation using hyperspectral remote sensing in the California Delta ecosystem. *Remote Sens. Environ.* **2008**, *112*, 4034–4047. [CrossRef]
13.   Kokaly, R.F.; Despain, D.G.; Clark, R.N.; Livo, K.E. Mapping vegetation in Yellowstone National Park using spectral feature analysis of AVIRIS data. *Remote Sens. Environ.* **2003**, *84*, 437–456. [CrossRef]
14.   Okin, G.S.; Roberts, D.A.; Murray, B.; Okin, W.J. Practical limits on hyperspectral vegetation discrimination in arid and semiarid environments. *Remote Sens. Environ.* **2001**, *77*, 212–225. [CrossRef]
15.   Lawrence, R.L.; Wood, S.D.; Sheley, R.L. Mapping invasive plants using hyperspectral imagery and Breiman Cutler classifications (randomForest). *Remote Sens. Environ.* **2006**, *100*, 356–362. [CrossRef]
16.   Pengra, B.W.; Johnston, C.A.; Loveland, T.R. Mapping an invasive plant, Phragmites australis, in coastal wetlands using the EO-1 Hyperion hyperspectral sensor. *Remote Sens. Environ.* **2007**, *108*, 74–81. [CrossRef]
17.   Lass, L.W.; Thill, D.C.; Shafii, B.; Prather, T.S. Detecting Spotted Knapweed (Centaurea maculosa) with Hyperspectral Remote Sensing. *Weed Technol.* **2002**, *16*, 426–432. [CrossRef]
18.   Skowronek, S.; Ewald, M.; Isermann, M.; Van De Kerchove, R.; Lenoir, J.; Aerts, R.; Warrie, J.; Hattab, T.; Honnay, O.; Schmidtlein, S.; et al. Mapping an invasive bryophyte species using hyperspectral remote sensing data. *Biol. Invasions* **2017**, *19*, 239–254. [CrossRef]
19.   Ishii, J.; Washitani, I. Early detection of the invasive alien plant Solidago altissima in moist tall grassland using hyperspectral imagery. *Int. J. Remote Sens.* **2013**, *34*, 5926–5936. [CrossRef]
20.   Atkinson, J.T.; Ismail, R.; Robertson, M. Mapping Bugweed (Solanum mauritianum) Infestations in Pinus patula Plantations Using Hyperspectral Imagery and Support Vector Machines. *IEEE J. Sel. Top. Appl. Earth Obs. Remote Sens.* **2014**, *7*, 17–28. [CrossRef]
21.   Sluiter, R.; Pebesma, E.J. Comparing techniques for vegetation classification using multi- and hyperspectral images and ancillary environmental data. *Int. J. Remote Sens.* **2010**, *31*, 6143–6161. [CrossRef]
22.   Vapnik, V.; Lerner, A. Pattern recognition using generalized portrait method. *Autom. Remote Control* **1963**, *24*, 774–780.
23.   Breiman, L. Random Forests. *Mach. Learn.* **2001**, *45*, 5–32. [CrossRef]
24.   Ghosh, A.; Fassnacht, F.E.; Joshi, P.K.; Koch, B. A framework for mapping tree species combining hyperspectral and LiDAR data: Role of selected classifiers and sensor across three spatial scales. *Int. J. Appl. Earth Obs. Geoinf.* **2014**, *26*, 49–63. [CrossRef]
25.   Camps-Valls, G.; Gomez-Chova, L.; Calpe-Maravilla, J.; Martin-Guerrero, J.D.; Soria-Olivas, E.; Alonso-Chorda, L.; Moreno, J. Robust support vector method for hyperspectral data classification and knowledge discovery. *IEEE Trans. Geosci. Remote Sens.* **2004**, *42*, 1530–1542. [CrossRef]
26.   Marcinkowska-Ochtyra, A.; Jarocińska, A.; Bzdęga, K.; Tokarska-Guzik, B. Classification of Expansive Grassland Species in Different Growth Stages Based on Hyperspectral and LiDAR Data. *Remote Sens.* **2018**, *10*, 2019. [CrossRef]
27.   Burai, P.; Deák, B.; Valkó, O.; Tomor, T. Classification of herbaceous vegetation using airborne hyperspectral imagery. *Remote Sens.* **2015**, *7*, 2046–2066. [CrossRef]
28.   McPheeters, K.D.; Skirvin, R.M.; Hall, H.K. Brambles (Rubus spp.). In *Crops II.*; Bajaj, Y.P.S., Ed.; Springer: Heidelberg, Germany, 1988; pp. 104–123. ISBN 3642735223.
29.   Grime, J.P.; Hodgson, J.G.; Hunt, R. *Comparative Plant Ecology*; Springer: Dordrecht, The Netherlands, 1988; Volume 51, ISBN 978-0-412-74170-8.

30. Balandier, P.; Marquier, A.; Casella, E.; Kiewitt, A.; Coll, L.; Wehrlen, L.; Harmer, R. Architecture, cover and light interception by bramble (Rubus fruticosus): a common understorey weed in temperate forests. *Forestry* **2013**, *86*, 39–46. [CrossRef]

31. Wolanin, M.M.; Wolanin, M.N.; Oklejewicz, K. Occurrence of brambles (Rubus L.) in young forest plantations on the Kolbuszowa Plateau. *For. Res. Pap.* **2017**, *78*, 179–186. [CrossRef]

32. Dehaan, R.; Louis, J.; Wilson, A.; Hall, A.; Rumbachs, R. Discrimination of blackberry (Rubus fruticosus sp. agg.) using hyperspectral imagery in Kosciuszko National Park, NSW, Australia. *ISPRS J. Photogramm. Remote Sens.* **2007**, *62*, 13–24. [CrossRef]

33. Pruchniewicz, D.; Żołnierz, L. The influence of Calamagrostis epigejos expansion on the species composition and soil properties of mountain mesic meadows. *Acta Soc. Bot. Pol.* **2016**, *86*, 1–11. [CrossRef]

34. Aiken, S.G.; Dore, W.G.; Lefkovitch, L.P.; Armstrong, K.C. Calamagrostis epigejos (Poaceae) in North America, especially Ontario. *Can. J. Bot.* **1989**, *67*, 3205–3218. [CrossRef]

35. Rebele, F.; Lehmann, C. Biological flora of central europe: Calamagrostis epigejos (L.) Roth. *Flora* **2001**, *196*, 325–344. [CrossRef]

36. Rebele, F. Competition and coexistence of rhizomatous perennial plants along a nutrient gradient. *Plant Ecol.* **2000**, *147*, 77–94. [CrossRef]

37. Kabuce, N.; Priede, A. NOBANIS-Invasive Alien Species Fact Sheet Solidago Canadensis. 2019. Available online: www.nobanis.org (accessed on 16 November 2019).

38. Guzikowa, M.; Maycock, P.F. The invasion and expansion of three North American species of goldenrod (Solidago canadensis L. sensu lato. S. gigantea Ait. and S. graminifolia (L) Salisb in Poland. *Acta Soc. Bot. Pol.* **2014**, *55*, 367–384. [CrossRef]

39. Weber, E. Current and Potential Ranges of Three Exotic Goldenrods (Solidago) in Europe. *Conserv. Biol.* **2001**, *15*, 122–128. [CrossRef]

40. Werner, P.A.; Gross, R.S.; Bradbury, I.K. The biology of Canadian Weeds. 45. Solidago canadensis L. *Can. J. Plant Sci.* **1980**, *60*, 1393–1409. [CrossRef]

41. Shui-Liang, G.; Fang, F. Physiological Adaptation of the Invasive Plant Solidago canadensis to Environments. *Chinese J. Plant Ecol.* **2003**, *27*, 47–52. [CrossRef]

42. Yang, R.-Y.; Tang, J.-J.; Yang, Y.-S.; Chen, X. Invasive and non-invasive plants differ in response to soil heavy metal lead contamination. *Bot. Stud.* **2007**, *48*, 453–458.

43. Marcinkowska-Ochtyra, A.; Zagajewski, B.; Raczko, E.; Ochtyra, A.; Jarocińska, A. Classification of High-Mountain Vegetation Communities within a Diverse Giant Mountains Ecosystem Using Airborne APEX Hyperspectral Imagery. *Remote Sens.* **2018**, *10*, 570. [CrossRef]

44. Melgani, F.; Bruzzone, L. Classification of hyperspectral remote sensing images with support vector machines. *IEEE Trans. Geosci. Remote Sens.* **2004**, *42*, 1778–1790. [CrossRef]

45. Lillesand, T.; Kiefer, R.; Chipman, J. *Remote Sensing and Image Interpretation*, 7th ed.; Wiley: Hoboken, NJ, USA, 2015; p. 736. ISBN 978-1-118-34328-9.

46. Cohen, J. A Coefficient of Agreement for Nominal Scales. *Educ. Psychol. Meas.* **1960**, *20*, 37–46. [CrossRef]

47. Sasaki, Y. The truth of the F-measure. *Teach. Tutor Mater.* **2007**, *1*, 1–5.

48. Van Rijsbergen, C.J. *Information Retrieval*, 2nd ed.; Butterworth-Heinemann: Newton, MA, USA, 1979; p. 208. ISBN 978-0-408-70929-3.

49. Mann, H.B.; Whitney, D.R. On a Test of Whether one of Two Random Variables is Stochastically Larger than the Other. *Ann. Math. Stat.* **1947**, *18*, 50–60. [CrossRef]

50. Marcinkowska-Ochtyra, A.; Gryguc, K.; Ochtyra, A.; Kopeć, D.; Jarocińska, A.; Sławik, Ł. Multitemporal Hyperspectral Data Fusion with Topographic Indices—Improving Classification of Natura 2000 Grassland Habitats. *Remote Sens.* **2019**, *11*, 2264. [CrossRef]

51. Fassnacht, F.E.; Neumann, C.; Forster, M.; Buddenbaum, H.; Ghosh, A.; Clasen, A.; Joshi, P.K.; Koch, B. Comparison of Feature Reduction Algorithms for Classifying Tree Species with Hyperspectral Data on Three Central European Test Sites. *IEEE J. Sel. Top. Appl. Earth Obs. Remote Sens.* **2014**, *7*, 2547–2561. [CrossRef]

52. Raczko, E.; Zagajewski, B. Tree Species Classification of the UNESCO Man and the Biosphere Karkonoski National Park (Poland) Using Artificial Neural Networks and APEX Hyperspectral Images. *Remote Sens.* **2018**, *10*, 1111. [CrossRef]

53. Raczko, E.; Zagajewski, B. Comparison of support vector machine, random forest and neural network classifiers for tree species classification on airborne hyperspectral APEX images. *Eur. J. Remote Sens.* **2017**, *50*, 144–154. [CrossRef]

54. Green, A.A.; Berman, M.; Switzer, P.; Craig, M.D. A Transformation for Ordering Multispectral Data in Terms of Image Quality with Implications for Noise Removal. *IEEE Trans. Geosci. Remote Sens.* **1988**, *26*, 65–74. [CrossRef]

55. Mather, P.M.; Koch, M. *Computer Processing of Remotely-Sensed Images*; John Wiley & Sons, Ltd: Chichester, UK, 2011; ISBN 9780470666517.

56. Plaza, A.; Benediktsson, J.A.; Boardman, J.W.; Brazile, J.; Bruzzone, L.; Camps-Valls, G.; Chanussot, J.; Fauvel, M.; Gamba, P.; Gualtieri, A.; et al. Recent advances in techniques for hyperspectral image processing. *Remote Sens. Environ.* **2009**, *113*, 110–122. [CrossRef]

57. Shen, G.; Sakai, K.; Hoshino, Y. High Spatial Resolution Hyperspectral Mapping for Forest Ecosystem at Tree Species Level. *Agric. Inf. Res.* **2010**, *19*, 71–78. [CrossRef]

58. Kopeć, D.; Zakrzewska, A.; Halladin-Dąbrowska, A.; Wylazłowska, J.; Kania, A.; Niedzielko, J. Using Airborne Hyperspectral Imaging Spectroscopy to Accurately Monitor Invasive and Expansive Herb Plants: Limitations and Requirements of the Method. *Sensors* **2019**, *19*, 2871. [CrossRef] [PubMed]

59. Schuster, C.; Schmidt, T.; Conrad, C.; Kleinschmit, B.; Förster, M. Grassland habitat mapping by intra-annual time series analysis -Comparison of RapidEye and TerraSAR-X satellite data. *Int. J. Appl. Earth Obs. Geoinf.* **2015**, *34*, 25–34. [CrossRef]

60. Marcinkowska-Ochtyra, A.; Zagajewski, B.; Ochtyra, A.; Jarocińska, A.; Wojtuń, B.; Rogass, C.; Mielke, C.; Lavender, S. Subalpine and alpine vegetation classification based on hyperspectral APEX and simulated EnMAP images. *Int. J. Remote Sens.* **2017**, *38*. [CrossRef]

61. Burai, P.; Laposi, R.; Enyedi, P.; Schmotzer, A.; Bognar, V.K. Mapping invasive vegetation using AISA Eagle airborne hyperspectral imagery in the Mid-Ipoly-Valley. In Proceedings of the 2011 3rd Workshop on Hyperspectral Image and Signal Processing: Evolution in Remote Sensing (WHISPERS), Lisbon, Portugal, 6–9 June 2011; pp. 1–4.

62. Chance, C.M.; Coops, N.C.; Plowright, A.A.; Tooke, T.R.; Christen, A.; Aven, N. Invasive Shrub Mapping in an Urban Environment from Hyperspectral and LiDAR-Derived Attributes. *Front. Plant Sci.* **2016**, *7*, 1–19. [CrossRef]

63. Rajah, P.; Odindi, J.; Mutanga, O. Evaluating the potential of freely available multispectral remotely sensed imagery in mapping American bramble (Rubus cuneifolius). *S. Afr. Geogr. J.* **2018**, *100*, 291–307. [CrossRef]

64. Zagajewski, B. *Assesment of Neural Networks and Imaging Spectroscopy for Vegetation Classification of the High Tatras*; Olędzki, J., Ed.; Klub Teledetekcji Środowiska Polskiego Towarzystwa Geograficznego: Warsaw, Poland, 2010; Volume 43, ISBN 0071-8076.

 *remote sensing*

Article

A Machine Learning Framework to Predict Nutrient Content in Valencia-Orange Leaf Hyperspectral Measurements

Lucas Prado Osco [1,*], Ana Paula Marques Ramos [2], Mayara Maezano Faita Pinheiro [2], Érika Akemi Saito Moriya [3], Nilton Nobuhiro Imai [3], Nayara Estrabis [1], Felipe Ianczyk [1], Fábio Fernando de Araújo [4], Veraldo Liesenberg [5], Lúcio André de Castro Jorge [6], Jonathan Li [7], Lingfei Ma [7], Wesley Nunes Gonçalves [1], José Marcato Junior [1] and José Eduardo Creste [4]

[1] Faculty of Engineering, Architecture, and Urbanism and Geography, Federal University of Mato Grosso do Sul (UFMS), 79070-900 Campo Grande, Brazil; nayara.estrabis@ufms.br (N.E.); felipe.ianczyk@ufms.br (F.I.); wesley.goncalves@ufms.br (W.N.G.); jose.marcato@ufms.br (J.M.J.)

[2] Environmental and Regional Development, University of Western São Paulo (UNOESTE), 19050-920 Presidente Prudente, Brazil; anaramos@unoeste.br (A.P.M.R.); mayarafaita@gmail.com (M.M.F.P.)

[3] Department of Cartographic Science, São Paulo State University (UNESP), 19060-900 Presidente Prudente, Brazil; eakemisaito@gmail.com (É.A.S.M.); nilton.imai@unesp.br (N.N.I.)

[4] Department of Agronomy, University of Western São Paulo (UNOESTE), 19050-920 Presidente Prudente, Brazil; fabio@unoeste.br (F.F.d.A.); jcreste@unoeste.br (J.E.C.)

[5] Forest Engineering Department, Santa Catarina State University (UDESC), 88520-000 Conta Dinheiro, Brazil; veraldo.liesenberg@udesc.br

[6] National Research Center of Development of Agricultural Instrumentation, Brazilian Agricultural Research Agency (EMBRAPA), 13560-970 São Carlos, Brazil; lucio.jorge@embrapa.br

[7] Department of Geography and Environmental Management and Department of Systems Design Engineering, University of Waterloo (UW), Waterloo, ON N2L 3G1, Canada; junli@uwaterloo.ca (J.L.); l53ma@uwaterloo.ca (L.M.)

* Correspondence: pradoosco@gmail.com

Received: 4 February 2020; Accepted: 7 March 2020; Published: 12 March 2020

Abstract: This paper presents a framework based on machine learning algorithms to predict nutrient content in leaf hyperspectral measurements. This is the first approach to evaluate macro- and micronutrient content with both machine learning and reflectance/first-derivative data. For this, citrus-leaves collected at a Valencia-orange orchard were used. Their spectral data was measured with a Fieldspec ASD FieldSpec® HandHeld 2 spectroradiometer and the surface reflectance and first-derivative spectra from the spectral range of 380 to 1020 nm (640 spectral bands) was evaluated. A total of 320 spectral signatures were collected, and the leaf-nutrient content (N, P, K, Mg, S, Cu, Fe, Mn, and Zn) was associated with them. For this, 204,800 (320 × 640) combinations were used. The following machine learning algorithms were used in this framework: k-Nearest Neighbor (kNN), Lasso Regression, Ridge Regression, Support Vector Machine (SVM), Artificial Neural Network (ANN), Decision Tree (DT), and Random Forest (RF). The training methods were assessed based on Cross-Validation and Leave-One-Out. The Relief-F metric of the algorithms' prediction was used to determine the most contributive wavelength or spectral region associated with each nutrient. This approach was able to return, with high predictions (R^2), nutrients like N (0.912), Mg (0.832), Cu (0.861), Mn (0.898), and Zn (0.855), and, to a lesser extent, P (0.771), K (0.763), and S (0.727). These accuracies were obtained with different algorithms, but RF was the most suitable to model most of them. The results indicate that, for the Valencia-orange leaves, surface reflectance data is more suitable to predict macronutrients, while first-derivative spectra is better linked to micronutrients. A final contribution of this study is the identification of the wavelengths responsible for contributing to these predictions.

Keywords: spectroscopy; proximal sensor; macronutrient; micronutrient; artificial intelligence

1. Introduction

Remote sensing techniques can be useful for the estimation of plant health conditions, including monitoring the nutritional status [1–4], the stress response [5–7], plant count [8,9], yield prediction [10–12], chlorophyll content [13–15], pest and disease identification [16,17], and biomass estimation [18], among others. Multisensory data is often used to accomplish this task, including the ones acquired by orbital sensors, aircraft or Unnamed Aerial Vehicle (UAV)-embedded cameras, terrestrial sensors, and field spectroradiometers, known as proximal sensors [19–23]. This type of sensor can measure the spectral response of a target at very-high resolutions while having a reductive amount of radiometric interference by being near the leaf sample.

The usage of proximal sensors for plant evaluation has assisted phenological studies of different species. Due to the high spectral resolution capability of these sensors, studies have been relatively successful in modeling phenomena, such as the ones previously stated, but at the leaf level, like plant stress, yield prediction, nutrient content, chlorophyll, and many other attributes [24–27]. They also have the advantage of helping to define, in detail, the appropriate spectral regions to estimate these phenomena. This definition is relatively important as it can guide future research towards the development of equipment specifically designed to measure these regions [23]. Another type of contribution is that it can assist in creating spectral vegetation indices or other simpler mathematical models that contribute to identifying the different characteristics of plants [13,28].

Currently, one of the most common problems in monitoring crops is knowing the proper amounts of fertilization rates. Traditional agronomic methods used to evaluate plant nutrients are done regularly, in key periods, to manage fertilization of agricultural fields [29]. These methods require the collection of a high number of leaves for the chemical analysis of the leaf tissue. However, this chemical analysis is a time-consuming, labor-intensive, and pollutive task [3,30,31]. Remote sensing, specifically proximal sensing, can provide an effective alternative in assisting nutritional analysis of plants more accurately. The use of proximal sensors has an advantage over traditional agronomic methods since it allows to infer vegetation conditions in a non-invasive and non-destructive manner [32–35].

Regarding the monitoring of plant and leaf nutritional conditions by remote sensing systems, recent research has made significant advances, especially in the estimation of nitrogen (N) content [1–4,21,25,28,31]. These studies were conducted at orbital, aerial, terrestrial, or proximal levels in different crops. N deficiency is linked to a characteristic chlorosis symptom, which is observable at the visible spectra [21,25,28]. Still, considerable research was also able to identify spectral bands and wavelengths in the near and short-wave infrared regions related to this nutrient [2,3,25,28,36]. Regardless, even though N is a pretty standard nutrient to be evaluated by remote sensing systems, the same cannot be said about others.

The evaluation of nutrients, other than N, by proximal sensors, is more unusual. One study was able to infer potassium (K) content by computing random two-band spectral indices calculated from hyperspectral data ranging from 350 to 2500 nm [37]. Others focused on evaluating a large group of macronutrients, magnesium (Mg), S, phosphor (P), K, and calcium (Ca), and found important associations between the spectral region of 470 to 800 nm with them [32] Lastly, one approach aimed to predict macronutrients like K, calcium (Ca), and magnesium (Mg), as well as micronutrients like manganese (Mn) and iron (Fe), by using near-infrared spectroscopy, but their method did not return satisfactory results for micronutrients [24]. Recent literature demonstrates how hyperspectral measurements are being linked to nutrients, specifically macro. However, there is a gap in terms of micronutrient prediction by spectral sensors that need to be addressed by new research, and few studies were conducted within this theme. In citrus, a study performed a Partial Least-Squares Regression (PLSR) evaluation on both macro- and micronutrients and archived interesting results using

Laser-Induced Breakdown Spectroscopy (LIBS) [38]. Similar research, focusing only on near-infrared spectroscopy, also returned high predictions for both classes of nutrients [39].

Another way to infer chemical components from hyperspectral measurements is by applying a derivative analysis. The derivation of the reflectance data allows highlighting absorption features of components that, in a traditional spectral curve (i.e., reflectance curve), may not be measured with the same accuracy or even be detected [40–42]. Studies that apply a derivation of the reflectance curves in plants have found good correlations with N [40,41] and cadmium (Cd) concentrations [42]. Since the gains of derivative analysis are known in the literature, there are also methods for data analysis in the remote sensing scenario that may benefit from it. The advantages proportionated by derivatives may assist in the evaluation of leaf nutritional content when combined with more robust techniques.

The aforementioned studies found high relationships with hyperspectral data by employing various statistical methods in their analysis. However, methods like Partial Least-Squares Regression (PLSR), Principal Component Analysis (PCA), Stepwise-Multiple Linear Regression (SMLR), among others, returned different accuracies even for the same cultures [15,24,32,37–40,42]. Some of these methods are also reductive, and the prediction may decrease if an increase occurs in the model complexity [32]. Since hyperspectral measurements produce high and complex amounts of data, one type of approach that could ideally deal with this is machine learning.

Machine learning algorithms are a robust and intelligent technique that can model different types of data [43,44]. These algorithms have the advantage of being non-parametric and non-linear while being able to analyze noised and imperfect data [45–47]. They are also capable to perform numerous combinations and calculations in a matter of seconds, achieving relative success in remote sensing applications regarding plant analysis [48,49]. Concerning hyperspectral measurements, among the applications evaluated, these algorithms were able to return state-of-the-art performances for many situations [5,16,23,50–52]. Even though, to date, no study evaluated the performance of machine learning algorithms in inferring both plant macro- and micro-nutritional content with only leaf hyperspectral measurements. Since these algorithms have returned good accuracies in different hyperspectral analyses, they could be appropriate to deal with the complexity imposed by this type of dataset in the described situation.

As previously stated, the first-derivative of the reflectance data has already been proved to be effective in associating with different chemical components. From this information, it could be assumed that both the reflectance data and its first-derivative could be of assistance in predicting different nutrients of the leaf tissue. Since the derivation of a reflectance curve can highlight hard-to-detect components at the first level, it is possible that, by integrating these curves with machine learning algorithms, one can create important information regarding proximal sensing and plant nutritional analysis. In this spirit, a framework adopting machine learning algorithms are proposed to predict macro- and micronutrient content in the leaf-tissue directly from its hyperspectral response. This is the first approach to evaluate different nutrient content combining machine learning methods and reflectance/first-derivative data.

In this study, citrus leaves—more specifically, from Valencia-orange trees—were selected to compose the experimental dataset. It is well known that a sufficient supply of both macro- and micronutrients is critical to the management and sustainability of these plants, and the balance of available nutrients is a key component to profitability [53]. Citrus plants are economically important to the agricultural sector of many countries and may benefit significantly from a rapid and indirect nutritional assessment, such as the one proposed here. In this manner, the aims of this work are to a) show a method to indicate the most suitable spectra (reflectance/first-derivative), in order to model the nutrient content according to the algorithms' performance; and b) determine the important wavelengths or spectral regions associated with each nutrient.

2. Materials and Methods

The framework proposed in this paper was divided into four main phases (Figure 1). In the first phase, the hyperspectral measurements of the leaf samples in a Valencia-orange orchard were performed. These measurements were conducted with a field spectroradiometer. In the second phase, the spectral measurements were corrected, and the data were pre-processed. These corrections aimed to convert the radiance signal to reflectance, as well as remove the noise and calculate their first-derivative. The third phase involved the data analysis by machine learning algorithms. In this phase, a fine-tuning to determine the most appropriate parameters to model the data was performed. The fourth and final phase consisted of the organization of the prediction values into a hyperspectral map, where it was identified as the most appropriate algorithm and wavelength (i.e., spectral window) to predict each nutrient.

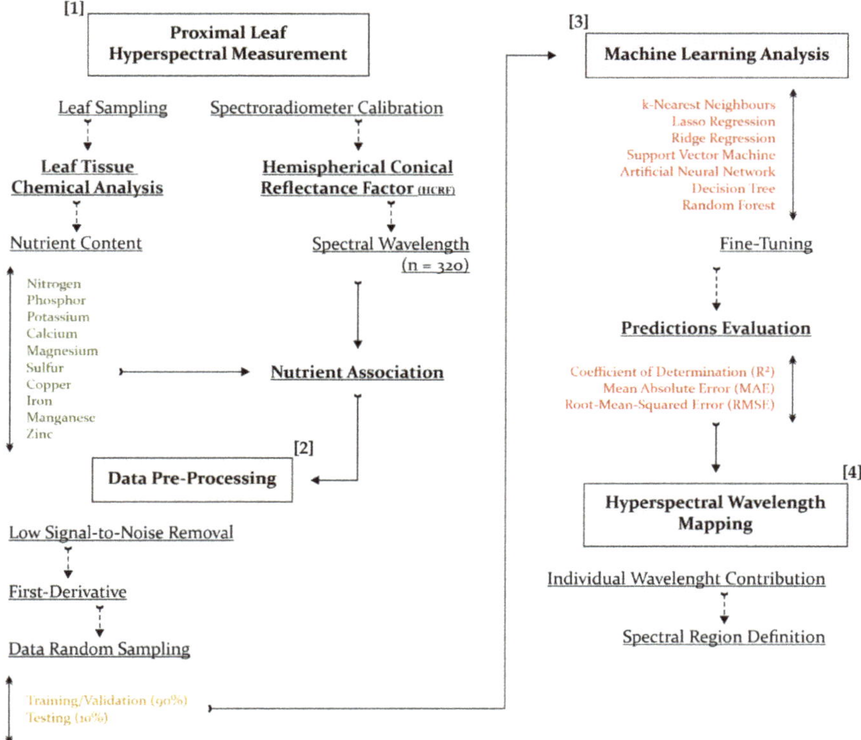

Figure 1. The workflow of the four main processes adopted for the proposed approach.

2.1. Study Area and Data Acquisition

As a study area, an open field of citrus trees, located on private property in the municipality of Ubirajara, São Paulo state, Brazil, was selected. The species analyzed were all of Valencia-orange (*Citrus sinensis* "Valencia"), planted on a Citrumelo–Swingle rootstock. During the evaluation, the trees were in their vegetative phase, with an adult size, measuring nearly 3 m in height (ground-related), with crown areas around 5.5 m². During the survey, the trees were at their maturing stage, which is five years from their initial planting. The area contains 752 trees per hectare, planted at approximately a 7 × 1.9 m spacing. Each plantation field was 250 m × 250 m in size, with some fields compensating for others accordingly to its location. The plantation fields were selected randomly inside this property and configured the different conditions of the treatments. Before the analysis, the soil was fertilized with 250 kg/ha of N in the form of urea, 125 kg/ha of phosphorus excreted, expressed as P_2O_5, and 167 kg/ha of potassium oxide (K_2O). The area is predominantly composed of red-yellow podzolic soil, situated in a Cwa Köppen [54] subtropical climate type unit.

This paper evaluated leaf samples from multiple orange tress scattered around different planting fields in an experimental portion of the orchard. In each field, the number of trees was selected according to the size of the planting field and trees planted per area. The selection was both to measure the leaf hyperspectral response and to collect them. A total of 320 samples collected in the field, with both spectra curves and later-known nutrient content, were gathered during this survey. The sampling method followed standard recommended agronomic procedures, guided by a field specialist. To represent the proper conditions of a citrus tree, only leaves at a medium canopy height and those visually healthy with no signs of diseases or damages were evaluated. A lift platform was used to elevate the person with the equipment. Since the chemical analysis of the leaf tissue recommends the 3rd or 4th leaf of a fruit branch to be sampled, the spectroradiometer equipment was directed as close as possible to the leaves that shared this description.

After measuring the spectral radiance, the leaves were extracted from their respective branches, separated, and identified them into plastic bags to be submitted to the laboratory. The leaf samples consisted of the same leaves that had their spectral radiance measured. They were conditioned at an appropriate temperature and transported accordingly. In the laboratory, the leaves were washed with a neutral detergent to remove any impurities. Later, they were dried in an oven, for 48 h, at 60–65 °C, and then crushed. From the crushed material, 100 mg was used for the N analysis. For that, the Kjeldahl titration method [55], divided into 3 phases, was followed: (1) digestion; (2) distillation in a nitrogen distiller; and (3) titration with sulfuric acid (H_2SO_4). The remaining material was separated and used for the analysis of the other macronutrient (P, K, Ca, Mg, and S) and the micronutrients (copper (Cu), Fe, Mn, and zinc (Zn)), following standard laboratory procedures of chemical analysis of the leaf-tissue [56].

2.2. Hyperspectral Measurement Processing

The spectral radiance of the Valencia-orange leaves was measured with a Fieldspec ASD FieldSpec® HandHeld 2 spectroradiometer. To record each target, the equipment was positioned at a 45° angle concerning the tree canopy. For that, a lift platform was used to ensure the correct height. This equipment operates at a spectral range of 325 nm to 1075 nm. In this study, a 10° aperture lens was adopted, and 10 readings/measurements were conducted in each leaf to produce one mean spectral signature. This procedure was important to reduce noise and variance for the same target. Before each spectral measurement, a Lambertian (Spectralon® plate) surface plate was registered. This Lambertian plate was used to calibrate the equipment and convert the digital number to a physical signal. As mentioned, the leaf-spectral response in-field was recorded into 320 measurements for this experiment.

The measured spectral curves consist of the radiance value of the target (i.e., leaf samples) spread along the electromagnetic spectrum. To produce the reflectance value (i.e., reflectance factor), the Hemispherical Conical Reflectance Factor (HCRF) was calculated as shown in Equation (1) [57]:

$$\mathrm{HCRF}(\omega_i \omega_r) = \frac{dL\,(\theta_r, \Phi_r)\,(\mathrm{target})}{dL\,(\theta_r, \Phi_r)\,(\mathrm{reference})}\, K\,(\theta_i, \Phi_i, \theta_r, \Phi_r) \tag{1}$$

where dL is the radiance; ω is the solid angle; θ and Φ are the zenith and azimuth angles, respectively; i is the incident flux; and r is the reflected energy flux. The K value is the calibration coefficient (i.e., correction factor specified for the equipment). The target corresponds to the radiance of the leaf and the reference is the radiance of the Lambertian surface plate. The HCRF represents the spectral signature of the recorded target.

After obtaining the reflectance factor of each leaf, a low signal-to-noise removal was performed by excluding wavelengths under 380 nm and above 1020 nm. After this, the first-derivative of all the HCRFs (n = 320) was calculated. The first-derivative calculation is a traditional method for modeling spectral data, and many approaches have discussed this issue. For this study, a linear least mean-squared smoothing filter [58] was firstly performed to reduce the random noise that may vary with the wavelengths and affect the derivative function. In most cases, noise can be assumed to be stationary with constant variance. It then can estimate a noise-free spectrum s(λ) in terms of the current value of the observed data. By knowing the correct signal of the spectrum giving a specific wavelength s(λ), it is possible to perform a final approximation to estimate derivatives by suitable difference schemes according to a finite band resolution: $\Delta\lambda$. Thus, the first-derivative was calculated according to [58]:

$$\left.\frac{d_s}{d_\lambda}\right| = \frac{s(\lambda_i) - s(\lambda_j)}{\Delta\lambda} \tag{2}$$

where $\Delta\lambda$ is calculated as $|\lambda j - \lambda i|$, assuming that the interval between the bands is constant. Additional tests involving further derivates, such as the second, third, and fourth, were also made in the experimental phase of this study. However, there were no indications of an improvement over the first-derivate for the used dataset during the machine learning analysis. For this reason, the proposed framework was limited only to the first-derivative, but future research using different leaf data to process additional derivatives is encouraged.

From the total leaf measurements (n = 320) used here, 10% (n = 32) was randomly separated and designed to compose the testing dataset (Figure 2). Wavelengths ranging from 380 to 1020 nm were used in the software as columns, while the leaf measurements (320) were used as rows. The 32 measurements were configured as an independent dataset, which belonged to the Valencia-orange trees located at different plantation fields, never before seen by the algorithms. The other 288 measurements configured the dependent dataset and belonged to trees with conditions or characteristics distinct from one another, observed during the field campaign. To indicate that, a descriptive statistical analysis was conducted with the nutrients' concentration from the chemical analysis of the leaf tissue, and the following parameters were calculated: minimum, maximum, mean, standard deviation, median, and coefficient of variation. They were important to demonstrate the discrepancy of the dependent (calibration/training) data used, and how representative it could be of the nutritional conditions of Valencia-orange leaf tissue in the analyzed period.

2.3. Machine Learning Analysis and Hyperspectral Mapping

In a computational environment, the nutrients were individually selected as the target variables. As input parameters, the reflectance and the first-derivative were used, and the performance of the algorithms in predicting these nutrients was evaluated. As stated, the curves were separated into three sets. The training dataset was used to set-up the hyperparametrization of the chosen algorithms. For that, the Random Search approach [59] was used. The same conjunction of training/validation data was adopted for all algorithms. This process was repeated with a fine-tuning until the reduction in the mean absolute error (MAE) did not result in any more practical gains, as the modification in the parameters impacted the processing time. Once the hyperparameters of each algorithm were defined, the testing dataset was used to verify its real performance.

To configure and run the algorithms, the open-source computer program RapidMiner 9.5 was used, which is based on a particular Python Library [60], while still permitting the development and implementation of different codes. The workstation for this task was equipped with an Intel(R) Core (TM) i7-8550U CPU 4.00 GHz, a Nvidia GeForce MX-150 4Gb GDDR5 64-bits 6008 MHz GPU, and 8GB RAM DDR4 2400MHz. The algorithms for the proposed framework were as follows: k-Nearest Neighbor (kNN), Lasso Regression, Ridge Regression, Support Vector Machine (SVM), Artificial Neural Network (ANN), Decision Tree (DT), and Random Forest (RF). The prediction metrics to evaluate these algorithms were the coefficient of determination (R^2), mean absolute error (MAE), and root-mean-squared error (RMSE). To ascertain the relationship between the measured data and the predicted data, the overall finest models were evaluated in a regression plot.

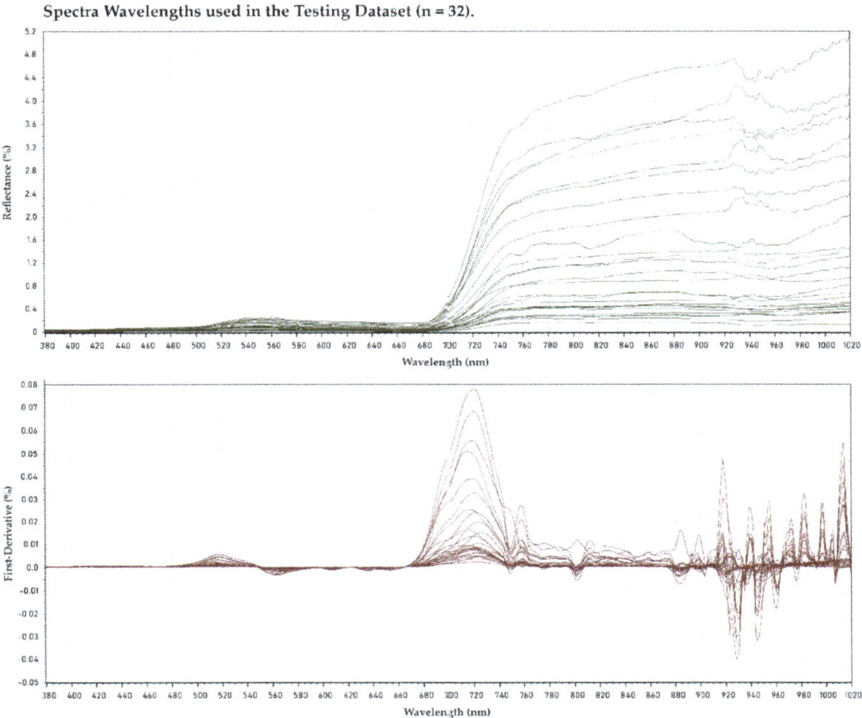

Figure 2. Spectral wavelengths used for testing the machine learning algorithms' performance. In green are the spectral reflectance, while in dark-red are their respective first-derivatives.

Regarding the configuration of each algorithm, the parameters of the used methods were set to the library default values, except those described in Table 1. For both the DT and RF algorithms an Extreme Gradient Boosting (XGBoost) model was used to increase their performances. This model adopts a forward-learning ensemble method [61], which obtains predictive results in gradually improved estimations. To illustrate the machine learning architecture regarding data inputs and outputs in the proposed analysis, a structure was organized in Figure 3.

Table 1. Detailed information regarding the algorithms adopted in the proposed framework.

Algorithm	Hyperparameter	Criteria
kNN	Distance Number of Neighbors	Euclidian k-Neighbors = 5
Lasso Regression (L1)	Strength (α) Elastic Net Mixing Proportion (L1–L2)	1.0 0.57:0.43
Ridge Regression (L2)	Strength (α) Elastic Net Mixing Proportion (L1–L2)	0.1 0.57:0.43
SVM	Radial Basis Function (RBF) Kernel exp(-g|x-y|^2)	g = automatic Regression Loss = 1.00 SVM Type Cost = 1 Tolerance = 0.001 Interaction (limit) = Unlimited
ANN	Activation: Logistic Function Adam Optimizer Regularization (α) = 0.0001	Neurons (First Hidden Layer) = 400 Neurons (Second Hidden Layer) = 200 Interactions = 10,000
DT	Number of Leaves Trees Depth	Leaves (minimal) = 2 Tree-depth (maximum) = 100
RF	Number of Trees Nodes	Trees = 100 Nodes (maximum) = 5

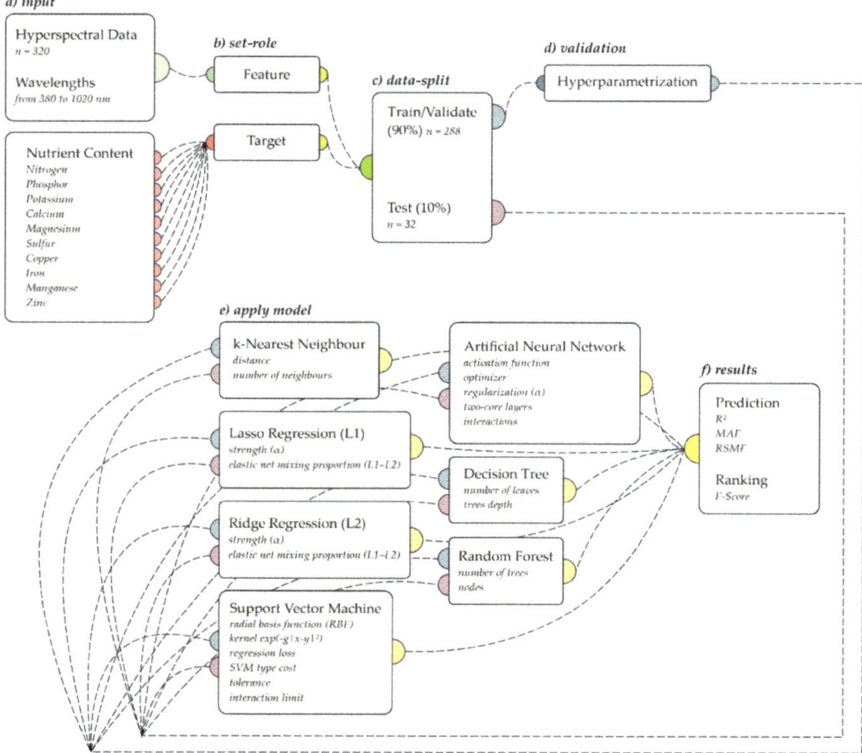

Figure 3. Structure of the machine learning architecture of the proposed framework.

It is also important to address that, although hyperspectral data is relatively easy to obtain, leaf tissue analysis can be limited. This is mostly because the chemical analysis can be highly cost if considering the amounts of data required to process the machine learning algorithms. Therefore, the appropriate number of samples is something to be observed in each case. In this study, the amount of data used to train and validate/test the used algorithms should be sufficient based on the literature. One study [62] compared different learners, such as RF, SVM, kNN, and others, in diverse settings. Between these settings, they evaluated the number of classes per problem (from 2 to 50) and the number of samples per class (from 5 to 100). This returned a variation of 10 to 5000 samples. Through their study, it was demonstrated that data curation could be modeled by these algorithms from a few to a high number of samples and still achieve appropriate results. In comparison with the proposed approach, other machine learning frameworks also adopted similar sample sizes, like 324 leaf measurements that were used to model the water-stress response from lettuce [23], 189 hyperspectral observations that were used to model grapevine water status [63], and 266 observations that were used as training to predict nitrogen content in rice fields [64].

As a discussion example, a recent paper collected 500 samples per class with 540 spectral bands and adopted a Cross-Validation method with a dataset considering 200 samples for each validation to demonstrate the importance of the feature selection methods [65]. Regardless, hyperspectral data have a characteristic distinct from most data, which is a high number of bands/wavelengths available to model a given problem. The used dataset was composed of 320 leaf measurement (in which 32 were separated as a test) and 640 spectral bands (380–1020 nm). This gives a total of 204,800 combinations to work with, which should be enough to configure a training/testing dataset. Although this high dimensionality could offer potential problems to hyperspectral data processes [65], studies already suggested that maintaining the original data could also outperform feature-selected subsets [66,67].

As mentioned, though the aforementioned studies did use similar sample sizes of data to train, validate, and test their learners, little information related to hyperspectral wavelengths and machine learning method sample size could be encountered in the literature [65]. Since there is no research focused on evaluating the impact of the training set to model spectral data, a previous comparison regarding two well-known sampling methods was performed. The first is the cross-validation method, which is more suitable to deal with the most common tasks in machine learning data curation [43]. The Cross-Validation method was performed with 10k folders. This model separates the data into 10(k) parts while using nine of them to train the algorithm and one to validate. This process is done sequentially, constantly changing the folder used for validation. In this manner, the chosen algorithm is always validated by data not used during the training phase. The second method used was the Leave-One-Out approach. This method is similar to Cross-Validation, but instead it only takes one data instance for validation each time. The method is considered a very time-intensive procedure, and it is only recommended for smaller datasets [43]. After applying the Random Search approach [59] to perform a fine-tuning, both training methods' results were compared (Table 2).

In the Cross-Validation method, from the 288 samples, 90% was used to train while 10% was used to validate, and was repeated 10 times randomly. In the Leave-One-Out method, 287 samples were used to train, while one sample was used to test it. This was repeated until all instances were used. The low difference between MAE predictions in each nutrient for both methods indicates that, even when adopting a more suitable training approach to model smaller datasets (Leave-One-Out), the training results are similar. Still, while the Leave-One-Out method is approximately unbiased, it could result in a high variance. Normally, the variance in fitting a model tends to be higher in small datasets since it is more sensitive to noise and artifacts in the used training sample. Because of that, a Cross-Validation method would also show signs of high variance, as well as a high bias if given a limited amount of data. This was not the case here, since both methods returned high predictions and similar metrics, thus indicating that, whatever the training method, the amount of data (204,800 combinations) was sufficient to model the given problem. Regardless, the Leave-One-Out method needed a higher computational cost, which is something to be considered when evaluating the amount of processed data. In the workstation, the Leave-One-Out-averaged processing time for all algorithms was 7.5 times slower than the Cross-Validation method. Because of that, the Cross-Validation method was adopted in this study, but future research should considerer both methods according to their respective dataset size and characteristics.

Lastly, the contribution of each spectral wavelength into the performance of the algorithm was computed by displaying their Relief-F value. Relief-F uses a kNN scoring to address noise data while handles incomplete data [68]. It is considered a reliable metric to inform a feature score and then be applied to rank top-scoring features. Here, the Relief-F values were used to map the hyperspectral response of each nutrient regarding the strength of the individual wavelengths to the performance of the evaluated algorithms. Aside from that, to help ascertain the hyperspectral relationship with the evaluated nutrients dataset, an analysis of each nutrient and a Shapiro–Wilk normality test at a 95% confidence interval was performed. As the normality test returned a p-value under 0.05, a non-parametric Spearman's correlation test in a pairwise comparison was executed to verify the association between each nutrient.

Table 2. MAE returned from both training methods: Cross-Validation and Leave-One-Out.

Method	N	P	K	Ca	Mg	S	Cu	Fe	Mn	Zn
kNN										
MAE (Ref.)—Cross-Validation	0.682	0.122	1.070	6.125	0.201	0.179	6.793	17.022	4.890	5.058
MAE (1st Der.)—Cross-Validation	0.884	0.174	1.235	6.358	0.357	0.154	6.589	16.789	3.012	4.246
MAE (Ref.)—Leave-One-Out	0.700	0.158	1.107	6.549	0.224	0.144	6.723	17.686	4.705	5.157
MAE (1st Der.)—Leave-One-Out	0.912	0.178	1.301	6.897	0.377	0.163	6.453	16.949	3.192	4.349
Lasso Regression										
MAE (Ref.)—Cross-Validation	1.986	0.083	1.322	7.010	0.523	0.118	17.552	31.849	8.004	7.128
MAE (1st Der.)—Cross-Validation	1.056	0.079	1.756	6.849	0.446	0.287	17.389	30.595	7.341	9.058
MAE (Ref.)—Leave-One-Out	1.898	0.095	1.352	7.101	0.538	0.107	17.515	32.256	8.058	7.185
MAE (1st Der.)—Leave-One-Out	1.285	0.087	1.975	6.942	0.455	0.284	17.395	30.112	7.578	9.088
Ridge Regression										
MAE (Ref.)—Cross-Validation	2.058	0.245	1.022	9.388	0.589	0.284	17.383	27.218	9.085	6.143
MAE (1st Der.)—Cross-Validation	1.359	0.234	1.766	6.238	0.687	0.235	17.287	33.185	7.898	8.897
MAE (Ref.)—Leave-One-Out	2.045	0.277	1.079	9.183	0.587	0.225	17.858	27.351	9.056	6.041
MAE (1st Der.)—Leave-One-Out	1.350	0.202	1.768	6.156	0.678	0.202	17.084	33.584	7.984	8.789
SVM										
MAE (Ref.)—Cross-Validation	0.789	0.134	1.012	5.888	0.456	0.179	18.894	20.170	3.374	7.071
MAE (1st Der.)—Cross-Validation	1.058	0.158	1.415	5.979	0.568	0.199	10.152	23.158	6.385	5.759
MAE (Ref.)—Leave-One-Out	0.798	0.159	1.028	5.978	0.456	0.178	18.265	20.588	3.318	7.052
MAE (1st Der.)—Leave-One-Out	1.158	0.178	1.456	5.899	0.589	0.158	10.128	23.480	6.318	5.878
ANN										
MAE (Ref.)—Cross-Validation	0.789	0.157	1.025	7.126	0.285	0.134	7.895	29.389	5.058	6.388
MAE (1st Der.)—Cross-Validation	1.058	0.193	1.453	6.087	0.456	0.146	9.185	19.241	4.358	5.268
MAE (Ref.)—Leave-One-Out	0.744	0.155	1.064	7.235	0.259	0.138	7.563	29.289	5.568	6.456
MAE (1st Der.)—Leave-One-Out	1.057	0.189	1.487	6.023	0.482	0.105	9.458	19.568	4.238	5.215
DT										
MAE (Ref.)—Cross-Validation	0.689	0.158	1.056	6.054	0.315	0.113	7.874	19.498	4.286	5.158
MAE (1st Der.)—Cross-Validation	1.055	0.123	1.126	6.586	0.467	0.205	6.894	31.218	4.878	4.238
MAE (Ref.)—Leave-One-Out	0.641	0.102	1.085	6.088	0.305	0.106	7.415	19.352	4.984	5.512
MAE (1st Der.)—Leave-One-Out	1.028	0.112	1.285	6.547	0.489	0.202	6.897	31.189	4.489	4.354
RF										
MAE (Ref.)—Cross-Validation	0.689	0.078	1.112	3.025	0.210	0.087	7.289	18.548	4.898	3.789
MAE (1st Der.)—Cross-Validation	0.638	0.101	1.207	6.238	0.389	0.179	6.046	16.189	3.874	1.789
MAE (Ref.)—Leave-One-Out	0.677	0.089	1.103	3.045	0.207	0.088	7.358	18.895	4.984	3.898
MAE (1st Der.)—Leave-One-Out	0.622	0.107	1.201	6.125	0.379	0.189	6.215	16.189	3.789	1.875

3. Results

The chemical analysis of the leaf tissue returned heterogeneous and non-parametric results for the nutrient content of the analyzed leaves (Table 3). Analysis has shown that the majority of the nutrients presented a high variability and uniform distribution. This behavior was most noticeable in nutrients, such as Ca, Fe, Mn, and Zn. Regardless, this condition is important to demonstrate the applicability of the proposed framework. Since this is a heterogeneous dataset, machine learning algorithms are advantageous for modeling data with such characteristics.

Table 3. Descriptive data from the chemical analysis of the Valencia-orange leaves.

Summary	Macronutrient (g/kg)						Micronutrient (mg/kg)			
	N	P	K	Ca	Mg	S	Cu	Fe	Mn	Zn
Mean	29.55	2.13	17.07	30.72	5.36	2.36	72.20	86.95	36.69	27.77
Std. Dev.	2.95	0.43	3.34	13.18	1.39	0.38	26.09	39.50	19.11	13.09
Median	29.45	2.17	16.70	28.85	5.25	2.35	69.90	78.35	33.10	.22.80
Min.	24.00	1.21	11.80	10.70	2.70	1.60	25.50	26.20	14.30	10.90
Max.	36.70	2.98	28.30	78.60	9.90	3.60	128.90	207.30	122.10	69.80
Coeff. Var.	9.98	20.39	19.60	42.90	25.97	16.37	36.14	45.44	52.105	47.16

All of the nutrients returned a p-value under 0.05 for the Shapiro–Wilk normality test at a 95% confidence interval.

The correlation between nutrients (Figure 4) is important information to characterize a dataset. The correlation coefficients indicated that, although significant, most nutrients have a low correlation value among themselves. Still, the pairwise comparison returned an expected behavior. Macronutrients, such as N, P, and K, showed positive correlations with each other while presenting negative correlation coefficients (N and P) with the other nutrients. The correlation coefficients between the nutrients varied around 0.5 or below, with the highest reaching 0.59 and lowest reaching −0.60. The low correlation value is also favorable for the proposed framework, as it helps to isolate the effects of the nutrient on the evaluated wavelengths.

	N	P	K	Ca	Mg	S	Cu	Fe	Mn	Zn
N		<0.001	<0.001	<0.001	0.361	<0.001	<0.001	<0.001	<0.001	<0.001
P	0.257		0.119	<0.001	<0.001	<0.001	<0.001	<0.001	<0.001	<0.001
K	0.335	−0.087		<0.001	<0.001	<0.001	<0.001	0.908	<0.001	<0.001
Ca	−0.205	−0.604	0.121		<0.001	<0.001	<0.001	<0.001	<0.001	<0.001
Mg	−0.051	−0.257	0.144	0.557		<0.001	0.423	<0.001	<0.001	<0.001
S	0.313	−0.380	0.331	0.596	0.300		<0.001	<0.001	<0.001	<0.001
Cu	−0.141	−0.533	0.270	0.390	0.044	0.198		<0.001	<0.001	<0.001
Fe	−0.397	−0.332	−0.006	0.507	0.272	0.315	0.329		<0.001	<0.001
Mn	−0.173	−0.439	0.265	0.535	0.135	0.256	0.687	0.551		<0.001
Zn	−0.241	−0.562	0.268	0.592	0.306	0.215	0.598	0.272	0.470	

Correlation test between the nutrient's content in the sampled citrus-leaves

Figure 4. Correlation between the measured nutrients for the Valencia-orange leaf samples.

For the machine learning algorithms used, the results were separated between the two datasets: reflectance (Table 4) and first-derivative (Table 5). The algorithms returned good performances ($R^2 > 0.80$) for the macronutrients with the spectral reflectance as predictors. When the first-derivative was used, the algorithms performed well on both macro- and micronutrients (some $R^2 > 0.80$), but all performances were improved for the micronutrients. This is an important discovery, as it highlights the importance of first-derivative measurements and their relationship with micronutrients in the Valencia-orange leaf tissue. In both datasets, algorithms like RF, ANN, and kNN returned better predictions than most linear ones, such as Lasso and Ridge Regressions and SVM. The MAE predictions returned here are similar to the predictions resulted from the training phase, which indicates how adjusted the sampling method was.

Table 4. The machine learning algorithms' accuracy performance for the reflectance data.

Method	N	P	K	Ca	Mg	S	Cu	Fe	Mn	Zn
kNN										
R^2	0.852	0.623	0.621	0.179	0.797	0.119	0.834	0.437	0.592	0.431
MAE	0.704	0.163	1.087	6.765	0.285	0.204	7.083	18.142	5.105	6.005
RMSE	1.245	0.278	2.041	13.905	0.445	0.416	11.362	36.248	11.707	10.157
Lasso Regression										
R^2	0.394	0.452	0.315	0.157	0.413	0.660	0.215	0.180	0.189	0.128
MAE	2.145	0.193	1.542	7.304	0.627	0.137	19.881	34.140	8.470	7.852
RMSE	2.526	0.335	2.745	14.091	0.757	0.258	24.744	43.751	16.513	12.573
Ridge Regression										
R^2	0.351	0.153	0.354	0.169	0.139	0.056	0.232	0.222	0.190	0.137
MAE	2.468	0.284	1.347	9.923	0.698	0.298	19.321	28.456	9.456	6.541
RMSE	2.912	0.417	2.597	13.989	0.916	0.431	24.485	42.158	16.502	12.982
SVM										
R^2	0.638	0.404	0.530	0.336	0.458	0.400	0.277	0.308	0.742	0.447
MAE	0.902	0.247	1.546	6.551	0.505	0.233	19.692	21.421	3.666	7.891
RMSE	1.952	0.349	2.275	12.501	0.752	0.344	23.754	40.201	9.309	9.741
ANN										
R^2	0.860	0.656	**0.762**	0.481	0.733	0.438	**0.841**	0.340	0.698	0.595
MAE	0.840	0.177	**1.265**	7.637	0.359	0.174	**8.377**	30.259	5.880	6.949
RMSE	1.211	0.265	**1.619**	11.052	0.510	0.332	**11.120**	39.251	10.078	8.567
DT										
R^2	0.743	0.661	0.613	0.576	0.759	0.452	0.731	0.453	0.730	0.640
MAE	0.787	0.178	1.434	6.375	0.345	0.166	8.835	20.016	5.090	5.681
RMSE	1.644	0.263	2.064	12.123	0.484	0.328	14.472	35.726	9.525	8.0811
RF										
R^2	**0.912**	**0.771**	0.699	**0.624**	**0.832**	**0.727**	0.754	**0.527**	**0.854**	**0.741**
MAE	**0.706**	**0.093**	1.146	**3.525**	**0.234**	**0.100**	7.828	**19.375**	5.093	**4.246**
RMSE	**1.059**	**0.216**	1.818	**9.404**	**0.405**	**0.231**	13.850	**33.233**	7.007	**6.846**

The bolded in the table are the scores representing the overall best performance of each nutrient.

Table 5. The machine learning algorithms' accuracy performance for first-derivative data.

Method	N	P	K	Ca	Mg	S	Cu	Fe	Mn	Zn
kNN										
R^2	0.669	0.453	0.329	0.311	0.512	0.172	0.752	0.512	**0.898**	0.587
MAE	0.944	0.180	1.341	6.994	0.398	0.184	7.456	17.846	**3.594**	4.948
RMSE	1.867	0.335	2.717	12.039	0.689	0.404	13.903	33.740	**5.859**	8.655
Lasso Regression										
R^2	0.257	0.401	0.161	0.287	0.401	0.158	0.292	0.190	0.234	0.130
MAE	1.489	0.197	1.986	7.048	0.486	0.304	19.513	33.146	7.934	9.588
RMSE	2.804	0.354	3.040	12.258	0.789	0.407	23.503	43.487	16.055	12.561
Ridge Regression										
R^2	0.210	0.310	0.157	0.302	0.129	0.222	0.273	0.183	0.300	0.158
MAE	1.650	0.265	1.997	6.978	0.789	0.248	19.212	34.240	8.242	9.047
RMSE	2.987	0.380	3.101	12.268	0.987	0.358	23.819	44.289	15.348	11.978
SVM										
R^2	0.373	0.323	0.330	0.531	0.270	0.459	0.679	0.423	0.649	0.513
MAE	1.357	0.250	1.714	6.745	0.678	0.240	11.870	24.653	6.676	6.159
RMSE	2.262	0.373	2.716	10.511	0.844	0.326	15.829	36.705	10.855	9.397
ANN										
R^2	0.721	0.554	0.445	**0.566**	**0.564**	**0.582**	0.800	0.444	0.838	0.731
MAE	1.287	0.219	1.680	**6.495**	**0.559**	**0.197**	10.030	20.466	4.617	5.605
RMSE	1.712	0.302	2.471	**10.113**	**0.652**	**0.287**	12.493	36.028	7.372	6.983
DT										
R^2	0.703	0.633	0.491	0.491	0.479	0.474	0.786	0.509	0.728	0.584
MAE	1.298	0.136	1.223	6.850	0.597	0.232	7.013	32.559	4.976	4.832
RMSE	2.767	0.274	2.368	13.015	0.832	0.314	25.209	33.861	15.036	11.391
RF										
R^2	**0.866**	**0.765**	**0.548**	0.501	0.507	0.453	**0.861**	**0.612**	0.879	**0.855**
MAE	**0.738**	**0.119**	**1.225**	6.668	0.424	0.209	**6.509**	**17.280**	4.050	**2.075**
RMSE	**1.185**	**0.219**	**2.231**	10.839	0.693	0.328	**10.389**	**31.640**	6.377	**5.121**

The bolded in the table are the scores representing the overall best performance of each nutrient.

To ascertain the relationship between each nutrient prediction, their regression values were plotted (Figure 5a,b). A quick analysis of the best-predicted values versus the laboratory-measured values demonstrates how the performance of the algorithms varied with the increase in the nutrient concentrations. Nutrients such as P and Ca did not show a closer resemblance with a 1:1 relationship (dashed-line—Figure 5a,b) as much as the other nutrients' predictions, even lower ones such as Fe. Regardless, most predictions were quite well related to the laboratory data, and on-site measurements of nutrients like N, K, Mg, S, Cu, Mn, and Zn may benefit from the advantage promulgated by the approach presented here.

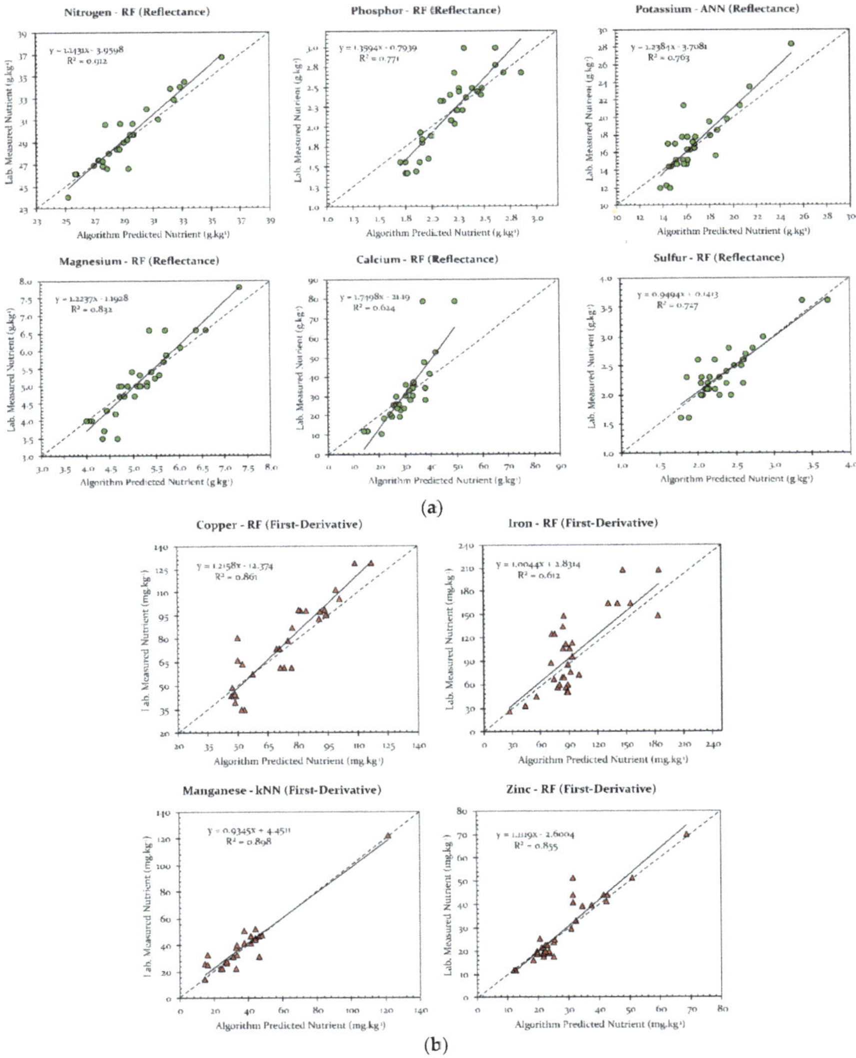

Figure 5. (**a**) Macronutrient prediction comparison against laboratory measurements for the best algorithms' results. (**b**) Micronutrient prediction comparison against laboratory measurements for the best algorithms' results.

The calculated Relief-F value showed the contribution of each wavelength to the algorithms' performance (Figure 6a,b). This contribution is important to isolate specific spectral regions and wavelengths of the electromagnetic spectrum most closely related to each nutrient. This relationship, however, is limited to the evaluated algorithm and its performance. Still, since most performances were relatively high (Tables 3 and 4) for most nutrients, this metric is an interesting parameter, as it shines some light on the spectral mapping of Valencia-orange leaf nutrients, as not much is known about their spectral behavior.

As mentioned, the Relief-F value calculated for each wavelength indicated important contributions in different ranges for each nutrient (Figure 5a,b). Because of that, certain bands of the electromagnetic spectrum contributed more than others. To summarize the information obtained from the proposed framework, a table (Table 6) indicating the nutrient and its class (macro or micro), the machine learning method most suitable to model it, its coefficient of determination (R^2), the spectral data which its prediction was calculated from, and the most contributive wavelengths or spectral regions to model the measured nutrient from the Relief-F value was created. These results demonstrate the potential of applying different machine learning algorithms for this task. So far, this is the first approach of its kind with nutrient content in leaf tissue analysis.

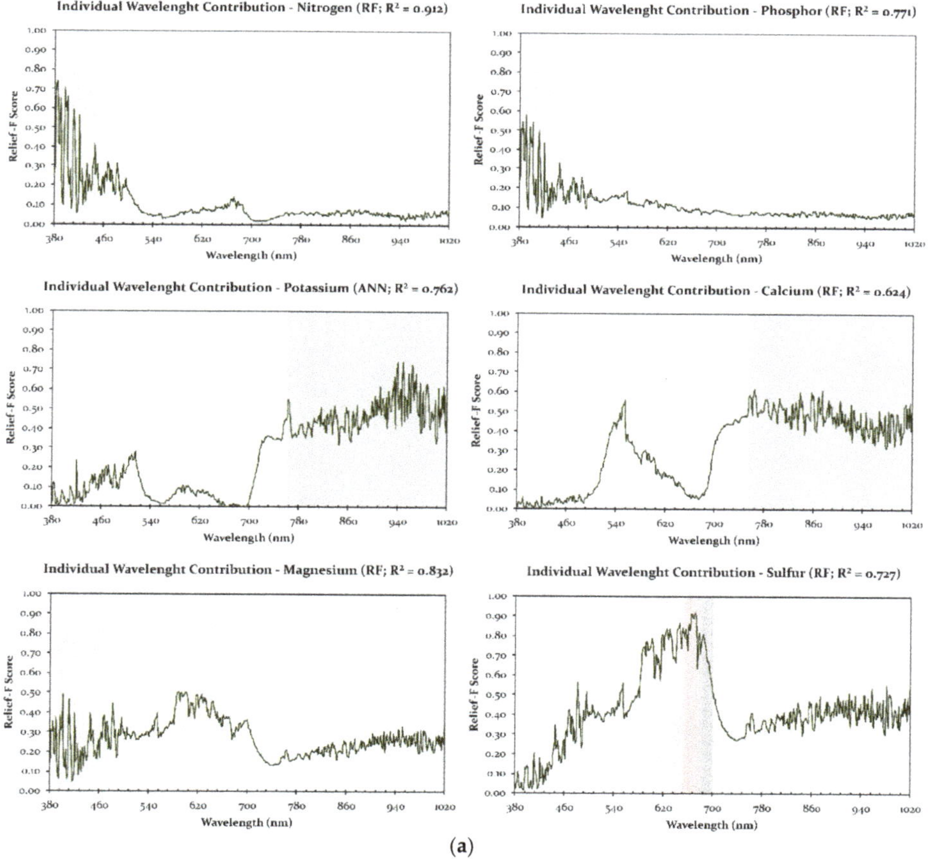

Figure 6. *Cont.*

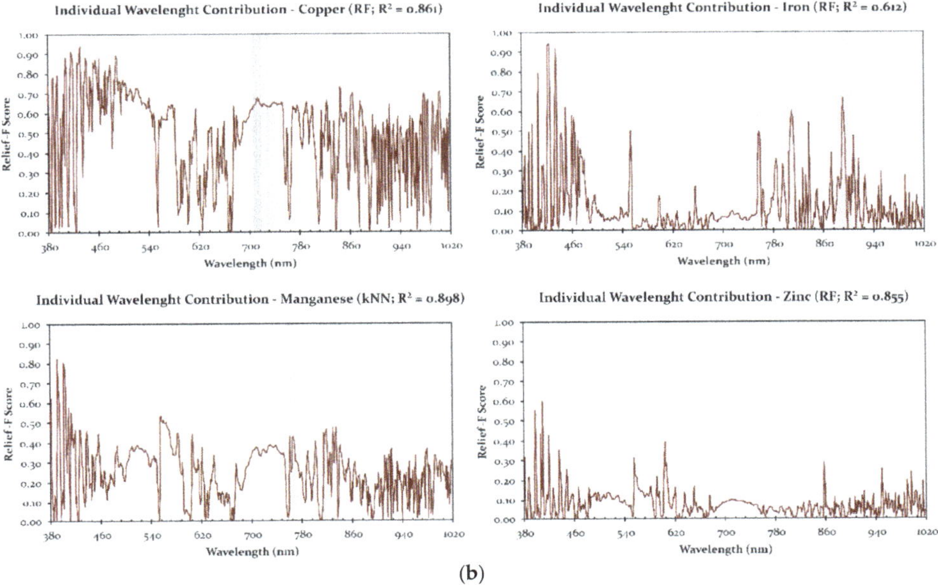

(b)

Figure 6. (**a**) The individual contribution of wavelengths for each macronutrients' prediction. (**b**) The individual contribution of wavelengths for each micronutrients' prediction.

Table 6. Summarized information on the results obtained by the proposed framework.

Nutrient	Class	Method	R^2	MAE	Spectral Data	Contributive Wavelengths/Spectral Regions (nm) *
N	Macro	RF	0.912	0.706	Reflectance	384–412; 421; 423; 432; 433; 435; 440–455; 464–472; 480–487
P	Macro	RF	0.771	0.093	Reflectance	385–411; 438–456; 472–477; 502; 521; 527; 544–555;
K	Macro	ANN	0.763	1.265	Reflectance	762–764; 816; 838; 857; 903 908; 915–925; 934–957; 973–1020
Ca	Macro	RF	0.624	3.525	Reflectance	545–551; 749–787; 843–888; 901–1020
Mg	Macro	RF	0.832	0.234	Reflectance	390; 411–412; 445; 496; 554; 586–630; 643; 656–669
S	Macro	RF	0.727	0.100	Reflectance	579; 590; 595; 609–612; 618; 624–632; 645–680; 684–689; 700
Cu	Micro	RF	0.861	6.509	First-Derivative	388; 394; 416–419; 430–432; 440; 452–456; 461; 475; 512; 523; 823; 863–865; 951; 977–979;
Fe	Micro	RF	0.612	17.280	First-Derivative	391–396; 405; 421–424; 433–436; 474–477; 552; 758; 810; 837; 890–892; 910; 926
Mn	Micro	kNN	0.898	3.594	First-Derivative	381; 392–410; 414; 438; 555–568; 582; 819; 607; 761–767; 823–835; 841
Zn	Micro	RF	0.855	2.075	First-Derivative	381; 398; 407–411; 420; 449; 555–559; 604–607; 858

* These wavelengths and regions were obtained by sorting the highest Relief-F values of each prediction.

4. Discussion

In the proposed framework, both reflectance data and their first-derivatives were used to predict macro- and micronutrients. This approach used a robust technique (machine learning) to model the hyperspectral data, which helped to ascertain some discoveries. The results demonstrated compelling performances to predict most of the nutrients (Tables 3–5). Nutrients like N, Mg, Cu, Mn, and Zn were predicted with an R^2 of 0.912, 0.832, 0.861, 0.898, and 0.855, respectively. Other nutrients like P, K, and S presented inferior performances with an R^2 of 0.771, 0.763, and 0.727, respectively. The worst performances were obtained for nutrients like Ca and Fe, with their R^2's equal to 0.624 and 0.612, respectively. In comparison to the literature, most of these performances, specifically those related to macronutrients, were similar or superior for other types of plants and methods. For N, predictions using visible to infrared data returned accuracies between 0.73 to 0.87 (R^2) [2,30,40]. For

K, a three-band combination index predicted the nutrient with an R^2 equal to 0.74 [37]. In nutrients like Mg, S, P, Ca, and others, the predictions (R^2) variated between lower values of 0.27 up to 0.98, depending on the method applied and the plant evaluated [24,26,30,32,38,39,50].

One important finding from this research is the relationship between nutrients, algorithms, and leaf-spectral curves. For macronutrients, the performance of the algorithms was superior when adopting the surface reflectance data. As for the micronutrients, the first-derivative of the spectral reflectance returned better performances for the algorithms (Figure 5a,b and Table 6). This finding can be related to information reported in the literature [39–41]. Since first-derivative spectra allow for highlighting absorption features of the original spectra, it could potentially be linked to different components not so easily observable in spectral reflectance data alone. This approach demonstrated a better relation for all micronutrients when linked to the algorithms with the first-derivative, so this could offer a possible explanation. As previously mentioned in Section 2.2, other derivatives of the dataset were evaluated, but could not find a significant difference over the first derivative. Still, further research should continue to explore the association between first-derivative spectra, second- and third-derivatives, and micronutrients.

Another contribution of the proposed framework is that, although, with a limitation in the accuracy of the algorithms, it is possible to identify the wavelengths and the spectral regions that most contributed to predicting each nutrient (Figure 6a,b and Table 6). While some nutrients show contributions from the same wavelengths, these contributions vary in value (Relief-F). Even so, most of the nutrients showcase particular wavelengths that could potentially be isolated or used in combination with others to ascertain their relationship with the prediction (Table 6). This finding could help to map the Valencia-orange leaf spectral behavior related to both macro- and micronutrients and promote the investigation of simpler mathematical models or spectral indices capable of modeling these nutrients by focusing on these wavelengths.

Machine learning algorithms have the advantage of modeling data in a non-linear and a non-parametric manner. Unlike many traditional statistical methods, these algorithms are built with the advantage of dealing with noisy, complex, and heterogeneous data [16,23,50–52]. These characteristics proved to be an advantage for this study, as the data used had higher variance, was not-normal (Table 3), and, while statistically significant, low-correlated in a pairwise manner (Figure 4). Previous research aimed to model nutrient content with similar characteristics by using multiple mathematical methods in the analysis of plant hyperspectral data, but it did not return the same accuracies [15,24,32,40,42]. Nonetheless, since machine learning methods can deal with most of the data inconsistencies, both in hyperspectral measurements and in nutrient content analysis, the proposed framework should be more appropriate to combine these features not requiring data modification while still returning good performances.

Finally, the different performances returned by the algorithms should be discussed. It is clear that regression models like Lasso, Ridge, and SVM were inferior to others (Tables 3 and 4) in both scenarios (reflectance/first-derivative). Although SVM is known to handle high dimensionality data and do well with a limited training dataset [45], it performed poorly in the used dataset in comparison with the rest. The DT algorithm, though not as inferior in performance as the aforementioned algorithms, achieved middling results in comparison with the remaining methods. For the DT, the XGBoost model was adopted to improve the prediction performance, which was also implemented in the RF base model. During the experimental phase, this boosting model proved to be of assistance in enhancing the performance of both algorithms. Still, DT did not return predictions as accurate as did the RF algorithm.

The highest performances were obtained by the RF, ANN, and kNN algorithms; both for macro- and micronutrients. While RF was better in almost all predictions, ANN and kNN performed well in only particular cases, especially for K (reflectance data) and Mn (first-derivative data), respectively. kNN is a simpler method than ANN and RF, being relatively faster. However, throughout the different nutrients, RF and ANN had better consistency. The ANN was constructed in a manner that could predict most of the nutrients, but performance rates were limited. The amount of available data for training the algorithms could also be a potential hindrance for deep learning networks to handle. While the ANN method benefited from a multilayer perceptron, with two hidden layers and a high number of neurons and interactions, the RF algorithm was boosted with the XGBoost model, which returned a continuous performance for the reflectance and first-derivative datasets. Regardless, as no machine learning algorithm is considered universally appropriate to deal with any task, a framework like the one proposed here is recommended since it makes uses of multiple algorithms because different data could potentially impact its performance.

5. Conclusions

The proposed approach uses leaf spectral data in the visible and near-infrared regions, and switches between reflectance and its first-derivative data to predict the amount of macro- and micronutrients measured in the laboratory. This method was able to return high predictions (R^2) for nutrients like N (0.912), Mg (0.832), Cu (0.861), Mn (0.898), and Zn (0.855), and, to a lesser extent, P (0.771), K (0.763), and S (0.727). These accuracies were obtained with the RF, ANN, and kNN algorithms, among which RF performed the best. Another discovery was that reflectance data is more suitable to model macronutrients, while the first-derivative of the reflectance data is better related to micronutrients. Another contribution also made by this study is the identification (by the Relief-F value) of the wavelengths most responsible for the prediction results. Each nutrient was better correlated to one or more spectral wavelengths. Because of it, future research should evaluate simpler models or spectral vegetation indices capable of modeling the nutrient content by focusing on these wavelengths. Although the presented method was used for evaluating the nutritional conditions of Valencia-orange leaves, it can be replicated for different plants and cultivars, with the possibility of even better performances being achievable. Furthermore, as an advantage of this approach, this framework may be implemented in hyperspectral data obtained with sensors embedded in UAV-based systems.

Author Contributions: Conceptualization, L.P.O., A.P.M.R., J.M.J., and J.E.C.; methodology, L.P.O., A.P.M.R., É.A.S.M., and J.E.C. formal analysis, L.P.O. and M.M.F.P.; resources, J.E.C., N.N.I., F.F.d.A. and J.M.J.; data curation, L.P.O., M.M.F.P., A.P.M.R., É.A.S.M., J.E.C., and J.M.J.; writing—original draft preparation, L.P.O.; writing—review and editing, A.P.M.R, É.A.S.M., N.N.I., F.F.d.A., N.E., F.I., V.L., L.A.d.C.J., J.L., L.M., W.N.G. and J.M.J.; supervision, A.P.M.R., J.E.C., N.N.I. and J.M.J; project administration, A.P.M.R., J.E.C and J.M.J.; funding acquisition, L.P.O., and J.E.C. All authors have read and agreed to the published version of the manuscript.

Funding: This study was financed in part by the Coordenação de Aperfeiçoamento de Pessoal de Nível Superior—Brasil (CAPES)—Finance Code 001 and supported by the Universidade Federal de Mato Grosso do Sul (UFMS). É.A.S. Moriya is supported by FUNDUNESP/Print (p: 3030/2019). V. Liesenberg is supported by FAPESC (2017TR1762) and CNPq (313887/2018-7). N.N. Imai is supported by CNPq (310128/2018-8). J. Marcato Junior is supported by CNPq (433783/2018-4, 303559/2019-5) and Fundect (59/300.066/2015).

Acknowledgments: The authors acknowledge Universidade Federal de Mato Grosso do Sul (UFMS) for supporting the research, and Fazenda Brasilia, located in Areia Branca, Ubirajara—SP (Brazil), for contributing to the experimental site.

Conflicts of Interest: The authors declare no conflict of interest. The funders had no role in the design of the study; in the collection, analyses, or interpretation of data; in the writing of the manuscript, or in the decision to publish the results.

References

1. Li, F.; Wang, L.; Liu, J.; Wang, Y.; Chang, Q. Evaluation of leaf N concentration in winter wheat based on discrete wavelet transform analysis. *Remote Sens.* **2019**, *11*, 1331. [CrossRef]
2. Li, Z.; Jin, X.; Yang, G.; Drummond, J.; Yang, H.; Clark, B.; Li, Z.; Zhao, C. Remote sensing of leaf and canopy nitrogen status in winter wheat (Triticum aestivum L.) based on N-PROSAIL model. *Remote Sens.* **2018**, *10*, 1463. [CrossRef]
3. Osco, L.P.; Marques Ramos, A.P.; Saito Moriya, É.A.; de Souza, M.; Marcato Junior, J.; Matsubara, E.T.; Imai, N.N.; Creste, J.E. Improvement of leaf nitrogen content inference in Valencia-orange trees applying spectral analysis algorithms in UAV mounted-sensor images. *Int. J. Appl. Earth Obs. Geoinf.* **2019**, *83*, 101907. [CrossRef]

4. Zheng, H.; Li, W.; Jiang, J.; Liu, Y.; Cheng, T.; Tian, Y.; Zhu, Y.; Cao, W.; Zhang, Y.; Yao, X. A comparative assessment of different modeling algorithms for estimating leaf nitrogen content in winter wheat using multispectral images from an unmanned aerial vehicle. *Remote Sens.* **2018**, *10*, 2026. [CrossRef]

5. Loggenberg, K.; Strever, A.; Greyling, B.; Poona, N. Modelling water stress in a Shiraz vineyard using hyperspectral imaging and machine learning. *Remote Sens.* **2018**, *10*, 202. [CrossRef]

6. Gerhards, M.; Schlerf, M.; Rascher, U.; Udelhoven, T.; Juszczak, R.; Alberti, G.; Miglietta, F.; Inoue, Y. Analysis of airborne optical and thermal imagery for detection of water stress symptoms. *Remote Sens.* **2018**, *10*, 1139. [CrossRef]

7. Johnson, K.; Sankaran, S.; Ehsani, R. Identification of Water Stress in Citrus Leaves Using Sensing Technologies. *Agronomy* **2013**, *3*, 747–756. [CrossRef]

8. Osco, L.P.; dos Santos de Arruda, M.; Marcato Junior, J.; da Silva, N.B.; Ramos, A.P.M.; Moryia, É.A.S.; Imai, N.N.; Pereira, D.R.; Creste, J.E.; Matsubara, E.T.; et al. A convolutional neural network approach for counting and geolocating citrus-trees in UAV multispectral imagery. *ISPRS J. Photogramm. Remote Sens.* **2020**, *160*, 97–106. [CrossRef]

9. Weinstein, B.G.; Marconi, S.; Bohlman, S.; Zare, A.; White, E. Individual tree-crown detection in rgb imagery using semi-supervised deep learning neural networks. *Remote Sens.* **2019**, *11*, 1309. [CrossRef]

10. Zhang, K.; Ge, X.; Shen, P.; Li, W.; Liu, X.; Cao, Q.; Zhu, Y.; Cao, W.; Tian, Y. Predicting rice grain yield based on dynamic changes in vegetation indexes during early to mid-growth stages. *Remote Sens.* **2019**, *11*, 387. [CrossRef]

11. Nevavuori, P.; Narra, N.; Lipping, T. Crop yield prediction with deep convolutional neural networks. *Comput. Electron. Agric.* **2019**, *163*, 104859. [CrossRef]

12. Hunt, M.L.; Blackburn, G.A.; Carrasco, L.; Redhead, J.W.; Rowland, C.S. High resolution wheat yield mapping using Sentinel-2. *Remote Sens. Environ.* **2019**, *233*, 111410. [CrossRef]

13. Cui, B.; Zhao, Q.; Huang, W.; Song, X.; Ye, H.; Zhou, X. A new integrated vegetation index for the estimation of winter wheat leaf chlorophyll content. *Remote Sens.* **2019**, *11*, 974. [CrossRef]

14. Guo, T.; Tan, C.; Li, Q.; Cui, G.; Li, H. Estimating leaf chlorophyll content in tobacco based on various canopy hyperspectral parameters. *J. Ambient Intell. Humaniz. Comput.* **2019**, *10*, 3239–3247. [CrossRef]

15. Peng, Z.; Guan, L.; Liao, Y.; Lian, S. Estimating total leaf chlorophyll content of gannan navel orange leaves using hyperspectral data based on partial least squares regression. *IEEE Access* **2019**, *7*, 155540–155551. [CrossRef]

16. Abdulridha, J.; Batuman, O.; Ampatzidis, Y. UAV-based remote sensing technique to detect citrus canker disease utilizing hyperspectral imaging and machine learning. *Remote Sens.* **2019**, *11*, 1373. [CrossRef]

17. Yao, Z.; Lei, Y.; He, D. Early visual detection of wheat stripe rust using visible/near-infrared hyperspectral imaging. *Sensors (Switzerland)* **2019**, *19*, 952. [CrossRef]

18. Pham, T.D.; Yokoya, N.; Bui, D.T.; Yoshino, K.; Friess, D.A. Remote sensing approaches for monitoring mangrove species, structure, and biomass: Opportunities and challenges. *Remote Sens.* **2019**, *11*, 230. [CrossRef]

19. Brinkhoff, J.; Dunn, B.W.; Robson, A.J.; Dunn, T.S.; Dehaan, R.L. Modeling mid-season rice nitrogen uptake using multispectral satellite data. *Remote Sens.* **2019**, *11*, 1837. [CrossRef]

20. Zhou, C.; Ye, H.; Xu, Z.; Hu, J.; Shi, X.; Hua, S.; Yue, J.; Yang, G. Estimating maize-leaf coverage in field conditions by applying a machine learning algorithm to UAV remote sensing images. *Appl. Sci.* **2019**, *9*, 2389. [CrossRef]

21. Delloye, C.; Weiss, M.; Defourny, P. Retrieval of the canopy chlorophyll content from Sentinel-2 spectral bands to estimate nitrogen uptake in intensive winter wheat cropping systems. *Remote Sens. Environ.* **2018**, *216*, 245–261. [CrossRef]

22. Vanbrabant, Y.; Tits, L.; Delalieux, S.; Pauly, K.; Verjans, W.; Somers, B. Multitemporal chlorophyll mapping in pome fruit orchards from remotely piloted aircraft systems. *Remote Sens.* **2019**, *11*, 1468. [CrossRef]

23. Osco, L.P.; Ramos, A.P.M.; Moriya, E.A.S.; Bavaresco, L.G.; Lima, B.C.; Estrabis, N.; Pereira, D.R.; Creste, J.E.; Marcato Junior, J.; Gonçalves, W.N.; et al. Modeling hyperspectral response of water-stress induced lettuce plants using artificial neural networks. *Remote Sens.* **2019**, *11*, 2797. [CrossRef]

24. de Oliveira, D.M.; Fontes, L.M.; Pasquini, C. Comparing laser induced breakdown spectroscopy, near infrared spectroscopy, and their integration for simultaneous multi-elemental determination of micro- and macronutrients in vegetable samples. *Anal. Chim. Acta* **2019**, *1062*, 28–36. [CrossRef]

25. Chen, J.; Li, F.; Wang, R.; Fan, Y.; Raza, M.A.; Liu, Q.; Wang, Z.; Cheng, Y.; Wu, X.; Yang, F.; et al. Estimation of nitrogen and carbon content from soybean leaf reflectance spectra using wavelet analysis under shade stress. *Comput. Electron. Agric.* **2019**, *156*, 482–489. [CrossRef]

26. Cuq, S.; Lemetter, V.; Kleiber, D.; Levasseur-Garcia, C. Assessing macro-element content in vine leaves and grape berries of vitis vinifera by using near-infrared spectroscopy and chemometrics. *Int. J. Environ. Anal. Chem.* **2019**. [CrossRef]

27. Santoso, H.; Tani, H.; Wang, X.; Segah, H. Predicting oil palm leaf nutrient contents in kalimantan, indonesia by measuring reflectance with a spectroradiometer. *Int. J. Remote Sens.* **2019**, *40*, 7581–7602. [CrossRef]

28. Osco, L.P.; Ramos, A.P.M.; Pereira, D.R.; Moriya, E.A.S.; Imai, N.N.; Matsubara, E.T. Predicting canopy nitrogen content in citrus-trees using random forest algorithm associated to spectral vegetation indices from UAV-Imagery. *Remote Sens.* **2019**, *11*, 2925. [CrossRef]

29. Allen, V.; Barker, D.J.P. *Handbook of Plant Nutrition*; Taylor et Francis, 2015. Available online: https://www.bokus.com/bok/9781439881989/handbook-of-plant-nutrition/ (accessed on 4 February 2020).

30. Malmir, M.; Tahmasbian, I.; Xu, Z.; Farrar, M.B.; Bai, S.H. Prediction of macronutrients in plant leaves using chemometric analysis and wavelength selection. *J. Soils Sediments* **2019**. [CrossRef]

31. Ye, X.; Abe, S.; Zhang, S. Estimation and mapping of nitrogen content in apple trees at leaf and canopy levels using hyperspectral imaging. *Precis. Agric.* **2019**. [CrossRef]

32. Ling, B.; Goodin, D.G.; Raynor, E.J.; Joern, A. Hyperspectral analysis of leaf pigments and nutritional elements in tallgrass prairie vegetation. *Front. Plant Sci.* **2019**, *10*, 1–13. [CrossRef] [PubMed]

33. Zhao, Y.; Yan, C.; Lu, S.; Wang, P.; Qiu, G.Y.; Li, R. Estimation of chlorophyll content in intertidal mangrove leaves with different thicknesses using hyperspectral data. *Ecol. Indic.* **2019**, *106*, 105511. [CrossRef]

34. Shi, J.; Li, W.; Zhai, X.; Guo, Z.; Holmes, M.; Elrasheid Tahir, H.; Zou, X. Nondestructive diagnostics of magnesium deficiency based on distribution features of chlorophyll concentrations map on cucumber leaf. *J. Plant Nutr.* **2019**, *42*, 2773–2783. [CrossRef]

35. Román, J.R.; Rodríguez-Caballero, E.; Rodríguez-Lozano, B.; Roncero-Ramos, B.; Chamizo, S.; Águila-Carricondo, P.; Cantón, Y. Spectral response analysis: An indirect and non-destructive methodology for the chlorophyll quantification of biocrusts. *Remote Sens.* **2019**, *11*, 1350. [CrossRef]

36. Bruning, B.; Liu, H.; Brien, C.; Berger, B.; Lewis, M.; Garnett, T. The Development of hyperspectral distribution maps to predict the content and distribution of nitrogen and water in wheat (Triticum aestivum). *Front. Plant Sci.* **2019**. [CrossRef]

37. Lu, J.; Yang, T.; Su, X.; Qi, H.; Yao, X.; Cheng, T.; Zhu, Y.; Cao, W.; Tian, Y. Monitoring leaf potassium content using hyperspectral vegetation indices in rice leaves. *Precis. Agric.* **2019**. [CrossRef]

38. Jull, H.; Künnemeyer, R.; Schaare, P. Nutrient quantification in fresh and dried mixtures of ryegrass and clover leaves using laser-induced breakdown spectroscopy. *Precis. Agric.* **2018**, *19*, 823–839. [CrossRef]

39. Galvez-Sola, L.; García-Sánchez, F.; Pérez-Pérez, J.G.; Gimeno, V.; Navarro, J.M.; Moral, R.; Martínez-Nicolás, J.J.; Nieves, M. Rapid estimation of nutritional elements on citrus leaves by near infrared reflectance spectroscopy. *Front. Plant Sci.* **2015**, *6*, 1–3. [CrossRef]

40. Yang, J.; Du, L.; Gong, W.; Shi, S.; Sun, J.; Chen, B. Analyzing the performance of the first-derivative fluorescence spectrum for estimating leaf nitrogen concentration. *Opt. Express* **2019**, *27*, 3978. [CrossRef]

41. Yang, J.; Cheng, Y.; Du, L.; Gong, W.; Shi, S.; Sun, J.; Chen, B. Selection of the optimal bands of first-derivative fluorescence characteristics for leaf nitrogen concentration estimation. *Appl. Opt.* **2019**, *58*, 5720–5727. [CrossRef]

42. Zhou, W.; Zhang, J.; Zou, M.; Liu, X.; Du, X.; Wang, Q.; Liu, Y.; Liu, Y.; Li, J. Prediction of cadmium concentration in brown rice before harvest by hyperspectral remote sensing. *Environ. Sci. Pollut. Res.* **2019**, *26*, 1848–1856. [CrossRef] [PubMed]

43. Mitchell, T.M. *Machine Learning*, 1st ed.; McGraw-Hill, Inc.: New York, NY, USA, 1997.

44. Ball, J.E.; Anderson, D.T.; Chan, C.S. Comprehensive survey of deep learning in remote sensing: Theories, tools, and challenges for the community. *J. Appl. Remote Sens.* **2017**, *11*, 1. [CrossRef]

45. Ma, L.; Liu, Y.; Zhang, X.; Ye, Y.; Yin, G.; Johnson, B.A. ISPRS Journal of Photogrammetry and Remote Sensing Deep learning in remote sensing applications: A meta-analysis and review. *ISPRS J. Photogramm. Remote Sens.* **2019**, *152*, 166–177. [CrossRef]

46. Gao, J.; Meng, B.; Liang, T.; Feng, Q.; Ge, J.; Yin, J. ISPRS Journal of Photogrammetry and Remote Sensing Modeling alpine grassland forage phosphorus based on hyperspectral remote sensing and a multi-factor machine learning algorithm in the east of Tibetan. *ISPRS J. Photogramm. Remote Sens.* **2019**, *147*, 104–117. [CrossRef]

47. Feng, P.; Wang, B.; Liu, D.L.; Yu, Q. Machine learning-based integration of remotely-sensed drought factors can improve the estimation of agricultural drought in South-Eastern Australia. *Agric. Syst.* **2019**, *173*, 303–316. [CrossRef]

48. Han, L.; Yang, G.; Dai, H.; Xu, B.; Yang, H.; Feng, H.; Li, Z.; Yang, X. Modeling maize above-ground biomass based on machine learning approaches using UAV remote-sensing data. *Plant Methods* **2019**, *15*, 10. [CrossRef]

49. Singhal, G.; Bansod, B.; Mathew, L.; Goswami, J.; Choudhury, B.U.; Raju, P.L.N. Chlorophyll estimation using multi-spectral unmanned aerial system based on machine learning techniques. *Remote Sens. Appl. Soc. Environ.* **2019**, *15*, 100235. [CrossRef]

50. Chanda, S.; Hazarika, A.K.; Choudhury, N.; Islam, S.A.; Manna, R.; Sabhapondit, S.; Tudu, B.; Bandyopadhyay, R. Support vector machine regression on selected wavelength regions for quantitative analysis of caffeine in tea leaves by near infrared spectroscopy. *J. Chemom.* **2019**, *33*, 10. [CrossRef]

51. Shah, S.H.; Angel, Y.; Houborg, R.; Ali, S.; McCabe, M.F. A random forest machine learning approach for the retrieval of leaf chlorophyll content in wheat. *Remote Sens.* **2019**, *11*, 920. [CrossRef]

52. Fu, P.; Meacham-Hensold, K.; Guan, K.; Bernacchi, C.J. Hyperspectral leaf reflectance as proxy for photosynthetic capacities: An ensemble approach based on multiple machine learning algorithms. *Front. Plant Sci.* **2019**, *10*. [CrossRef]

53. Obreza, T.A.; Morgan, K.T. (Eds.) *Nutrition of Florida Citrus Trees*, 2nd ed.; IFAS Extension; University of Florida: Gainesville, FL, USA, 2008.

54. Köppen, W.; Volken, E.; Brönnimann, S. The thermal zones of the Earth according to the duration of hot, moderate and cold periods and to the impact of heat on the organic world. *Meteorol. Zeitschrift* **2011**, *20*, 351–360. [CrossRef] [PubMed]

55. Nitrogen Determination by Kjeldahl Method PanReac AppliChem ITW Reagents. Available online: https://www.itwreagents.com/uploads/20180114/A173_EN.pdf (accessed on 27 February 2019).

56. Malavolta, E.; Vitti, G.C.; Oliveira, S.A. *Evaluation of Nutritional Status of Plants: Principles and Perspectives*, 2nd ed.; POTAFOS: Piracicaba, SP, Brazil, 1997; 319p.

57. Anderson, K.; Rossini, M.; Labrador, J.P.; Balzarolo, M.; Arthur, A.; Fava, F.; Julitta, T.; Vescovo, L. Inter-comparison of hemispherical conical reflectance factors (HCRF) measured with four fiber-based spectrometers. *Opt. Express.* **2013**, *21*, 605–617. [CrossRef] [PubMed]

58. Tsai, F.; Philpot, W. Derivative Analysis of Hyperspectral Data. *Remote Sens. Environ.* **1998**, *66*, 41–51. [CrossRef]

59. Bergstra, J.; Bengio, Y. Random search for hyper-parameter optimization. *J. Mach. Learn. Res.* **2012**, *13*, 281–305.

60. RapidMiner. RapidMiner Python Package. Available online: https://github.com/rapidminer/pythonrapidminer (accessed on 5 December 2019).

61. XGBoost. eXtreme Gradient Boosting. Available online: https://github.com/dmlc/xgboost (accessed on 5 December 2019).

62. Amancio, D.R.; Comin, C.H.; Casanova, D.; Travieso, G.; Bruno, O.M.; Rodrigues, F.A.; Da Fontoura Costa, L. A systematic comparison of supervised classifiers. *PLoS ONE* **2014**, *9*, 1–14. [CrossRef]

63. Pôças, I.; Tosin, R.; Gonçalves, I.; Cunha, M. Toward a generalized predictive model of grapevine water status in Douro region from hyperspectral data. *Agric. For. Meteorol.* **2020**, *280*, 107793. [CrossRef]

64. Zha, H.; Miao, Y.; Wang, T.; Li, Y.; Zhang, J.; Sun, W. Sensing-Based Rice Nitrogen Nutrition Index Prediction with Machine Learning. *Remote Sens.* **2020**, *12*, 215. [CrossRef]

65. Hennessy, A.; Clarke, K.; Lewis, M. Hyperspectral Classification of Plants: A Review of Waveband Selection Generalisability. *Remote Sens.* **2020**, *12*, 113. [CrossRef]

66. Chan, J.C.W.; Paelinckx, D. Evaluation of Random Forest and Adaboost tree-based ensemble classification and spectral band selection for ecotope mapping using airborne hyperspectral imagery. *Remote Sens. Environ.* **2008**, *112*, 2999–3011. [CrossRef]

67. Alonzo, M.; Bookhagen, B.; Roberts, D.A. Urban tree species mapping using hyperspectral and lidar data fusion. *Remote Sens. Environ.* **2014**, *148*, 70–83. [CrossRef]
68. Urbanowicz, R.J.; Meeker, M.; La Cava, W.; Olson, R.S.; Moore, J.H. Relief-based feature selection: Introduction and review. *J. Biomed. Inform.* **2018**, *85*, 189–203. [CrossRef] [PubMed]

 *remote sensing*

Review

Recent Advances of Hyperspectral Imaging Technology and Applications in Agriculture

Bing Lu [1], Phuong D. Dao [1,2], Jiangui Liu [3], Yuhong He [1,*] and Jiali Shang [3]

[1] Department of Geography, Geomatics and Environment, University of Toronto Mississauga, 3359 Mississauga Road, Mississauga, ON L5L 1C6, Canada; bing.lu@mail.utoronto.ca (B.L.); phuong.dao@mail.utoronto.ca (P.D.D.)

[2] School of the Environment, University of Toronto, 33 Willcocks Street, Toronto, ON M5S 3E8, Canada

[3] Agriculture and Agri-Food Canada, 960 Carling Avenue, Ottawa, ON K1A 0C6, Canada; jiangui.liu@canada.ca (J.L.); jiali.shang@Canada.ca (J.S.)

* Correspondence: yuhong.he@utoronto.ca

Received: 12 July 2020; Accepted: 16 August 2020; Published: 18 August 2020

Abstract: Remote sensing is a useful tool for monitoring spatio-temporal variations of crop morphological and physiological status and supporting practices in precision farming. In comparison with multispectral imaging, hyperspectral imaging is a more advanced technique that is capable of acquiring a detailed spectral response of target features. Due to limited accessibility outside of the scientific community, hyperspectral images have not been widely used in precision agriculture. In recent years, different mini-sized and low-cost airborne hyperspectral sensors (e.g., Headwall Micro-Hyperspec, Cubert UHD 185-Firefly) have been developed, and advanced spaceborne hyperspectral sensors have also been or will be launched (e.g., PRISMA, DESIS, EnMAP, HyspIRI). Hyperspectral imaging is becoming more widely available to agricultural applications. Meanwhile, the acquisition, processing, and analysis of hyperspectral imagery still remain a challenging research topic (e.g., large data volume, high data dimensionality, and complex information analysis). It is hence beneficial to conduct a thorough and in-depth review of the hyperspectral imaging technology (e.g., different platforms and sensors), methods available for processing and analyzing hyperspectral information, and recent advances of hyperspectral imaging in agricultural applications. Publications over the past 30 years in hyperspectral imaging technology and applications in agriculture were thus reviewed. The imaging platforms and sensors, together with analytic methods used in the literature, were discussed. Performances of hyperspectral imaging for different applications (e.g., crop biophysical and biochemical properties' mapping, soil characteristics, and crop classification) were also evaluated. This review is intended to assist agricultural researchers and practitioners to better understand the strengths and limitations of hyperspectral imaging to agricultural applications and promote the adoption of this valuable technology. Recommendations for future hyperspectral imaging research for precision agriculture are also presented.

Keywords: precision agriculture; remote sensing; hyperspectral imaging; platforms and sensors; analytical methods; crop properties; soil characteristics; classification of agricultural features

---

1. Introduction

The global agricultural sector is facing increasing challenges posed by a range of stressors, including a rapidly growing population, the depletion of natural resources, environmental pollution, crop diseases, and climate change. Precision agriculture is a promising approach to address these challenges through improving farming practices, e.g., adaptive inputs (e.g., water and fertilizer), ensured outputs (e.g., crop yield and biomass), and reduced environmental impacts. Remote sensing is capable of identifying within-field variability of soils and crops and providing useful information for site-specific management practices [1,2]. There are two types of remote sensing technologies given the source of energy, passive (e.g., optical) and active remote sensing (e.g., LiDAR and Radar). Passive optical remote sensing is usually further divided into two groups based on the spectral resolutions of sensors, multispectral and hyperspectral remote sensing [3]. Multispectral imaging is facilitated by collecting spectral signals in a few discrete bands, each spanning a broad spectral range from tens to hundreds of nanometers. In contrast, hyperspectral imaging detects spectral signals in a series of continuous channels with a narrow spectral bandwidth (e.g., typically below 10 nm); therefore, it can capture fine-scale spectral features of targets that otherwise could be compromised [4].

Multispectral images (e.g., Landsat, Sentinel 2, and SPOT images) have been widely used in agricultural studies to retrieve various crop and soil attributes, such as crop chlorophyll content, biomass, yield, and soil degradation [5–10]. However, due to the limitations in spectral resolution, the accuracy of the retrieved variables is often limited, and early signals of crop stresses (e.g., nutrient deficiency, crop disease) cannot be effectively detected in a timely manner [11]. Hyperspectral images (e.g., Hyperion, CASI, and Headwall Micro-Hyperspec) with hundreds of bands can capture more detailed spectral responses; hence, it is more capable of detecting subtle variations of ground covers and their changes over time. Therefore, hyperspectral imagery can be used to address the aforementioned challenges and facilitate more accurate and timely detection of crop physiological status [12,13]. Previous studies have also demonstrated the superior performance of hyperspectral over multispectral images in monitoring vegetation properties, such as estimating the leaf area index (LAI) [14], discriminating crop types [15], retrieving crop biomass [16], and assessing leaf nitrogen content [17]. Despite its outstanding performance, hyperspectral imaging has been utilized comparatively less in operational agricultural applications in the past few decades due to the high cost of the sensors and imaging missions, and various technical challenges (e.g., low signal-to-noise ratio and large data volume) [18–21]. Although ground-based hyperspectral reflectance data can be quickly measured using a spectroradiometer (e.g., ASD Field Spec, Analytical Spectral Devices Inc., Boulder, CO, USA) and have been widely used for observing canopy- and leaf-level spectral features [22–24], such ground-based measurements are limited to a few numbers of field sites, and they cannot capture spatial variability across large areas. In contrast, hyperspectral imaging sensors are more convenient to acquire spatial variability of spectral information across a region.

In recent years, a wide range of mini-sized and low-cost hyperspectral sensors have been developed and are available for commercial use, such as Micro- and Nano-Hyperspec (Headwall Photonics Inc., Boston, MA, USA), HySpex VNIR (HySpex, Skedsmo, Skjetten, Norway), and FireflEYE (Cubert GmbH, Ulm, Germany) [11,25]. These sensors can be mounted on manned or unmanned airborne platforms (e.g., airplanes, helicopters, and unmanned aerial vehicles (UAVs)) for acquiring hyperspectral images and supporting various monitoring missions [13,26,27]. In addition, new spaceborne hyperspectral sensors have been launched recently, such as the DESIS—launched in 2018 [28]—and PRISMA— launched in 2019 [29]—or will be launched in the next few years, such as EnMAP, with scheduled launching in 2020 [30,31]. Overall, increasingly more airborne or spaceborne hyperspectral images have become available, bringing unprecedented opportunities for better monitoring of ground targets, especially for better investigation of crop and soil variabilities and supporting precision agriculture. Therefore, a literature search was performed to examine if more research in using hyperspectral imaging for agricultural purposes had been published in recent years. Both Web of Science and Google Scholar were used for conducting the literature search with topics or keywords, including hyperspectral, imaging, agriculture, or farming, and publication over a 30-year time span (1990 to 2020). The searched results were further verified to ensure that each publication falls within the scope of hyperspectral imaging for agriculture applications. It was found that there was an increasing number of publications in recent years that used hyperspectral imaging for agricultural applications (Figure 1). Substantially more studies have been published in the recent decade (e.g., 245 articles published in 2011–2020) than that in the previous one (e.g., 97 published in 2001–2010).

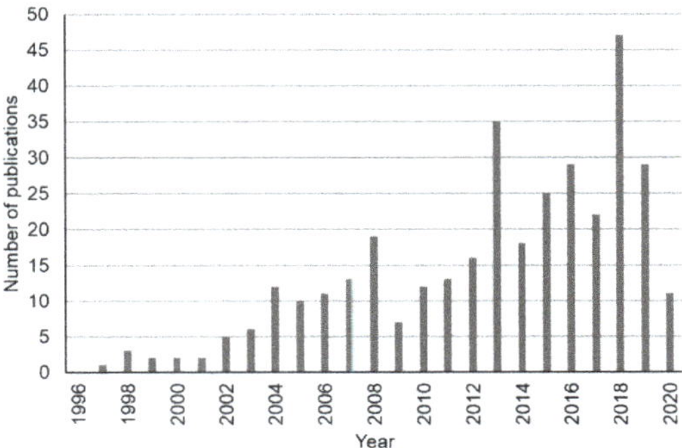

Figure 1. The number of publications that utilized hyperspectral imaging for agriculture applications (by May 2020).

This review is designed to focus on the acquisition, processing, and analysis of hyperspectral imagery for different agricultural applications. The review is organized in the following main aspects: (1) Hyperspectral imaging platforms and sensors, (2) methods for processing and analyzing hyperspectral images, and (3) hyperspectral applications in agriculture (Table 1). Regarding imaging platforms, different types, including satellites, airplanes, helicopters, fixed-wing UAVs, multi-rotor UAVs, and close-range platforms (e.g., ground or lab based), have been used. These platforms acquire images with different spatial coverage, spatial resolution, temporal resolution, operational complexity, and mission cost. It will be beneficial to summarize various platforms in terms of these features to support the selection of the appropriate one(s) for different monitoring purposes. After raw hyperspectral imagery is acquired, pre-processing is the step for obtaining accurate spectral information. Several procedures need to be carried out during pre-processing (usually implemented in a specialized remote sensing software), including radiometric calibration, spectral correction, atmospheric correction, and geometric correction. Although these are standard processing steps for most satellite imagery, it still can be challenging to perform on many airborne hyperspectral images due to different technical issues (e.g., the requirement of high-accuracy Global Positioning System (GPS) signals for proper geometric correction, the measurement of real-time solar radiance for accurate spectral correction). There are no standardized protocols for all sensors due to the limited availability of hyperspectral imaging in the past and the fact that the new mini-sized and low-cost hyperspectral sensors in the market are from different manufacturers with varying sensor configurations. Various approaches have been used in previous studies to address these challenges [12,19,32,33]. Therefore, it is essential to review these approaches to support other researchers for more accurate and efficient hyperspectral image processing. After pre-preprocessing, such as calibration and correction, spectral information extraction (e.g., band selection and dimension reduction) can be performed to further improve the usability of the hyperspectral image. Techniques for these procedures are reviewed in this study.

Table 1. Topics reviewed in this article.

Procedures of Applying Hyperspectral Imagery	Image Acquisition	Image Processing and Analysis	Image Applications
Review Focuses	**Platforms:** - Satellites - Airplanes - UAVs - Close-range platforms **Sensors:** - EO-1 Hyperion - AVIRIS - CASI - Headwall Hyperspec etc.	**Pre-processing:** - Geometric and radiometric correction etc. - Dimension reduction - Band selection **Analytical Methods:** - Empirical regression - Radiative transfer modelling - Machine learning and deep learning	**Specific Applications:** - Estimating crop biochemical and biophysical properties - Evaluating crop nutrient status - Classifying imagery to identify crop types, growing stages, weeds/invasive species, stress/disease - Retrieving soil moisture, fertility, and other physical or chemical properties

With pre-processed hyperspectral images, a robust and efficient analytical method is required for analyzing the tremendous amount of information contained in the images (e.g., spectral, spatial, and textural features) and extracting target properties (e.g., crop and soil characteristics). Previous studies have used a suite of analytical methods, including empirical regression (e.g., linear regression, partial least square regression (PLSR), and multi-variable regression (MLR)), radiative transfer modelling (RTM, e.g., PROSPECT and PROSAIL), machine learning (e.g., random forest (RF)), and deep learning (e.g., convolutional neural network (CNN)) [34–37]. These methods have been developed based on different theories and have different operational complexity, computation efficiency, and performance accuracy. Therefore, it is essential to review the strengths and limitations of these methods and help to choose the appropriate one(s) for specific research purposes. Using hyperspectral information, researchers have investigated a wide range of agricultural features. Some popular ones include crop water content, LAI, chlorophyll and nitrogen contents, pests and disease, plant height, phenological information, soil moisture, and soil organic matter content [11,38]. It will also be valuable to review the performances of hyperspectral imaging in these studies and further explore the potential of this technology for monitoring other agricultural features. Lastly, challenges of using hyperspectral imaging for precision agriculture, together with future research directions, are discussed. A few previous review articles have discussed some of these topics to some extent [11,38,39]. More details and contributions of this review will be discussed in each specific section. Overall, this review aims to examine the main procedures in collecting and utilizing hyperspectral images for different agricultural applications, to further understand the strengths and limitations of hyperspectral technology, and to promote the faster adoption of this valuable technology in precision farming.

2. Hyperspectral Imaging Platforms and Sensors

Hyperspectral sensors can be mounted on different platforms, such as satellites, airplanes, UAVs, and close-range platforms, to acquire images with different spatial and temporal resolutions. Platforms used in the literature were identified and summarized over the publication years, aiming to find, if any, the platforms that had been used more frequently in a specific time period, and the results are shown in Figure 2. Airplanes have been the most widely used platforms for hyperspectral imaging in agriculture (Figure 2). Approximately 30 articles that used airplanes were published every five years starting from 2001 (e.g., 27 publications in 2001–2005 and 38 in 2006–2010). In comparison, satellite-based hyperspectral imaging has been used less frequently; approximately 20 or fewer articles were published in all five-year periods. UAVs are popular platforms for remote sensing and have been widely used in the last decade for hyperspectral imaging in agriculture (e.g., more than 20 publications in 2011–2015 and 2016–2020). Close-range platforms have been the most widely used in the last five years (i.e., 2016–2020), with 49 publications (Figure 2). The review in this section is structured based on different platforms, including satellites, airplanes, UAVs, and close-range platforms. In contrast to previous articles reviewing hyperspectral platforms [20,38,39], the review in this section focuses more on recent advancements of imaging platforms (e.g., UAVs, helicopters, and close range) and their applications to precision farming (e.g., weed classification, fine-scale evaluation of crop health, pests, and disease).

2.1. Satellite-Based Hyperspectral Imaging

Compared with a large number of satellite-based multispectral sensors (e.g., Landsat, SPOT, WorldView, QuickBird, Sentinel-2), there are significantly fewer hyperspectral sensors. EO-1 Hyperion, PROBA-CHRIS, and TianGong-1 [40] are a few examples of the available satellite hyperspectral sensors [20]. EO-1 Hyperion is the most widely used satellite-based hyperspectral sensor for agriculture (e.g., more than 40 publications). It collects data in the visible, near-infrared, and shortwave infrared ranges with a spectral resolution of 10 nm and a spatial resolution of 30 m. More sensor specifications of EO-1 Hyperion are given in Table 2. The sensor was in operation from 2000 to 2017, which corresponds to the period having more publications using satellite-based hyperspectral imaging (e.g., 2006 to 2020 in Figure 2). The use of Hyperion data has been reported in a variety of agricultural studies for monitoring different crop and soil properties, including detecting crop disease [41,42], estimating crop properties (e.g., chlorophyll, LAI, biomass) [43–45], assessing crop residues [46,47], classifying crop types [48], and investigating soil features [49,50]. A few featured ones include Wu et al. [45], who estimated vegetation chlorophyll content and LAI in a mixed agricultural field using Hyperion data and evaluated spectral bands that are sensitive to these vegetation properties. Camacho Velasco et al. [48] used Hyperion hyperspectral imagery and different classification algorithms (e.g., spectral angle mapper and adaptive coherence estimator) for identifying five types of crops (e.g., oil palm, rubber, grass for grazing, citrus, and sugar cane) in Colombia. Gomez et al. [49] predicted soil organic carbon (SOC) using both spectroradiometer data and a Hyperion hyperspectral image, and they found that using Hyperion data resulted in a lower accuracy compared with results derived from spectroradiometer data.

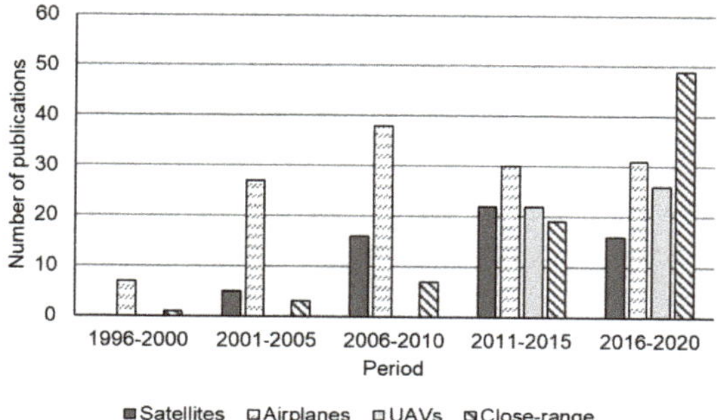

Figure 2. Number of publications that used different hyperspectral imaging platforms over time.

Studies have also been conducted to compare the performances of Hyperion hyperspectral imagery with multispectral imagery for estimating crop properties or classifying crop types. For instance, Mariotto et al. [15] compared Hyperion hyperspectral imagery with Landsat multispectral imagery for the estimation of crop productivity and the classification of crop types. The authors reported better performances of using hyperspectral imagery than using Landsat imagery for both research purposes. Similarly, Bostan et al. [51] compared Hyperion hyperspectral imagery with Landsat multispectral imagery for crop classification and also found that higher classification accuracy can be achieved by using hyperspectral imagery.

Table 2. Specifications of commonly used hyperspectral sensors [11,20,52–56].

	Satellite-Based			Airplane-Based				UAV-Based *	
Sensor	Hyperion	PROBA-CHRIS		AVIRIS	CASI	AISA	HyMap	Headwall Hyperspec	UHD 185-Firefly
Spectral range (nm)	357–2576	415–1050		400–2500	380–1050 (CASI-1500)	400–970 (Eagle)	440–2500	400–1000 (VNIR)	450–950
Number of spectral bands	220	19	63	224	288	244	128	270 (Nano) 324 (Micro)	138
Spectral Resolution (nm)	10	34	17	10	<3.5	3.3	15	6 (Nano) 2.5 (Micro)	4
Operational altitudes (km)	705 (swath 7.7 km)	830 (swath 14 km)		1–20				<0.15	
Spatial resolution (m)	30	17	36	1–20				0.01–0.5	
Temporal resolution (days)	16–30	8		Depends on flight operations (hours to days)					
Organization	NASA, USA	ESA, UK		Jet Propulsion Laboratory, USA	Itres, Canada	Specim, Finland	Integrated Spectronics, Australia	Headwall Photonics, USA	Cubert GmbH, Germany
Number of publications	41	9		18	22	20	12	9	6

* UAV-based sensors typically can also be mounted on airplanes for imaging.

PROBA-CHRIS is another commonly used satellite-based hyperspectral sensor that was launched in 2001. Specific studies, such as Verger et al. [57], utilized PROBA-CHRIS data for retrieving LAI, the fraction of vegetation cover (fCover), and the fraction of absorbed photosynthetically active radiation (FAPAR) in an agricultural field. Antony et al. [58] identified three growth stages of wheat using multi-angle PROBA-CHRIS images and found the optimal view angles for the identification. Casa et al. [59] evaluated the performance of airborne Multispectral Infrared Visible Imaging Spectrometer (MIVIS) data and spaceborne PROBA-CHRIS data for investigating soil texture, and they found that these two data have similar performances, although the PROBA-CHRIS data have a lower spatial resolution.

There are a few other satellite-based hyperspectral sensors that have not been commonly used in an agricultural environment. For instance, Hyperspectral Imager (HySI) is a hyperspectral sensor equipped on the Indian Microsatellite-1 (IMS-1) launched in 2008 [60]. It collects spectral signals in the range of 400–950 nm with a spatial resolution of 550 m at nadir [61]. HySI imagery has been used to map different agricultural features, such as soil moisture and soil salinity [62]. It has also been used for crop classification [63]. However, this data has not been widely used in precision farming, which is probably due to the low spatial resolution and limited data availability. The Hyperspectral Imager for the Coastal Ocean (HICO) is another spaceborne hyperspectral sensor that takes images with a spectral range from 380 to 960 nm at a spatial resolution of 90 m [64]. This sensor was mainly designed to sample the coastal ocean and operated from 2009 to 2015.

In recent years, several spaceborne hyperspectral sensors have been launched or scheduled for launching in the next few years. For instance, the German Aerospace Center (DLR) Earth Sensing Imaging Spectrometer (DESIS), a hyperspectral sensor mounted on the International Space Station, was launched in 2018 [65]. This sensor acquires images in the range from 400 to 1000 nm with a spectral resolution of 2.5 nm and a spatial resolution of 30 m. The Hyperspectral Imager Suite (HISUI) is a Japanese hyperspectral sensor that is also onboard the International Space Station [66]. It was launched in 2019 and collects data in the range from 400 to 2500 nm with a spatial resolution of 20 m and a temporal resolution of 2 to 60 days [20]. Hyperspectral Precursor and Application Mission (PRISMA) is an Italian hyperspectral mission with the sensor launched in March 2019. Its spectral resolution is 12 nm in the range of 400-2500 nm (~250 bands in visible to shortwave infrared). Its hyperspectral imagery has a spatial resolution of 30 and 5 m for the panchromatic band [67]. The Environmental Mapping and Analysis Program (EnMAP) is a German hyperspectral satellite mission that is still in the development and production phase [68]. The EnMAP sensor will collect data from the visible to the shortwave infrared range with a spatial resolution of 30 m. It is planned to be launched in 2020. The Spaceborne Hyperspectral Applicative Land and Ocean Mission (SHALOM) is a joint mission by Israeli and Italian space agencies, and the satellite is scheduled to be launched in 2022 [69]. This sensor will collect hyperspectral images with a spatial resolution of 10 m in the spectral range of 400–2500 nm and panchromatic images with a spatial resolution of 2.5 m [70]. HyspIRI is another hyperspectral mission that is also at the study stage [71]. This sensor will collect data in the 380 to 2500 nm range with an interval of 10 nm and a spatial resolution of 60 m.

Although the actual PRISMA, EnMAP, and HyspIRI data are not yet available, researchers have simulated the images using other data and tested the performance of the simulated images for investigating different vegetation and soil features. For instance, Malec et al. [72], Siegmann et al. [73], and Locherer et al. [74] simulated EnMAP imagery using different airborne or spaceborne images and applied the simulated images for investigating different crop and soil properties. Bachmann et al. [75] produced an image using the EnMAP's end-to-end simulation tool and examined the uncertainties associated with spectral and radiometric calibration. Castaldi et al. [76] simulated data of four current (EO-1 ALI and Hyperion, Landsat 8 Operational Land Imager (OLI), Sentinel-2 MultiSpectral Instrument (MSI)) and three forthcoming (EnMAP, PRISMA, and HyspIRI) sensors using a soil spectral library and compared their performance for estimating soil properties. Castaldi et al. [77] used PRISMA data that were simulated with lab-measured spectral data for estimating clay content and attempted to reduce the influence of soil moisture on the estimation of clay.

Previous studies have confirmed the good performance of satellite-based hyperspectral sensors for studying agricultural features; however, several factors could potentially affect the broad applications of these data in precision farming, including the spatial resolution, temporal resolution, and data quality. The detection and monitoring of many agricultural features, such as crop disease, pest infestation, and nutrient status, require high spatial and temporal resolution. Most of the satellite-based hyperspectral sensors have medium spatial resolutions, such as 17 or 36 m for PROBA-CHRIS; 30 m for Hyperion, PRISMA, and EnMAP, DESIS; and 60 m for HyspIRI. Previous studies have indicated that such spatial resolutions are not sufficient for precision farming applications [20,49]. To overcome such limitations, researchers have attempted to pansharpen hyperspectral images, aiming to improve spatial resolution [73,78–80]. Loncan et al. [81] also reviewed different pansharpening methods for generating high-spatial resolution hyperspectral images.

Temporal resolution is another factor that could potentially limit the applications of satellite-based hyperspectral images to precision agriculture. Most of the satellite-based sensors have a long revisit cycle (e.g., typically around two weeks), and thus early signals of crop stress (e.g., disease and pest) may be missed. This limitation can be further aggravated by unfavorable weather conditions (e.g., cloud contamination). Lastly, low data quality is also an issue that can affect the performance of satellite-based hyperspectral imaging for investigating agricultural features. A low signal-to-noise ratio is a well-known issue of Hyperion data (e.g., in the shortwave infrared (SWIR) range), which has affected the accuracy of retrieving different agricultural features [20]. For instance, Asner and Heidebrecht [82], Gomez et al. [49], and Weng et al. [83] found that the low signal-to-noise ratio influenced the accuracies of estimating non-photosynthetic vegetation and soil cover, soil organic matter, and soil salinity, respectively. Future satellite-based hyperspectral missions are expected to solve the data quality issue.

2.2. Airplane-Based Hyperspectral Imaging

Airborne hyperspectral imaging has been widely used to collect hyperspectral imagery for different monitoring purposes (e.g., for agriculture or forestry). The first hyperspectral sensor was an airborne visible/infrared imaging spectrometer (AVIRIS) that was developed and utilized in 1987 [84]. It collects spectral signals in 224 bands in the visible to SWIR range (Table 2). Researchers have applied AVIRIS data to help understand a wide range of agricultural features, such as investigating vegetation properties (e.g., yield, LAI, chlorophyll, and water content) [85–88], analyzing soil properties [89], evaluating crop health or identifying pest infestation [90–92], and mapping crop area or agricultural tillage practices [93,94].

Besides AVIRIS, the Compact Airborne Spectrographic Imager (CASI), Hyperspectral Mapper (HyMap), and AISA Eagle are also widely used airborne hyperspectral sensors (Table 2). For instance, CASI images have been used for estimating crop chlorophyll content [95], investigating crop cover fraction [96], classifying weeds [97], and delineating management zones [2]. The HyMap imagery has been applied to examining crop biophysical and biochemical variables (e.g., LAI, chlorophyll and water content) [98–100], detecting plant stress signals [101], and investigating the spatial patterns of SOC [102]. Regarding AISA Eagle imagery, Ryu et al. [35] and Cilia et al. [103] used this data for estimating crop nitrogen content, and Ambrus et al. [104] used it for estimating biomass.

Several other airborne hyperspectral sensors have also been used in previous studies. For instance, AVIS images were used for investigating a range of vegetation characteristics (e.g., biomass and chlorophyll) [105], Probe-1 hyperspectral images were used for investigating crop residues [106], RDACS-H4 hyperspectral images were used for detecting crop disease [34], AHS-160 hyperspectral sensor was used for mapping SOC [107], the SWIR Hyper Spectral Imaging (HSI) sensor was used for estimating soil moisture [108], the Pushbroom Hyperspectral Imager (PHI) was used for estimating winter wheat LAI [109], and airborne prism experiment (APEX) data were used for studying the relationship between SOC in croplands and the spectral signals [110].

Most of the aforementioned airborne hyperspectral images have been acquired by airplanes at medium to high altitude (e.g., 1–4 km altitude for CASI, 20 km for AVIRIS), and the acquired images generally having high to medium spatial resolution, such as 4 m for CASI imagery, 5 m for HyMap, and 20 m for AVIRIS [111–113]. Such spatial resolutions are appropriate for mapping many crop and soil features. However, image acquisition usually needs to be scheduled months or even years in advance, and flight missions are expensive [19]. Furthermore, for some specific applications, such as investigating species-level or community-level features (e.g., identification of weeds or early signal of crop disease), images with very high spatial resolutions (e.g., sub-meter) are preferred [114,115]. In addition, due to the unstable nature of airplanes as imaging platforms, a gimbal or high-accuracy inertial measurement unit (IMU) will be required to compensate for the orientation change of the airplanes or recording the orientation information for subsequent image correction, respectively. These factors limited the full application of airborne hyperspectral imaging in precision agriculture. Manned helicopters have also been used as platforms for hyperspectral imaging and investigation of vegetation features [27,116]. Helicopters have more flexible flight heights (e.g., 100 m–2 km) than airplanes and are capable of acquiring high-spatial-resolution images (e.g., sub-meter) over large areas. An aviation company with a manned helicopter is generally needed for the imaging task, which requires extra funding support and far advanced pre-scheduling.

2.3. UAV-Based Hyperspectral Imaging

UAV has become a popular platform in recent years for remote sensing data acquisition, especially for multispectral imaging using digital cameras or multispectral sensors. With the increased availability of lightweight hyperspectral sensors, researchers have experimented on mounting these sensors on UAVs to acquire high-spatial-resolution hyperspectral imagery [19,117]. Different types of UAVs, including multi-rotors, helicopters, and fixed wings, have been utilized in previous studies (Figure 3). Compared with manned airplanes and helicopters, UAVs are capable of acquiring high-spatial-resolution images with a much lower cost and have high flexibility in terms of scheduling a flight mission [118]. Several specific agricultural applications of UAV-based hyperspectral imaging are summarized in Table 3.

Figure 3. Hyperspectral UAV systems used in previous agricultural studies. Figures were reproduced with permission from the corresponding publishers: (**a**) MDPI [119], (**b**) MDPI [120], (**c**) MDPI [121], and (**d**) SPIE [122].

Table 3. Example applications of UAV-based hyperspectral imaging in agriculture.

Applications	Previous Studies	Research Focuses
Estimating LAI and chlorophyll	Yu et al. [37]	Estimated a range of vegetation phenotyping variables (e.g., LAI and leaf chlorophyll) using UAV-based hyperspectral imagery and radiative transfer modelling.
Estimating biomass	Honkavaara et al. [123]	Mounted a hyperspectral sensor and a consumer-level camera on a UAV for estimating biomass in a wheat and a barley field.
	Yue et al. [124]	Utilized UAV-based hyperspectral images for estimating winter wheat above-ground biomass.
Estimating nitrogen content	Pölönen et al. [125]	Used lightweight UAVs for collecting hyperspectral images and estimated crop biomass and nitrogen content.
	Kaivosoja et al. [126]	Applied UAV-based hyperspectral imagery to investigate biomass and nitrogen contents in a wheat field.
	Akhtman et al. [127]	Utilized UAV-based hyperspectral images for estimating nitrogen content and phytomass in corn and wheat fields and monitored temporal variations of these properties.
Estimating water content	Izzo et al. [128]	Evaluated water content in the commercial vineyard using UAV-based hyperspectral images and determined wavelengths sensitive to canopy water content.
Classifying weeds	Scherrer et al. [129]	Classified herbicide-resistant weeds in different crop fields (e.g., barley, corn, and dry pea) using both ground- and UAV-based hyperspectral imagery.
Detecting disease	Bohnenkamp et al. [119]	Used both ground- and UAV-based hyperspectral images for detecting yellow rust in wheat.

Various lightweight hyperspectral sensors have been developed in recent years and can be mounted on UAVs. Examples of sensors include the widely-used Headwall Micro- and Nano-Hyperspec VNIR [12,13,26,128], UHD 185-Firefly [53,130], the PIKA II sensor [19,32], and the HySpex VNIR [25,131]. These hyperspectral sensors contain more than 100 bands in the visible-near infrared spectral range (Table 2). These sensors are small and compact (1–2 kg), thus they can be deployed quickly on various manned or unmanned remote sensing platforms. Previous studies conducted by Adão et al. [11] and Lodhi et al. [52] also compared and summarized various lightweight hyperspectral sensors.

A large number of factors need to be considered in the application of UAV-based hyperspectral imaging, ranging from sensor setup and data collection, to image processing. Saari et al. [122] tested the feasibility of a UAV-based hyperspectral imaging system for agricultural and forest applications and discussed several challenges regarding the imaging technology (e.g., hardware requirements and system settings). Aasen et al. [132] focused on the calibration of images collected with a frame-based sensor and discussed several challenges related to the use of UAV-based hyperspectral imaging for vegetation and crop investigation (e.g., the payload of UAV, signal-to-noise ratio, and spectral calibration). Habib et al. [120] attempted to perform orthorectification of UAV-acquired pushbroom-based hyperspectral imagery with frame-based RGB images over an agricultural field. Adão et al. [11] reviewed applications of UAV-based hyperspectral imaging in agriculture and forestry and listed several hyperspectral sensors that can be mounted on UAVs. The authors also discussed several challenges in collecting and analyzing UAV-based hyperspectral imagery, such as radiometric noise, the low quality of UAV georeferencing, and a low signal-to-noise ratio.

UAV-based hyperspectral imaging has become more popular in recent years; therefore, it is critical to review its strengths and limitations. To explore more features of this technology, this section of the review is not limited to agricultural applications alone. Different types of UAVs have been used as hyperspectral imaging platforms, with the two most widely used as multi-rotors [130,133,134] and fixed-wing planes [33,120,135]. Slow flights at low altitudes are preferred to achieve high-spatial-resolution hyperspectral imagery with a high signal-to-noise ratio. Thus, a multi-rotor is more competitive than fixed-wing planes for hyperspectral imaging in terms of flight operation. Specifically, the multi-rotor allows for a low flight altitude, flexible flight speed, and vertical takeoff and landing, while the fixed wing requires a minimum flight altitude, speed, and, sometimes, accessories for takeoff and landing (e.g., runway, launcher, and parachute). A hyperspectral imaging system, which consists of a hyperspectral sensor, a data processing unit, a GPS, and an IMU, has a considerable weight (e.g., 1–3 kg), thus bringing challenges to the payload capacity of the UAV system and its battery endurance. The multi-rotors are generally powered by high-performance batteries (e.g., LiPo), and most have a short endurance (e.g., less than 20 min). The endurance can be as short as 3 min [12]. In contrast, many fixed-wing UAVs are powered by fuel, thus having a much longer endurance (e.g., 1–10 h) [19,135]. However, these fixed-wing planes are mostly large and heavy (e.g., a 5 m wingspan and 14 kg take-off weight) [135], and thus bring challenges to the flight operation. Using UAV, researchers need to consider the UAV SWaP (size, weight, and power), geographical coverage, time aloft, altitude, and other variables. In addition to the challenges in building a UAV system and performing flight operations, researchers likely need to apply for flight permission from an aviation authority (e.g., Special Flight Operations Certificate (SFOC) from Transport Canada), and purchase suitable UAV flight insurance [136]. UAV size and weight are essential parameters to consider in these processes. Furthermore, the UAVs are required to be visible during flight missions, so that the pilot can maintain constant visual contact with the aircraft. This could create a major challenge when flying over a large area, a hilly area, or an area with forests.

2.4. Close-Range (Ground- or Lab-Based) Hyperspectral Imaging

Close-range hyperspectral imaging, including ground (Figure 4a–c) or lab based (Figure 4d,e), is an emerging technology in recent years, and it is capable of acquiring super-high-spatial-resolution (e.g., cm or sub-cm level) hyperspectral imagery [137–139]. Therefore, this imaging technology can be used for investigating fine-scale (e.g., leaf and canopy level) vegetation features and thus greatly support the investigation of crop growing status and detection of early signs of crop stress (e.g., disease, weeds, or nutrition deficiency). Sensors are mounted on moving or static platforms (e.g., linear stages, scaffolds, or trucks) that can be deployed indoors or outdoors for collecting images. Lamps (e.g., halogen lamp) or the sun are used as light sources in these platforms, respectively.

Researchers have utilized different types of platforms and hyperspectral sensors for collecting super-high-spatial-resolution hyperspectral imagery to study different agricultural features, as shown in Table 4.

Table 4. Example applications of close-range hyperspectral imaging in previous studies.

Applications	Previous Studies	Research Focuses
Investigating biochemical components	Feng et al. [140]	Designed a hyperspectral imaging system that consists of a Headwall hyperspectral camera, a halogen lamp, a computer, and a translation stage and used this system for taking images of rice leaves to study leaf chlorophyll distribution.
	Mohd Asaari et al. [141]	Mounted a visible and near-infrared HIS camera in a high-throughput plant phenotyping platform for evaluating plant water status and detecting early stage signs of plant drought stress.
	Zhu et al. [142]	Installed a hyperspectral camera and halogen lamp on a moving stage and used this imaging system for estimating sugar and nitrogen contents in tomato leaves.
Detecting crop disease	Morel et al. [143]	Used a HySpex hyperspectral camera installed in a close-range imaging system for investigating black leaf streak disease in banana leaves.
	Nagasubramanian et al. [144]	Integrated a Pika XC hyperspectral line imaging scanner and halogen illumination lamps for taking images of soybeans and monitoring fungal disease.

Table 4. *Cont.*

Applications	Previous Studies	Research Focuses
Identifying vegetation species or weeds	Eddy et al. [139]	Mounted a hyperspectral sensor on a boom arm that was installed on a truck for acquiring images at 1 m above the ground and applied the hyperspectral images to classifying weeds in different crop fields.
	Lopatin et al. [145]	Installed an AISA Eagle imaging spectrometer on a scaffold at the height of 2.5 m above ground, aiming to collect hyperspectral imagery in a grassland area for classifying grassland species.
Phenotyping	Behmann et al. [146]	Utilized hyperspectral cameras and a close-range 3D laser scanner that were mounted on a linear stage for collecting hyperspectral images and 3D point models, respectively, and used these two datasets for generating hyperspectral 3D plant models for better monitoring plant phenotyping features.
Monitoring soil properties	Antonucci et al. [147]	Attempted to estimate copper concentration in contaminated soils using hyperspectral images that were acquired from a lab-based spectral scanner.
	Malmir et al. [137]	Collected close-range soil images using Pika XC2 hyperspectral camera that was mounted on a linear stage and used the hyperspectral imagery for investigating soil macro- and micro-elements.

Overall, the close-range hyperspectral imaging platform is capable of acquiring super-high-spatial-resolution hyperspectral imagery that is critical for investigating fine-scale crop or soil features. These features provide detailed information about the plant's biophysical and biochemical processes and how plants respond to environmental stresses and diseases. However, the image collection and processing also suffer from different issues, such as uninformative variability caused by the interaction of light with the plant structure (i.e., illumination effects), influences of shadows, and expanding applications of the platform to a large scale [141,146]. Further research in these areas is warranted.

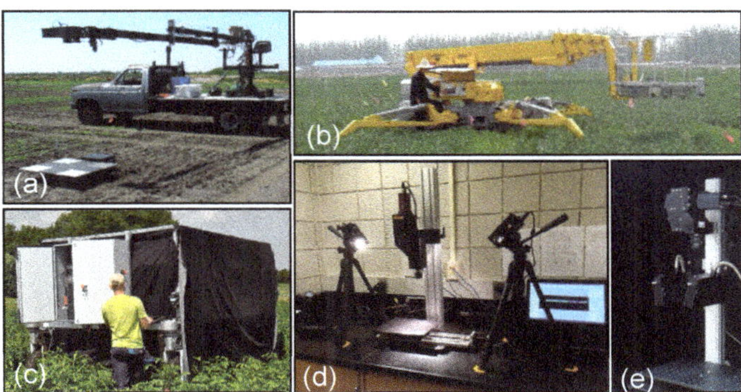

Figure 4. Close-range imaging platforms used in previous studies. Figures were reproduced with permission from corresponding publishers: (**a**) American Society for Photogrammetry and Remote Sensing (ASPRS), Bethesda, Maryland, asprs.org [139]; (**b**) SPIE [148]; (**c**) Elsevier [138]; (**d**) Springer Nature [144]; (**e**) Elsevier [149].

In summary, different hyperspectral imaging platforms, including satellites, airplanes, helicopters, UAVs, and close-range, have different advantages and disadvantages for applications in precision agriculture. Detailed comparisons of these platforms for agricultural applications are shown in Table 5. In brief, satellite-based systems provide images covering large areas but suffer from medium spatial resolution and limited data availability (e.g., a limited number of operating sensors and long revisit time). Airplane- and helicopter-based imaging platforms acquire data with suitable spatial coverage and resolution for most of the agricultural applications. However, they are limited by a high mission cost and scheduling challenges and thus are not suitable for repeated monitoring. UAV-based systems are capable of acquiring high-spatial resolution images repeatedly and have high flexibility. However, they can only cover a small area due to the limited battery endurance and aviation regulations. The close-range imaging systems are capable of obtaining super-high-spatial-resolution images, but they can only be used at leaf or canopy levels. Therefore, the following factors should be taken into consideration when selecting a platform for a specific research project: spatial resolution needed for the study, flight area and flight endurance, weight of the imaging system, platform payload capacity, flight safety and regulations, operation flexibility, and cost.

Table 5. Comparison of hyperspectral imaging platforms.

	Satellites	Airplanes	Helicopters	Fixed-Wing UAVs	Multi-Rotor UAVs	Close-Range Platforms
Example Photos	(Photo: Swales Aerospace)					(Photo: ASPRS)
Operational Altitudes	400–700 km	1–20 km	100 m–2 km	<150 m		<10 m
Spatial Coverage	Very large	Medium—large	Medium	Small—medium	Small	Very small
	e.g., one Hyperion scene covers 42 km × 7.7 km	A 10-min flight/operation covers				
		~100 km^2	~10 km^2	~5 km^2	~0.5 km^2	~0.005 km^2
Spatial Resolution	20–60 m	1–20 m	0.1–1 m	0.01–0.5 m		0.0001–0.01 m
Temporal Resolution	Days to weeks	Depends on flight operations (hours to days)				
Flexibility	Low (e.g., fixed repeating cycles)	Medium (e.g., limited by the availability of aviation company)		High		
Operational Complexity	Low (Final data provided to users)	Medium (Depends on who operates the sensor, users or data vendors)		High (users typically operate sensors and need to set up hardware and software properly)		
Applicable Scales	Regional—global	Landscape—regional		Canopy—landscape		Leaf—canopy
Major Limiting Factors	Weather (e.g., rain and clouds)	Unfavorable flight height/speed, unstable illumination conditions		Short battery endurance (e.g., 10–30 min), flight regulations		Platform design and operation
Image Acquisition Cost	Low to medium	High (typically requires hiring an aviation company to fly)		High (If need to cover a large area)		
Number of publications *	59	133	3	4	38	79

* The number of publications was counted based on which specific platform was used in each of the literature reviewed.

3. Methods for Processing and Analyzing Hyperspectral Images

Hyperspectral images acquired by different platforms and sensors are typically provided in a raw format (e.g., digital numbers) that needs to be pre-processed (e.g., atmospheric, radiometric, and spectral corrections) to retrieve accurate spectral information. Afterward, different approaches can be used for analyzing the hyperspectral information and investigating various agricultural features (e.g., crop and soil properties). A few commonly used methods include linear regression, advanced regression (e.g., PLSR), machine learning and deep learning (e.g., RF, CNN), and radiative transfer modelling (e.g., PROSPECT and PROSAIL). Researchers have used one or more of these methods for investigations of different agricultural features. In this section, the review is arranged based on the different methods used in the studies.

3.1. Pre-Processing of Hyperspectral Images

Typical processing of hyperspectral imagery includes geometric correction, orthorectification, radiometric correction, and atmospheric correction. For satellite- and airplane-based hyperspectral images, the geometric and orthorectification correction are generally performed by data providers, and the radiometric and atmospheric corrections can be done following standard image processing steps available in remote sensing software. For UAV-based images, in contrast, the users need to conduct these processing steps and decide on appropriate processing methods and associated parameters. For instance, a digital elevation model (DEM) and ground control points (GCPs) are usually needed for performing the orthorectification and geometric correction [12]. If the sensor mounted on UAV is pushbroom based, accurate sensor orientation information recorded by an IMU will be needed for these corrections, and the IMU needs to be integrated into the UAV and well-calibrated [12,27]. Software packages commonly used in previous studies for performing these corrections on UAV-based hyperspectral images include ENVI (Exelis Visual Information Solutions, Boulder, CO, USA) and PARGE (ReSe Applications Schläpfer, Wil, Switzerland) [12,26,117].

Radiometric correction is conducted to convert image digital numbers to radiance using calibration coefficients that are provided by the sensor manufacturer [11]. These coefficients may need to be updated over time due to the degradation of spectral materials used to construct the hyperspectral sensors. Regarding atmospheric correction, although the UAVs are flown at low altitudes, the signals acquired are still subjective to the influence of various atmospheric absorptions and scatterings, such as oxygen absorption at 760nm; water absorption near 820, 940, 1140, 1380, and 1880 nm; and carbon dioxide absorption at 2010 and 2060 nm [12,13,26,150]. Therefore, atmospheric correction is critical for obtaining good-quality spectral information. However, Adão et al. [11] suggest that this process might be skipped if the UAVs are operated close to the ground. Therefore, the application of atmospheric correction will depend on specific flight missions and research purposes (e.g., flight altitudes, if atmosphere-influenced spectral bands are needed). Software or methods commonly used in previous studies for performing atmospheric correction on UAV-based hyperspectral images include the MODTRAN model (Spectral Sciences Inc.), ENVI FLAASH (L3Harris Geospatial), PCI Geomatica (PCI Geomatics Corporate), SMARTS model (Solar Consulting Services), and empirical line correction [12,19,27,32,33,116].

Hyperspectral images typically have hundreds of bands, and many of them are highly correlated. Therefore, dimension reduction is also an essential procedure to consider in the pre-processing of hyperspectral imagery. Many previous studies using hyperspectral imagery have discussed the challenges of data redundancy and have used different methods for dimension reduction. For instance, Miglani et al. [151] performed principal component analysis (PCA) on hyperspectral images and indicated that 99% of the information could be explained in the first 10 principal components. Amato et al. [152] discussed a few previous methods of dimension reduction, such as PCA, minimum noise fraction (MNF), and singular value decomposition (SVD), and proposed a dimension reduction algorithm based on discriminant analysis for supervised classification. Teke et al. [38] reviewed several dimension reduction methods and summarized them based on transformation techniques. Thenkabail et al. [153] discussed the problems of high dimensionality and listed a number of spectral bands that are more important for investigating crop features. Sahoo et al. [4] reviewed different methods for dimension reduction, such as PCA, uniform feature design (UMD), wavelet transforms, and artificial neural networks (ANNs), and discussed their features of operation. Wang et al. [154] proposed an auto-encoder-based dimensionality reduction method that is a deep learning-based approach. Of these different methods, the wavelet transform is one of the most widely used ones for dimension reduction. This technique decomposes a signal into a series of scaled versions of the mother wavelet function and allows the variation of the wavelet based on the frequency information to extract localized features (e.g., local spectral variation) [155,156]. It has also been successfully used for image fusion, feature extraction, and image classification [156–158].

In addition to dimensionality reduction, band sensitivity analysis and band selection have also been widely used in hyperspectral remote sensing to reduce the data size by selecting only the bands that are sensitive to the object of interest. Different algorithms have been proposed in previous studies for band selection, such as a fast volume-gradient-based method that is an unsupervised method and removes the most redundant band successively based on the gradient of volume [159], a column subset selection-based method that maximizes the volume of the selected subset of columns (i.e., bands) and is robust to noisy bands [160], and a manifold ranking-based salient band selection method that puts band vectors in manifold space and selects a band-based ranking that can tackle the problem of inappropriate measurement of the band difference [161]. With the sensitivity analysis, previous studies have identified spectral bands that are sensitive to different crop properties, for instance, ~515, ~550, ~570, ~670, 700–740, ~800, and ~855 nm for investigating chlorophyll content; ~405, ~515, ~570, ~705, and ~720 nm for evaluating nitrogen status; ~970, ~1180, ~1245, ~1450, and ~1950 nm for assessing water content; ~682, ~855, ~910, ~970, ~1075, ~1245, ~1518, ~1725, and ~2260 nm for estimating biomass; and ~550, ~682, ~855, ~1075, ~1180, ~1450, and ~1725 nm for crop classification [36,44,153,162]. Overall, pre-processing is an essential step for improving the quality of hyperspectral images and preparing for further data analysis. After the pre-processing, the analytical methods to be discussed below can be used for analyzing the hyperspectral information and investigating various agricultural features on the ground.

3.2. Empirical Relationships

Linear regression is a widely used method for analyzing hyperspectral imagery and retrieving target information (e.g., crop and soil properties). Both spectral reflectance and vegetation indices can be used as predictor variables in establishing a linear relationship. For instance, using spectral bands, Finn et al. [108] built linear regressions between field-measured soil moisture data and the spectral reflectance of collected hyperspectral imagery and identified bands that have stronger correlations with soil moisture. More studies have used vegetation indices in the regression for a better performance as some indices can enhance the signal of targeted features and minimize the background noise. Some of the previous studies are shown in Table 6.

Table 6. Selected previous studies utilized linear regression and hyperspectral vegetation indices for investigating agricultural features.

Applications	Previous Studies	Research Focuses
	Oppelt and Mauser [105]	Utilized the Chlorophyll Absorption Integral (CAI), Optimized Soil-Adjusted Vegetation Index (OSAVI), and hyperspectral Normalized Difference Vegetation Index (h NDVI) for estimating leaf chlorophyll and nitrogen content from hyperspectral imagery and evaluated the performance of each of the indices.
Estimating leaf chlorophyll and nitrogen content	Wu et al. [45]	Tested a range of vegetation indices (e.g., NDVI, Simple Ratio (SR), and Triangular Vegetation Index (TVI)) for retrieving vegetation chlorophyll content and LAI from Hyperion images and determined the indices that produced high accuracies.
	Cilia et al. [103]	Utilized the Double-peak Canopy Nitrogen Index (DCNI) and Modified Chlorophyll Absorption Ratio Index/Modified Triangular Vegetation Index 2 (MCARI/MTVI2) for estimating nitrogen content, as well as the Transformed Chlorophyll Absorption in Reflectance Index (TCARI), MERIS Terrestrial Chlorophyll Index (MTCI) and Triangular Chlorophyll Index (TCI) for estimating leaf pigments.

Table 6. *Cont.*

Applications	Previous Studies	Research Focuses
Estimating LAI and biomass	Xie et al. [109]	Evaluated a range of vegetation indices, such as the modified simple ratio index (MSR), NDVI, a newly proposed index NDVI-like (which resembles NDVI), modified triangular vegetation index (MTVI2), and modified soil adjusted vegetation index (MSAVI) for estimating winter wheat LAI from hyperspectral images.
	Ambrus et al. [104]	Tested the NDVI and Red Edge Position (REP) for estimating field-scale winter wheat biomass.
	Richter et al. [98]	Examined a range of techniques (e.g., index-based empirical regression, radiative transfer modelling, and artificial neural network) for estimating crop biophysical variables (e.g., LAI and water content) in terms of operational agricultural applications with airborne Hymap data and discussed the unique features of each technique.
Estimating nitrogen content	Nevalainen et al. [163]	Utilized 28 published vegetation indices (e.g., Chlorophyll Absorption Ratio Index (CARI) and Normalized Difference Red Edge (NDRE)) for estimating oat nitrogen and identified the best-performing one.
Detecting crop disease	Huang et al. [164]	Examined the performance of the photochemical reflectance index (PRI) for estimating the disease index of wheat yellow rust using canopy reflectance data and then applied the regression on an airborne hyperspectral imagery for mapping the disease-affected areas.
	Copenhaver et al. [34]	Calculated a range of vegetation indices (e.g., NDVI and red edge position index) for detecting crop disease and compared the effectiveness of these indices.
Estimating crop residue cover	Galloza and Crawford [47]	Utilized the Normalized Difference Tillage Index (NDTI) and Cellulose Absorption Index (CAI), together with ALI, Hyperion, and airborne hyperspectral (SpecTIR) data, for estimating crop residue cover for conservation tillage application.
Crop classification	Thenkabail et al. [44]	Utilized both spectral bands and vegetation indices for classifying different crop types and estimating vegetation properties and evaluated the performance difference of using various bands or indices.

Overall, linear regression has been commonly used for estimating a wide range of crop or soil properties. It is easy to establish, and most of the index-based regressions generated satisfactory accuracies. However, there are several potential issues associated with this approach, such as the large number of indices available and it is unknown which performs better, regression may be very sensitive to data size and quality, and the saturation problem of indices [36,165]. It is thus critical to consider these potential issues and adopt appropriate solutions when establishing linear regressions with hyperspectral data. For instance, selecting appropriate vegetation indices with targeted crop or soil variables is recommended. Researchers have evaluated a wide range of hyperspectral vegetation indices for different research purposes. Haboudane et al. [166] examined 11 hyperspectral vegetation indices for estimating crop chlorophyll content. Main et al. [167] investigated 73 vegetation indices for estimating chlorophyll content in crop and savanna tree species. Peng and Gitelson [168] tested 10 multispectral indices and 4 hyperspectral indices for quantifying crop gross primary productivity. Croft et al. [169] analyzed 47 hyperspectral indices for estimating the leaf chlorophyll content of different tree species. Zhou et al. [170] evaluated eight hyperspectral indices for estimating the canopy-level wheat nitrogen content. Tong and He [165] evaluated 21 multispectral and 123 hyperspectral vegetation indices for calculating the grass chlorophyll content at both the leaf and canopy scales. Yue et al. [171] examined 54 hyperspectral vegetation indices for estimating winter wheat biomass. Indices performed differently in these studies; thus, it is suggested to evaluate the top-performed ones in these studies and select the one that generates the highest accuracy.

To deal with issues of linear regression, advanced regression, such as MLR and PLSR, has also been commonly used in previous research for estimating crop and soil properties [172,173]. Compared with linear regression, the advanced regression models mostly use multiple predictor variables in the model to achieve a higher accuracy. PLSR is one of the most widely used models for investigating crop properties using hyperspectral images, such as Ryu et al. [35], Jarmer [99], Siegmann et al. [73], and Yue et al. [124] used PLSR and hyperspectral images for estimating different crop biophysical and biochemical variables (e.g., LAI, biomass, chlorophyll, content, fresh matter, and nitrogen contents). Thomas et al. [100] examined PLSR for retrieving the biogas potential from hyperspectral images and evaluated the influence of imaging time on retrieval accuracy. Regarding soil features, Gomez et al. [49], Van Wesemael et al. [107], Hbirkou et al. [102], and Castaldi et al. [110] built a PLSR model for estimating the SOC content using hyperspectral images. Zhang et al. [50] used PLSR for estimating a wide range of soil properties (e.g., soil moisture, soil organic matter, clay, total carbon, phosphorus, and nitrogen content) from hyperspectral imagery and identified factors that may affect the model accuracy (e.g., low signal-to-noise ratio, spectral overlap of different soil features). Casa et al. [59] used the PLSR model and different hyperspectral imagery for investigating soil textural features and evaluated various factors (e.g., spectral range and resolution, soil moisture, geolocation error) influencing the model performance.

The PLSR model is implemented in Python and R [174,175] and is widely used in many research areas, including forests [176], grasslands [177], and waters [178]. This model performed well in different studies owning to its strengths in dealing with a large number of inter-correlated predictor variables (i.e., by converting them to a few non-correlated latent variables), addressing the data noise challenge, and tackling the over-fitting problem [171,179]. Different techniques have also been confirmed to be efficient for improving the accuracy of the PLSR model, such as incorporating different types of predictor variables in the model (e.g., spectral bands, indices, textural variables), utilizing predicted residual error sum of squares (PRESS) statistics for determining the optimal number of latent variables, and feature evaluation for selecting more important predictor variables in the model [36]. It is thus critical to carefully examine these techniques for achieving the optimal model accuracy.

3.3. Radiative Transfer Modelling

Radiative transfer modelling is a physically based approach that uses physical laws to simulate the interaction of electromagnetic radiation with vegetation (e.g., reflection, transmission, and absorption) [180]. The RTMs simulate vegetation spectra (e.g., leaf reflectance and transmittance) using vegetation biophysical and biochemical properties (e.g., chlorophyll and water contents) in the forward mode, and for inversion of these variables from spectral measurements in the inverse mode [181]. PROSAIL is one of the most widely used RTMs. This model is an integration of the leaf-level PROSPECT model and canopy-level SAIL model and is capable of simulating canopy reflectance using leaf properties (e.g., chlorophyll and water contents), canopy structural parameters (e.g., LAI and leaf angle), and soil reflectance [18].

PROSAIL has also been used in agricultural environments for investigating crop and soil properties. For instance, Casa and Jones [182] inverted PROSAIL and a ray-tracing canopy model with spectroradiometer-measured hyperspectral reflectance data and imaging spectrometer-acquired hyperspectral image data, respectively, for estimating canopy LAI and evaluated factors influencing the estimation accuracy (e.g., the non-homogeneous surface caused by the crop row structure). Richter et al. [98] utilized PROSAIL for estimating LAI, fCover, canopy chlorophyll, and water content from hyperspectral images and compared its performance to other methods (e.g., artificial neural network). Richter et al. [183] applied PROSAIL to investigate similar vegetation variables and analyzed the accuracy and efficiency of this method. Wu et al. [184] examined the sensitivity of vegetation indices to vegetation chlorophyll content using simulated results from the PROSPECT model and suggested a few well-performed indices. Locherer et al. [74] attempted to estimate vegetation LAI using the PROSAIL model and multi-source hyperspectral images and tested several techniques (e.g., different cost functions and types of averaging methods) used for the inversion process. Yu et al. [37] estimated a range of vegetation phenotyping variables (e.g., LAI and leaf chlorophyll) using hyperspectral imagery and PROSAIL and examined the sensitivity of different spectral ranges to the parameters in the PROSAIL model.

Compared with the regression models discussed in previous sections, the RTMs have been less used in the literature for investigating agricultural features due mainly to their high model complexity and computational intensity. For instance, a wide range of parameters need to be considered in RTM (e.g., chlorophyll, carotenoids, water contents, leaf area index, leaf angles, solar angles, and soil reflectance, along with other parameters, in the PROSAIL model) and the users need to use different techniques (e.g., merit function, look-up table) to facilitate the forward and inversion operations of the model. In addition, it costs much more computing time than the regression models to achieve the predictions of target vegetation variables. However, it is also well known that the regression models tend to be site and time specific and are not readily transferable to other geographical regions or different times over the site [166]. In contrast, RTM is a more transferable approach owning to the fact that it is established based on physical laws and does not require training data for rebuilding the model. In addition, RTM is capable of estimating a range of vegetation properties in one model, while regression models typically can only estimate one variable [36,185].

3.4. Machine Learning and Deep Learning

Machine learning algorithms, including support vector machine regression (SVM) and RF, are powerful tools for analyzing hyperspectral information since they can process a large number of variables (e.g., spectral reflectance and vegetation indices) efficiently [186]. Machine learning has been widely used in the remote sensing field for estimating properties of ground features or classifying different ground covers [36,114,187]. Researchers have also used different machine learning algorithms and hyperspectral images for agricultural applications. SVM has been a commonly used algorithm in previous research for prediction or classification purposes. For instance, Honkavaara et al. [123] estimated crop biomass using SVM and UAV-acquired hyperspectral imagery. Bostan et al. [51] utilized SVM for classifying different crop types and achieved high classification accuracy. Ran et al. [93] used KNN and SVM classifiers for investigating tillage practices in agricultural fields and compared their performances. RF is another commonly used algorithm for investigating agricultural features with hyperspectral imagery. For instance, Gao et al. [188] successfully classified weed and maize using RF and lab-based hyperspectral images. Using ground-based hyperspectral reflectance data acquired by an ASD spectroradiometer, Siegmann and Jarmer [189] evaluated the performance of RF, SVM, and PLSR for estimating crop LAI and confirmed the good performance of RF. Similarly, using hyperspectral reflectance, Adam et al. [190] attempted to detect maize disease with the RF model. Overall, machine learning models generally have robust performances for investigating agricultural features using hyperspectral imagery.

Deep learning is a subset of machine learning and extends machine learning by adding more "depth" (i.e., hierarchical representation of the dataset) in the model [191,192]. It is a popular approach in recent years for recognizing patterns in remote sensing images and thus for investigating various ground features. Deep learning has been commonly used in the remote sensing field for image classification, such as land cover classification [193–195] and the identification of ground features (e.g., buildings) [196]. Deep learning has also been applied to precision farming to solve complicated issues. Existing studies are, for example, investigating the estimation of crop yield using CNN and multispectral images together with climate data [197], plant disease detection using CNN and smartphone-acquired images [198], crop classification using 3-D CNN and multi-temporal multispectral images [199], and classification of agricultural land cover using deep recurrent neural network and multi-temporal SAR images [200]. Kamilaris and Prenafeta-Boldú [191] reviewed applications of deep learning in agriculture and food production, although not all studies used remote sensing images. Singh et al. [201] reviewed a range of deep learning methods and their applications, specifically in plant phenotyping. Up to now, deep learning has not been well explored for processing and analyzing remote sensing images, especially hyperspectral images, for agricultural applications. Considering the capacity of deep learning for studying feature patterns in images and the rich information in hyperspectral imagery, the integration of the two has a wide range of agricultural applications (e.g., crop classification, weed monitoring, crop disease detection, and plant stress evaluation). Further research in these areas is warranted.

Machine learning or deep learning is capable of processing multi-source and multi-type data [202]. For instance, besides multi-type remote sensing images (e.g., optical, thermal, LiDAR, and Radar), other sources of data, such as weather, irrigation, and historical yield information, can also be incorporated in the modelling process for a possibly better evaluation of targeted agricultural features [203]. Although machine learning and deep learning models are powerful, it is also critical to keep in mind that these models require large-quantity and high-quality training samples to achieve robust performances [202]. Insufficient training datasets or data with issues (e.g., data incompleteness, noise, and biases) may cause undesired model performances.

In summary, different analytical methods (e.g., linear regression, advanced regression, machine learning and deep learning, and RTM) have different levels of complexity, performance, and transferability. More detailed comparisons on these methods are listed in Table 7. Overall, linear regression is the easiest method to use, and its performance is generally acceptable, although this method can be highly influenced by the choice of predictor variables and quality of the sample data. The advanced regression (e.g., PLSR) mostly performs better than the linear regression since it involves multiple variables in the model and is less sensitive to data noise. RTM (e.g., PROSAIL) is capable of producing multiple data products (e.g., chlorophyll, water, and LAI) with reasonably high accuracies. One essential advantage of this method is its high transferability. However, this method has the highest complexity as it requires a wide range of parameters and extensive programming. In terms of machine learning, many algorithms, such as RF and SVM, are well established and mostly performed well in previous studies. Some programming and model adjustments are needed for this method to achieve optimal performance. Deep learning is a relatively new method and is increasingly popular in recent years. Appropriate model design and programming are critical for this approach. It also requires a substantial amount of training data and computing resources to achieve a good model performance.

Table 7. Comparison of different analytical methods.

Methods	Linear Regression	Advanced Regression	Radiative Transfer Modelling	Machine Learning	Deep Learning
Parameters typically used in the model	- One predictor variable (e.g., reflectance or vegetation index) - Response variable (e.g., chlorophyll)	- Multiple predictor variables - Response variable - Parameters in the model (e.g., the number of latent variables in PLSR)	- A wide range of predictor variables (e.g., leaf biophysical and biochemical properties) - Parameters in the model (e.g., absorption coefficients, the refractive index of leaf material in PROSAIL)	- Multiple predictor variables - Response variable - Parameters in the model (e.g., number of trees in the RF model)	- Predictor variables as input layers - Sizes and weights of layers - Number of layers for calculating
Model complexity	Low	Medium	High	Medium	High
Model performance	Low—high (depend on predictor variable used)	Medium—high	Medium—high	Medium—high	Medium—high
Transferability in time and geographical location	Low	Low	High	Low	High
Typical agricultural applications	Prediction of agricultural variables (e.g., yield, LAI)				Prediction of agricultural variables Classification of agricultural features
Application recommendations	- Test a range of predictor variables and identify the best performed one - Check data noise in the training samples	- Involve different types of variables (e.g., spectral and textural) - Check contributions of variables to the model - Tuning model parameters to achieve optimal performance	- Collect a set of vegetation biophysical and biochemical parameters - Adjust the model to improve calculating efficiency	- Involve different types of variables (e.g., spectral and textural) - Tuning model parameters to achieve optimal performance	- Optimize model configurations - Large size of training samples

4. Hyperspectral Applications in Agriculture

Hyperspectral imaging has been used in agriculture for a wide range of purposes, including estimating crop biochemical properties (e.g., chlorophyll, carotenoids, and water contents) and biophysical properties (e.g., LAI, biomass) for understanding vegetation physiological status and predicting yield, evaluating crop nutrient status (e.g., nitrogen deficiency), monitoring crop disease, and investigating soil properties (e.g., soil moisture, soil organic matter, and soil carbon). Previous studies have also summarized some of the above-mentioned applications of hyperspectral remote sensing in precision agriculture [4,84]. In this section, we will thus focus more on recent hyperspectral studies and summarize these studies according to specific applications.

4.1. Estimation of Crop Biochemical and Biophysical Properties

One important hyperspectral application in agriculture is monitoring crop conditions through the retrieval of crop biochemical and biophysical properties [8,99]. For instance, the leaf chlorophyll content is an essential biochemical property influencing the vegetation photosynthetic capacity and controlling crop productivity [99]. In previous studies, Oppelt and Mauser [105] collected AVIS data to retrieve the chlorophyll and nitrogen contents in a winter wheat field. Similarly, Moharana and Dutta [43] used Hyperion data to estimate the contents of these two biochemical components in a rice field. LAI, on the other hand, is a fundamental vegetation biophysical parameter and is highly related to crop biomass and yield [98]. Previous studies have used hyperspectral remote sensing to estimate the LAI of different crops, and some of the example studies are shown in Table 8.

Table 8. Selected previous studies estimating LAI for different crop types using hyperspectral images.

Crops	Previous Studies	Research Focuses
Winter wheat	Xie et al. [109]	Estimated canopy LAI in a winter wheat field using airborne hyperspectral imagery and proposed a new vegetation index for improved estimation accuracy.
	Siegmann et al. [73]	Retrieved LAI of two wheat fields using EnMAP images and attempted to pan-sharp the images aiming to improve the spatial resolution of LAI products.
Barley	Jarmer [99]	Retrieved a range of canopy variables from barley, including LAI, chlorophyll, water, and fresh matter content using HyMap data and established an efficient approach for monitoring the spatial patterns of crop variables.
Rice	Yu et al. [37]	Investigated LAI, leaf chlorophyll content, canopy water content, and dry matter content using UAV-based hyperspectral imagery, aiming to understand the growing status of rice.
Mixed agricultural fields	Richter et al. [98]	Estimated crop LAI and water content with airborne HyMap data aiming to support operational agricultural practices (e.g., irrigation management and crop stress detection) in the context of the EnMap hyperspectral mission.
	Wu et al. [45]	Estimated chlorophyll content and LAI in a mixed agricultural field (e.g., corns, chestnuts trees, and tea plants) using Hyperion data and identified spectral bands and vegetation indices that generated the highest accuracy.
	Verger et al. [57]	Estimated LAI, fCover, and FAPAR in an agricultural site with different crops using PROBA-CHRIS data.
	Locherer et al. [74]	Estimated LAI in mixed crop fields using EnMAP data and compared the result accuracy to that of LAI estimation with airborne data.

In addition to the above-mentioned vegetation biochemical and biophysical properties, crop water content is a critical parameter for revealing water stress. Richter et al. [98] attempted to estimate the water content in maize, sugar beet, and winter wheat using airborne HyMap data. Moharana and Dutta [204] investigated the water stress in a rice field and its variations using Hyperion images and indicated that the remote sensing-estimated water content matched well with field-observed data. Izzo et al. [128] evaluated the water status in a commercial vineyard using UAV-based hyperspectral data and determined wavelengths sensitive to the canopy water content. Sahoo et al. [4] discussed the applications of hyperspectral remote sensing data for evaluating water features in crops and listed several vegetation indices for calculating the water content.

It can be found from the literature review that many previous studies have focused on estimating the crop chlorophyll content, LAI, and water content using hyperspectral imagery, while other important crop properties, such as carotenoids, that are sensitive to plant stress are less explored. In addition, crop production is influenced by all of these vegetation properties (e.g., chlorophyll, water, and LAI). Besides investigating the spatial and temporal variations of each property, it is also critical to evaluate the relationships between these properties and further understand how they affect crop growth and crop production.

Estimating crop biomass and forecasting yield are also important applications of remote sensing, as they will contribute to the understanding of crop productivity and implementing suitable management measures [126]. Yue et al. [124] utilized UAV-based hyperspectral images for estimating the above-ground biomass of winter wheat. Yang [205] and Mariotto et al. [15] utilized both multispectral and hyperspectral data to estimate crop yield and found that the hyperspectral imagery-based model performed better. In addition, crop residues left in the field are critical materials protecting soil from water and wind erosion and influencing soil biochemical processes. Previous studies, such as Bannari et al. [106], Galloza and Crawford [47], Bannari et al. [46], have used different hyperspectral images for the estimation of crop residues on farmlands

Beyond the estimation of crop biomass and residue, one further research topic is investigating bioenergy (e.g., biogas), which can be generated from the crop biomass. Thomas et al. [100] attempted to estimate the amount of biogas that can be generated per unit of biomass using airborne HyMap data and achieved satisfactory results. Overall, hyperspectral imagery has contributed greatly to the estimation of crop biomass, yield, and other related features (e.g., bioenergy, crop residues). Since crop biomass and yield are highly affected by agricultural practices (e.g., watering and nutrition treatment), involving these practice data, together with hyperspectral imagery, in the model can potentially generate better results. More research in this area is warranted.

4.2. Evaluating Crop Nutrient Status

Precision farming involves evaluating the crop nutrient status and providing recommendations on site-specific resource management according to crop needs [206]. Such an approach is critical for improving the resource use efficiency and reducing environmental impacts [4,103]. Previous studies have used hyperspectral images for estimating the nitrogen content of different crop types, as shown in Table 9.

Table 9. Selected previous studies estimating the nitrogen content for different crop types using hyperspectral images.

Crop types	Previous Studies	Research Focuses
Corn	Akhtman et al. [127]	Used UAV-based hyperspectral images for estimating nitrogen content and phytomass in corn and wheat fields and monitored the temporal variation of these properties.
	Goel et al. [207]	Collected hyperspectral images in a cornfield with different nitrogen treatments and weed controls aiming to evaluate to what extent the spectral signals can identify different nitrogen treatments, weed controls, or their interactions.

Table 9. *Cont.*

Crop types	Previous Studies	Research Focuses
	Cilia et al. [103]	Estimated nitrogen concentration and dry mass in an experimental maize field using airborne hyperspectral imagery, aiming to quantify the nitrogen deficit and provide a variable rate fertilization map. The authors also suggested a way to evaluate the minimum amount of nitrogen to apply without reducing crop yield and avoid excessive fertilization.
	Quemada et al. [208]	Evaluated plant nitrogen status in a maize field using airborne hyperspectral images and developed nitrogen fertilizer recommendations.
Wheat	Koppe et al. [209]	Attempted to investigate wheat nitrogen status and aboveground biomass using hyperspectral and radar images and to evaluate spectral signatures of wheats under different nitrogen treatments.
	Kaivosoja et al. [126]	Used UAV-based hyperspectral imagery to investigate nitrogen content and absolute biomass in a wheat field and evaluated the degree of nitrogen shortage on the date of image acquisition. In this research, historical farming data, including a yield map and a spring fertilization map, were used for estimating the optimal amount of fertilizer to be applied in different areas of the field.
	Castaldi et al. [210]	Estimated nitrogen content in wheat using multi-temporal satellite-based multispectral and hyperspectral images and found that the band selection affected estimation accuracy at different phenological stages.
	Moharana and Dutta [43]	Collected Hyperion images for monitoring nitrogen and chlorophyll contents in rice and investigated the performance of different spectral indices.
	Ryu et al. [35]	Used airborne hyperspectral images and multivariable analysis to estimate nitrogen content in rice at the heading stage.
Rice	Zheng et al. [211]	Tried to monitor rice nitrogen status using UAV-based hyperspectral images and tested the performance of different vegetation indices for estimating the nitrogen content.
	Zhou et al. [212]	Estimated leaf nitrogen concentration of rice using close-range hyperspectral images and tested if the variations of the spatial resolution of the imagery affect the estimation accuracy.
	Nasi et al. [213]	Evaluated the performance of using airborne hyperspectral images and photogrammetric features for estimating crop nitrogen content and biomass in a barley field and a grassland site, and examined if the integration of spectral and plant height information can improve the estimation results.
	Nigon et al. [214]	Examined nitrogen stress in potato fields using airborne hyperspectral imagery and identified spectral indices that are sensitive to nitrogen content.
Other crops (i.e., barley, potato, cabbage, tomato, sugarcane, and cacao)	Chen et al. [215]	Estimated nitrogen content in cabbage seedlings using close-range hyperspectral images and identified sensitive wavelengths for the estimation.
	Zhu et al. [142]	Investigated soluble sugar, total nitrogen, and their ratio in tomato leaves using close-range hyperspectral images and tested data fusion analysis techniques for improving the investigation accuracy.
	Miphokasap and Wannasiri [216]	Collected Hyperion images for investigating spatial variations of sugarcane canopy nitrogen concentration and attempted to identify the nutrient deficient areas for corresponding treatments.
	Malmir et al. [217]	Attempted to evaluate nutrient status (e.g., nitrogen, phosphorus, and potassium) of cacao leaves using close-range hyperspectral images and examined influences of band selection on the evaluation accuracy.

Overall, owing to the large amount of spectral information in hyperspectral imagery, crop nutrient status can be evaluated with high accuracies, and a corresponding fertilizer treatment plan can be proposed to achieve optimal crop productions. However, it is also essential to keep in mind that there is a wide range of factors, such as soil moisture, soil type, and topographic conditions, that can impact crop growth and production. A more comprehensive treatment plan that takes into consideration both the crop nutrient status and other influencing factors can make a greater contribution to crop production.

4.3. Classifying Imagery to Identify Crop Types, Growing Stages, Weeds/Invasive Species, and Stress/Disease

Besides quantifying crop properties, hyperspectral images have also been used for classification purposes, such as differentiating crop types, identifying crop growing stages, classifying weeds or invasive species, and detecting disease [218]. Examples of previous studies are shown in Table 10. Different agricultural land covers or crop types have different spectral characteristics; hence, hyperspectral images can contribute greatly to the classification of these agricultural features.

Table 10. Selected previous studies for the classification of agricultural features using hyperspectral images.

Applications	Previous Studies	Research Focuses
Classification of crop types	Camacho Velasco et al. [48]	Utilized Hyperion data and different classification algorithms (e.g., spectral angle mapper and adaptive coherence estimator) for identifying five types of crops (e.g., oil palm, rubber, grass for grazing, citrus, and sugar cane) in Colombia.
	Bostan et al. [51]	Classified different crop and land cover types (e.g., maize, cotton, urban, water, barren rock, and other crop types) using Landsat 8 multispectral and EO-1 Hyperion hyperspectral images and indicated that hyperspectral imagery performed better than the multispectral imagery.
	Amato et al. [152]	Assessed the potential of PRISMA data for classifying different agricultural land uses (e.g., soybean, corn, and sugar beet) and evaluated the contribution of spectral bands to image segmentation and classification.
	Nigam et al. [91]	Performed crop classification over homogeneous and heterogeneous agriculture and horticulture areas with airborne AVIRIS images and assessed crop health at the field scale.
	Sahoo et al. [4]	Reviewed a few previous studies that used hyperspectral images for classification purposes and indicated the robustness of hyperspectral imagery for classifying different crop types and different crop phonological stages.
Other classifications (e.g., growth stages and agricultural tillage practices)	Antony et al. [58]	Applied multi-angle PROBA-CHRIS data for classifying different growth stages of wheat.
	Ran et al. [93]	Attempted to detect agricultural tillage practices using hyperspectral imagery with different classification models and identified the best performing one.
	Teke et al. [38]	Discussed the application of spectral libraries for classification purposes and listed several spectral libraries available worldwide. The authors also indicated the limitations of using a spectral library, such as the spectral varieties within the same species or land cover, and highlighted the importance of having geographically specific libraries

Weed infestation is a severe issue in agricultural fields and could substantially affect crop growth and yield. Identifying and mapping weeds in agricultural fields using remote sensing will contribute greatly to variable rate treatment in the fields [219]. Researchers have utilized different remote sensing data and methods for weed mapping, as shown in Table 11. Overall, the identification of weeds typically requires a high spatial resolution since many weeds are small in size and mixed with crops.

UAV-based and close-range hyperspectral imaging is capable of acquiring high-spatial-resolution images, and thus has high potential to contribute to weed detection.

Table 11. Selected previous studies for detecting weeds using different hyperspectral imaging platforms.

Platforms	Previous Studies	Research Focuses
Airborne	Goel et al. [97]	Attempted to detect weed infestation in a cornfield that had different nitrogen treatments using airborne hyperspectral imagery and found the different nitrogen treatments affected the classification accuracy of weed.
	Karimi et al. [220]	Performed combinations of different nitrogen treatment rates and weed management practices in a cornfield and tried to classify these combinations with airborne hyperspectral images.
Close range	Zhang et al. [221]	Developed a close-range weed sensing system using hyperspectral images for classifying tomato and weeds and tested its performance in different environments.
	Eddy et al. [139]	Used a ground-based hyperspectral imaging system for classifying weeds in canola, pea, and wheat crops and evaluated the applicability of this approach for real-time detection of weeds in the field.
	Eddy et al. [222]	Used hyperspectral image data as well as secondary products with reduced bands to classify weeds and achieved good accuracy.
	Liu et al. [223]	Classified carrot and weeds using a ground-based hyperspectral imaging system and evaluated the number of spectral bands needed to achieve a good classification accuracy.
Multiple platforms	Scherrer et al. [129]	Attempted to classify herbicide-resistant weeds in different crop fields (e.g., barley, corn, and dry pea) using both ground- and UAV-based hyperspectral imagery and discussed factors influencing classification accuracy (e.g., crop type, plant age, and illumination condition).
Review studies	LÓPEZ-Granados [224]	Discussed the high potential of hyperspectral remote sensing images for mapping weeds but also indicated the limitations of this technology due to the high cost of data collection.

Monitoring crop disease is highly important to growers trying to reduce economic and yield losses [38]. Hyperspectral imaging collects signals at fine spectral resolutions (e.g., less than 10-nm intervals), and thus can possibly detect early symptoms of crop disease and support timely interventions [225]. Previous studies have used hyperspectral images for detecting diseases in different types of groups (Table 12). Overall, hyperspectral signals are sensitive to the variations of crop growth status (e.g., caused by disease or stress) and thus can indicate the occurrence of crop disease or stress. However, considering that crop status can be affected by other factors (e.g., nutrient deficiency), repeat imaging and analysis together with robust modelling would be critical for accurate and timely detection of crop disease or stress.

Table 12. Selected previous studies for detecting disease in different crops using hyperspectral images.

Crops	Previous Studies	Research Focuses
Wheat	Bohnenkamp et al. [119]	Used both ground- and UAV-based hyperspectral imaging platforms for detecting yellow rust in wheat and evaluated factors influencing the detection (e.g., measurement distance, spectral features to use).
	Bauriegel et al. [226]	Targeted the infestation of wheat by *Fusarium* and attempted to detect this disease using hyperspectral remote sensing data, and consequently suggested that farmers need to deal with infected crops separately from healthy crops.
	Zhang et al. [227]	Attempted to detect the *Fusarium* head blight in winter wheat similarly using close-range hyperspectral imaging and suggested that this is a stable and feasible way to monitor this disease using low-altitude remote sensing.
Corn	Copenhaver et al. [34]	Used airborne hyperspectral images to detect the signal of *Ostrinia nubilalis* in a cornfield (e.g., via monitoring rate of plant senescence) and tested the performance of this approach throughout the growing season.
Soybean	Nagasubramanian et al. [144]	Tried to detect charcoal rot in soybeans using close-range hyperspectral imaging and identified wavelength ranges that are sensitive to this disease.
Sugarcane	Apan et al. [41]	Detected sugarcane areas affected by orange rust disease using Hyperion data and developed specific vegetation indices that are sensitive to the disease.
Mustard	Dutta et al. [42]	Delineated mustard areas influenced by diseases using Hyperion images and evaluated the performance of different indices.
Review studies	Lowe et al. [218]	Focused on hyperspectral imaging and reviewed some of its applications in detecting and classifying crop disease and stress.
	Thomas et al. [225]	Reviewed the contributions of hyperspectral imaging to the detection of plant disease and discussed different factors (e.g., light and wind) that may limit its wide applications.
	Mahlein et al. [228]	Reviewed previous studies using remote sensing for detecting plant disease, but not limited to hyperspectral imaging.

4.4. Retrieving Soil Moisture, Fertility, and Other Physical or Chemical Properties

Agricultural soil properties, including soil moisture, soil organic matter, soil salinity, and roughness, are important factors influencing crop growth and final production [7]. Hyperspectral remote sensing can contribute greatly to the investigation of these factors. For instance, estimating soil moisture is one of the most popular research topics. Finn et al. [108] estimated soil moisture at three different depths using airborne hyperspectral images and linear regression and discussed the contributions and limitations of hyperspectral remote sensing for soil moisture studies. Casa et al. [229] investigated soil water, clay, and sand contents using a fusion of CHRIS-PROBA images and soil geophysical data. Shoshany et al. [7] summarized four main approaches for estimating soil moisture content: (1) Radar techniques; (2) radiation balance and surface temperature calculations; (3) reflectance in the visible, NIR, and SWIR ranges; and (4) integrative methods using multiple spectral ranges. Although soil moisture can be estimated using optical remote sensing data, it is often affected by the plant ground cover. Integrating multi-type remote sensing data, e.g., SAR and thermal data, can possibly generate more accurate estimates.

SOC is a critical component of soil fertility, which highly controls both the growth and yield of crops. Hyperspectral data provide fine spectral details that are critical for the estimation of SOC content. Previous studies have used hyperspectral images collected by different platforms for investigating SOC (Table 13). Overall, hyperspectral imagery has a high potential for the estimation of soil organic matter and carbon. However, similar to the evaluation of soil moisture, the investigation of soil organic matter and carbon can be highly influenced by vegetation cover. Therefore, collecting hyperspectral images in non-growing seasons could be a solution.

Table 13. Selected previous studies for estimating soil organic carbon using hyperspectral images acquired by different platforms.

Platforms	Previous Studies	Research Focuses
Satellites	Zhang et al. [50]	Utilized EO-1 Hyperion images for estimating several soil properties, including soil moisture, soil organic matter, total carbon, total phosphorus, total nitrogen, and clay content. The authors also found the influence of spectral resolution on the performance of retrieval models.
	Casa et al. [230]	Assessed soil organic matter and soil texture at the field scale using CHRIS-PROBA images and produced uniform soil zones for supporting irrigation management.
Airplanes	Hbirkou et al. [102]	Attempted to estimate SOC in agricultural fields using airborne HyMap images and tested the influences of soil surface conditions on the estimation, aiming to support soil management in precision farming.
	Gedminas and Martin [231]	Tried to map soil organic matter using airborne hyperspectral imagery in combination with topographic information extracted from LiDAR image and evaluated the correlation between soil organic matter and various spectral bands.
	Castaldi et al. [110]	Investigated the relationship between SOC in croplands and spectral signals using a soil database and then estimated SOC in their study sites using airborne hyperspectral imagery. With this approach, the authors attempted to reduce the amount of new data collection in the field or lab.
	Van Wesemael et al. [107]	Discussed the impacts of vegetation cover on soil and the estimation of SOC from remote sensing data and attempted to use spectral unmixing techniques to estimate the fraction of vegetation cover and then estimate the soil carbon content using the residue soil spectra.
Multiple platforms	Gomez et al. [49]	Estimated SOC using both lab-based hyperspectral reflectance data and Hyperion image data and found that using the lab-acquired reflectance data can generate more accurate results than using the Hyperion data. At the same time, the Hyperion data can generate a SOC map that matches field observations and thus can also be used for prediction.

Hyperspectral remote sensing data have also been used for estimating other soil features, as shown in Table 14. It can be found from these studies that hyperspectral images can be used for studying a wide range of soil features. Different soil features influence the spectral signals in different bands and with different magnitudes, while some of these influences may be spectrally overlapped. Therefore, when investigating a specific soil feature, it is critical to collect a suitable number of soil samples with other soil features generally controlled.

Table 14. Selected previous studies for investigating different soil features using hyperspectral images.

Soil Features	Previous Studies	Research Focuses
Soil texture	Casa et al. [59]	Investigated soil texture using airborne MIVIS and spaceborne PROBA-CHRIS hyperspectral images and discussed their performance and limitation (e.g., lack of SWIR band).
Soil nitrogen	Song et al. [232]	Used airborne hyperspectral images for evaluating the impact of soil nitrogen applications and variable-rate fertilization on winter wheat growth. The authors also indicated that the variable-rate fertilization in the field could reduce the growing difference of winter wheat caused by the spatial variations of soil nitrogen.
Copper concentration	Antonucci et al. [147]	Attempted to estimate in soil using lab-based hyperspectral measurement and achieved good accuracy.
Potassium content	Wang et al. [233]	Evaluated potassium content in cinnamon soil using close-range hyperspectral imaging aiming to better understand soil fertility and indicated the good performance of this approach when the potassium content is high (i.e., $\geq$ 100 mg/kg).
CO_2 leaks	McCann et al. [234]	Detected CO_2 leaks from the soil by monitoring vegetation stress signals using multi-temporal hyperspectral images.

In summary, hyperspectral imaging has been successfully applied to a wide range of agricultural applications, as reviewed above, and summarized in Table 15. Future research directions are also suggested.

Table 15. Hyperspectral applications in agriculture.

	Previous Focuses	Suggested Future Research Directions
Crop biochemical and biophysical properties	- Leaf area index - Chlorophyll content - Water content - Fraction of vegetation cover - Fresh/dry biomass, crop residue - Yield	- Vegetation properties related to crop stress (e.g., carotenoids) - Relationships between different properties and how they affect crop growth
Crop nutrient status	- Nitrogen content	- Other nutrients (e.g., phosphorus, magnesium, and boron etc.) that may limit crop growth - Optimized treatment plan targeting different limiting factors
Classification	Classification of: - Crop types - Soil types - Growing stages (i.e., crop phenological features) Classification and detection of stressors: - Weeds or invasive species - Disease/stress affected areas	- Improvement of classification methods (e.g., advanced algorithms) for target features - Fusion and application of multi-type and multi-temporal remote sensing data - Further exploration of UAV and close-range imaging for better identification of fine-scale signals
Soil properties	- Soil moisture - Soil organic matter - Soil salinity - Soil roughness	- Separation of spectral signals from soil and vegetation for better assessing soil features - Fusion and application of multi-type remote sensing data to capture different soil information - Further exploration of close-range sensing for investigating soil properties.
Agro-ecosystem	- Less explored using hyperspectral image	- Ecosystem services - Biodiversity - Adverse effects of agricultural practices on the environments

5. Conclusions and Recommendations

Hyperspectral imaging has great potential for applications in agriculture, particularly precision agriculture, owing to ample spectral information sensitive to different plant and soil biophysical and biochemical properties. Multiple platforms, including satellites, airplanes, UAVs, and close-range platforms, have become more widely available in recent years for collecting hyperspectral images with different spatial, temporal, and spectral resolutions. These platforms also have different strengths and limitations in terms of spatial coverage, flight endurance, flexibility, operational complexity, and cost. These factors need to be considered when choosing imaging platform(s) for specific research purposes. Further technological developments are also needed to overcome some of the limitations, such as the short battery endurance in UAV operations and high cost of hyperspectral sensors.

Different analytical methods, such as linear regression, advanced regression, machine learning, deep learning, and RTM, have been explored in previous studies for analyzing the tremendous amount of information in hyperspectral images for investigating different agricultural features. Previous studies have mainly used the regression approach, while more physically based methods, such as RTM, have been less explored. Deep learning and effective big-data analytics are powerful tools for recognizing patterns in remote sensing data. Together with hyperspectral imagery, deep learning models have high potential to support the monitoring of a wide range of agricultural features. Different analytical methods have different advantages and disadvantages, and thus it is critical to compare these methods for specific research (e.g., requirements of accuracy and computing efficiency) and choose an optimal approach. In addition, image spectral information has been commonly used as variables for prediction or classification tasks, while other information, such as texture, has been less explored. Further, some other sources of data, such as weather, irrigation records, and historical yield information, can also be used in some of the analytical methods (e.g., machine learning and deep learning) for better monitoring of crop features. More research in these fields is also warranted.

Hyperspectral imaging has been successfully applied in a wide range of agricultural applications, including estimating crop biochemical and biophysical properties; evaluating crop nutrient and stress status; classifying or detecting crop types, weeds, and diseases; and investigating soil characteristics. Previous studies have focused on discussing one or two of the many factors impacting crop growth performance and productivity, and thus cannot evaluate crop status and growth-limiting factors comprehensively. It is important to integrate these factors to achieve a better understanding of their inter-relationships for optimal crop production and environmental protection. Besides, previous studies using hyperspectral imaging have mainly targeted investigating crop growth, aiming to improve crop yield, while less research has focused on understanding the ecosystem side of crop production (e.g., ecosystem services and biodiversity). Further research in these areas is warranted.

Author Contributions: Conceptualization, J.S., J.L., Y.H., B.L. and P.D.D.; methodology, B.L., P.D.D. and Y.H.; investigation, B.L.; writing—original draft preparation, B.L.; writing—review and editing, P.D.D., J.S., J.L. and Y.H.; project administration, J.S., J.L. and Y.H.; funding acquisition, Y.H. All authors have read and agreed to the published version of the manuscript.

Funding: This work was funded by the Natural Sciences and Engineering Research Council of Canada (NSERC) under Discovery Grant RGPIN-386183 to Professor Yuhong He.

Conflicts of Interest: The authors declare no conflict of interest.

Abbreviations

ALI	Advanced Land Imager
APEX	Airborne Prism Experiment
AVIS	Airborne Visible Near-Infrared Imaging Spectrometer
AVIS	Airborne Visible Near-Infrared Imaging Spectrometer
AVIRIS	Airborne Visible/Infrared Imaging Spectrometer
ANN	Artificial Neural Networks
CAI	Cellulose Absorption Index

CAI	Chlorophyll Absorption Integral
CARI	Chlorophyll Absorption Ratio Index
CASI	Compact Airborne Spectrographic Imager
CHRIS	Compact High Resolution Imaging Spectrometer
CNN	Convolutional Neural Network
DEM	Digital Elevation Model
DESIS	Dlr Earth Sensing Imaging Spectrometer
DCNI	Double-Peak Canopy Nitrogen Index
EnMAP	Environmental Mapping And Analysis Program
FAPAR	Fraction Of Absorbed Photosynthetically Active Radiation
fCover	Fraction Of Vegetation Cover
GCPs	Ground Control Points
HSI	Hyper Spectral Imaging
HySI	Hyperspectral Imager
HICO	Hyperspectral Imager For The Coastal Ocean
HISUI	Hyperspectral Imager Suite
HyspIRI	Hyperspectral Infrared Imager
HyMap	Hyperspectral Mapper
h NDVI	Hyperspectral Normalized Difference Vegetation Index
PRISMA	Hyperspectral Precursor And Application Mission
IMU	Inertial Measurement Unit
LAI	Leaf Area Index
MTCI	Meris Terrestrial Chlorophyll Index
MNF	Minimum Noise Fraction
MCARI/MTVI2	Modified Chlorophyll Absorption Ratio Index/Modified Triangular Vegetation Index 2
MSR	Modified Simple Ratio Index
MSAVI	Modified Soil Adjusted Vegetation Index
MTVI2	Modified Triangular Vegetation Index
MIVIS	Multispectral Infrared Visible Imaging Spectrometer
MSI	Multispectral Instrument
MLR	Multi-Variable Regression
NDRE	Normalized Difference Red Edge
NDTI	Normalized Difference Tillage Index
OLI	Operational Land Imager
OSAVI	Optimized Soil-Adjusted Vegetation Index
PLSR	Partial Least Square Regression
PRI	Photochemical Reflectance Index
PRESS	Predicted Residual Error Sum Of Squares
PCA	Principal Component Analysis
PHI	Pushbroom Hyperspectral Imager
RTM	Radiative Transfer Modelling
RF	Random Forest
REP	Red Edge Position
SWIR	Shortwave Infrared
SR	Simple Ratio
SVD	Singular Value Decomposition
SOC	Soil Organic Carbon
SHALOM	Spaceborne Hyperspectral Applicative Land And Ocean Mission
SFOC	Special Flight Operations Certificate
SVM	Support Vector Machine Regression
TCARI	Transformed Chlorophyll Absorption In Reflectance Index
TCI	Triangular Chlorophyll Index
TVI	Triangular Vegetation Index
UMD	Uniform Feature Design
UAV	Unmanned Aerial Vehicle

References

1. Weiss, M.; Jacob, F.; Duveiller, G. Remote sensing for agricultural applications: A meta-review. *Remote Sens. Environ.* **2020**, *236*, 111402. [CrossRef]
2. Liu, J.; Miller, J.R.; Haboudane, D.; Pattey, E.; Nolin, M.C. Variability of seasonal CASI image data products and potential application for management zone delineation for precision agriculture. *Can. J. Remote Sens.* **2005**, *31*, 400–411. [CrossRef]
3. Jensen, J.R. *Remote Sensing of the Environment: An Earth Resource Perspective*; Prentice Hall: Upper Saddle River, NJ, USA, 2006.
4. Sahoo, R.N.; Ray, S.S.; Manjunath, K.R. Hyperspectral remote sensing of agriculture. *Curr. Sci.* **2015**, *108*, 848–859.
5. Alonso, F.G.; Soria, S.L.; Gozalo, J.C. Comparing two methodologies for crop area estimation in Spain using Landsat TM images and ground-gathered data. *Remote Sens. Environ.* **1991**, *35*, 29–35. [CrossRef]
6. McNairn, H.; Champagne, C.; Shang, J.; Holmstrom, D.; Reichert, G. Integration of optical and Synthetic Aperture Radar (SAR) imagery for delivering operational annual crop inventories. *ISPRS J. Photogramm.* **2009**, *64*, 434–449. [CrossRef]
7. Shoshany, M.; Goldshleger, N.; Chudnovsky, A. Monitoring of agricultural soil degradation by remote-sensing methods: A review. *Int. J. Remote Sens.* **2013**, *34*, 6152–6181. [CrossRef]
8. Hunt, E.R.; Daughtry, C.S.T. What good are unmanned aircraft systems for agricultural remote sensing and precision agriculture? *Int. J. Remote Sens.* **2018**, *39*, 5345–5376. [CrossRef]
9. Thenkabail, P.S. Biophysical and yield information for precision farming from near-real-time and historical Landsat TM images. *Int. J. Remote Sens.* **2003**, *24*, 2879–2904. [CrossRef]
10. Shang, J.; Liu, J.; Ma, B.; Zhao, T.; Jiao, X.; Geng, X.; Huffman, T.; Kovacs, J.M.; Walters, D. Mapping spatial variability of crop growth conditions using RapidEye data in Northern Ontario, Canada. *Remote Sens. Environ.* **2015**, *168*, 113–125. [CrossRef]
11. Adão, T.; Hruška, J.; Pádua, L.; Bessa, J.; Peres, E.; Morais, R.; Sousa, J. Hyperspectral Imaging: A Review on UAV-Based Sensors, Data Processing and Applications for Agriculture and Forestry. *Remote Sens.* **2017**, *9*, 1110. [CrossRef]
12. Lucieer, A.; Malenovský, Z.; Veness, T.; Wallace, L. HyperUAS-imaging spectroscopy from a multirotor unmanned aircraft system. *J. Field Robot.* **2014**, *31*, 571–590. [CrossRef]
13. Gonzalez-Dugo, V.; Hernandez, P.; Solis, I.; Zarco-Tejada, P. Using High-Resolution Hyperspectral and Thermal Airborne Imagery to Assess Physiological Condition in the Context of Wheat Phenotyping. *Remote Sens.* **2015**, *7*, 13586–13605. [CrossRef]
14. Lee, K.; Cohen, W.B.; Kennedy, R.E.; Maiersperger, T.K.; Gower, S.T. Hyperspectral versus multispectral data for estimating leaf area index in four different biomes. *Remote Sens. Environ.* **2004**, *91*, 508–520. [CrossRef]
15. Mariotto, I.; Thenkabail, P.S.; Huete, A.; Slonecker, E.T.; Platonov, A. Hyperspectral versus multispectral crop-productivity modeling and type discrimination for the HyspIRI mission. *Remote Sens. Environ.* **2013**, *139*, 291–305. [CrossRef]
16. Marshall, M.; Thenkabail, P. Advantage of hyperspectral EO-1 Hyperion over multispectral IKONOS, GeoEye-1, WorldView-2, Landsat ETM+, and MODIS vegetation indices in crop biomass estimation. *ISPRS J. Photogramm.* **2015**, *108*, 205–218. [CrossRef]
17. Sun, J.; Yang, J.; Shi, S.; Chen, B.; Du, L.; Gong, W.; Song, S. Estimating Rice Leaf Nitrogen Concentration: Influence of Regression Algorithms Based on Passive and Active Leaf Reflectance. *Remote Sens.* **2017**, *9*, 951. [CrossRef]
18. Darvishzadeh, R.; Matkan, A.A.; Ahangar, A.D. Inversion of a radiative transfer model for estimation of rice canopy chlorophyll content using a lookup-table approach. *IEEE J.-STARS* **2012**, *5*, 1222–1230. [CrossRef]
19. Hruska, R.; Mitchell, J.; Anderson, M.; Glenn, N.F. Radiometric and geometric analysis of hyperspectral imagery acquired from an unmanned aerial vehicle. *Remote Sens.* **2012**, *4*, 2736–2752. [CrossRef]
20. Transon, J.; d'Andrimont, R.; Maugnard, A.; Defourny, P. Survey of Hyperspectral Earth Observation Applications from Space in the Sentinel-2 Context. *Remote Sens.* **2018**, *10*, 157. [CrossRef]
21. Lodhi, V.; Chakravarty, D.; Mitra, P. Hyperspectral Imaging System: Development Aspects and Recent Trends. *Sens. Imaging* **2019**, *20*, 1–24. [CrossRef]

22. Hatfield, J.L.; Prueger, J.H. Value of Using Different Vegetative Indices to Quantify Agricultural Crop Characteristics at Different Growth Stages under Varying Management Practices. *Remote Sens.* **2010**, *2*, 562–578. [CrossRef]

23. Zhang, H.; Lan, Y.; Suh, C.P.C.; Westbrook, J.; Clint Hoffmann, W.; Yang, C.; Huang, Y. Fusion of remotely sensed data from airborne and ground-based sensors to enhance detection of cotton plants. *Comput. Electron. Agric.* **2013**, *93*, 55–59. [CrossRef]

24. Mahajan, G.R.; Pandey, R.N.; Sahoo, R.N.; Gupta, V.K.; Datta, S.C.; Kumar, D. Monitoring nitrogen, phosphorus and sulphur in hybrid rice (Oryza sativa L.) using hyperspectral remote sensing. *Precis. Agric.* **2017**, *18*, 736–761. [CrossRef]

25. Skauli, T.; Goa, P.E.; Baarstad, I.; Loke, T. A compact combined hyperspectral and polarimetric imager. In Proceedings of the Society of Photo-Optical Instrumentation Engineers, Stockholm, Sweden, 5 October 2006; Driggers, R.G., Huckridge, D.A., Eds.; SPIE-INT SOC Optical Engineering: Bellingham, WA, USA, 2016; Volume 6395, pp. 44–51.

26. Zarco-Tejada, P.J.; Suarez, L.; Gonzalez-Dugo, V. Spatial resolution effects on chlorophyll fluorescence retrieval in a heterogeneous canopy using hyperspectral imagery and radiative transfer simulation. *IEEE Geosci. Remote Soc.* **2013**, *10*, 937–941. [CrossRef]

27. Lu, B.; He, Y.; Dao, P.D. Comparing the Performance of Multispectral and Hyperspectral Images for Estimating Vegetation Properties. *IEEE J. STARS* **2019**, *12*, 1784–1797. [CrossRef]

28. ISS Utilization: MUSES-DESIS (Multi-User System for Earth Sensing) with DESIS instrument. Available online: https://directory.eoportal.org/web/eoportal/satellite-missions/content/-/article/iss-muses (accessed on 3 August 2020).

29. PRISMA (Hyperspectral Precursor and Application Mission). Available online: https://directory.eoportal.org/web/eoportal/satellite-missions/p/prisma-hyperspectral#launch (accessed on 3 August 2020).

30. Satellite Missions Database. Available online: https://directory.eoportal.org/web/eoportal/satellite-missions (accessed on 10 November 2019).

31. EnMAP (Environmental Monitoring and Analysis Program). Available online: https://directory.eoportal.org/web/eoportal/satellite-missions/e/enmap (accessed on 3 August 2020).

32. Mitchell, J.J.; Glenn, N.F.; Anderson, M.O.; Hruska, R.C.; Halford, A.; Baun, C.; Nydegger, N. Unmanned Aerial Vehicle (UAV) hyperspectral remote sensing for dryland vegetation monitoring. In Proceedings of the 2012 4th Workshop on Hyperspectral Image and Signal Processing: Evolution in Remote Sensing (WHISPERS), Shanghai, China, 4–7 June 2012; pp. 1–10.

33. Zarco-Tejada, P.J.; Guillén-Climent, M.L.; Hernández-Clemente, R.; Catalina, A.; González, M.R.; Martín, P. Estimating leaf carotenoid content in vineyards using high resolution hyperspectral imagery acquired from an unmanned aerial vehicle (UAV). *Agric. Forest Meteorol.* **2013**, *171*, 281–294. [CrossRef]

34. Copenhaver, K.; Hellmich, R.; Hunt, T.; Glaser, J.; Sappington, T.; Calvin, D.; Carroll, M.; Fridgen, J. Use of spectral vegetation indices derived from airborne hyperspectral imagery for detection of European corn borer infestation in Iowa corn plots. *J. Econ. Entomol.* **2008**, *101*, 1614–1623.

35. Ryu, C.; Suguri, M.; Umeda, M. Multivariate analysis of nitrogen content for rice at the heading stage using reflectance of airborne hyperspectral remote sensing. *Field Crops Res.* **2011**, *122*, 214–224. [CrossRef]

36. Lu, B.; He, Y. Evaluating Empirical Regression, Machine Learning, and Radiative Transfer Modelling for Estimating Vegetation Chlorophyll Content Using Bi-Seasonal Hyperspectral Images. *Remote Sens.* **2019**, *11*, 1979. [CrossRef]

37. Yu, F.; Xu, T.; Du, W.; Ma, H.; Zhang, G.; Chen, C. Radiative transfer models (RTMs) for field phenotyping inversion of rice based on UAV hyperspectral remote sensing. *Int. J. Agric. Biol. Eng.* **2017**, *10*, 150–157.

38. Teke, M.; Deveci, H.S.; Haliloglu, O.; Gurbuz, S.Z.; Sakarya, U. A short survey of hyperspectral remote sensing applications in agriculture. In Proceedings of the 2013 6th International Conference on Recent Advances in Space Technologies (RAST), Istanbul, Turkey, 12–14 June 2013; IEEE: New York, NY, USA, 2013; pp. 171–176.

39. Dale, L.M.; Thewis, A.; Boudry, C.; Rotar, I.; Dardenne, P.; Baeten, V.; Pierna, J.A.F. Hyperspectral Imaging Applications in Agriculture and Agro-Food Product Quality and Safety Control: A Review. *Appl. Spectrosc. Rev.* **2013**, *48*, 142–159. [CrossRef]

40. Tiangong/Shenzhou: China's Human Spaceflight Program/Tianzhou Cargo Spaceship. Available online: https://directory.eoportal.org/web/eoportal/satellite-missions/t/tiangong (accessed on 3 August 2020).

41. Apan, A.; Held, A.; Phinn, S.; Markley, J. Detecting sugarcane 'orange rust' disease using EO-1 Hyperion hyperspectral imagery. *Int. J. Remote Sens.* **2004**, *25*, 489–498. [CrossRef]

42. Dutta, S.; Bhattacharya, B.K.; Rajak, D.R.; Chattopadhayay, C.; Patel, N.K.; Parihar, J.S. Disease detection in mustard crop using eo-1 hyperion satellite data. *J. Indian Soc. Remote* **2006**, *34*, 325–330. [CrossRef]

43. Moharana, S.; Dutta, S. Spatial variability of chlorophyll and nitrogen content of rice from hyperspectral imagery. *ISPRS J. Photogramm.* **2016**, *122*, 17–29. [CrossRef]

44. Thenkabail, P.S.; Mariotto, I.; Gumma, M.K.; Middleton, E.M.; Landis, D.R.; Huemmrich, K.F. Selection of Hyperspectral Narrowbands (HNBs) and Composition of Hyperspectral Twoband Vegetation Indices (HVIs) for Biophysical Characterization and Discrimination of Crop Types Using Field Reflectance and Hyperion/EO-1 Data. *IEEE J. STARS* **2013**, *6*, 427–439. [CrossRef]

45. Wu, C.; Han, X.; Niu, Z.; Dong, J. An evaluation of EO-1 hyperspectral Hyperion data for chlorophyll content and leaf area index estimation. *Int. J. Remote Sens.* **2010**, *31*, 1079–1086. [CrossRef]

46. Bannari, A.; Staenz, K.; Champagne, C.; Khurshid, K. Spatial Variability Mapping of Crop Residue Using Hyperion (EO-1) Hyperspectral Data. *Remote Sens.* **2015**, *7*, 8107–8127. [CrossRef]

47. Galloza, M.S.; Crawford, M. Exploiting multisensor spectral data to improve crop residue cover estimates for management of agricultural water quality. In Proceedings of the IEEE Geoscience and Remote Sensing Society Symposium, Vancouver, BC, Canada, 24–29 July 2011; IEEE: New York, NY, USA, 2011; pp. 3668–3671.

48. Camacho Velasco, A.; Vargas García, C.A.; Arguello Fuentes, H. A comparative study of target detection algorithms in hyperspectral imagery applied to agricultural crops in Colombia. *Revista Tecnura* **2016**, *20*, 86–99. [CrossRef]

49. Gomez, C.; Rossel, R.A.V.; McBratney, A.B. Soil organic carbon prediction by hyperspectral remote sensing and field vis-NIR spectroscopy: An Australian case study. *Geoderma* **2008**, *146*, 403–411. [CrossRef]

50. Zhang, T.; Li, L.; Zheng, B. Estimation of agricultural soil properties with imaging and laboratory spectroscopy. *J. Appl. Remote Sens.* **2013**, *7*, 73587. [CrossRef]

51. Bostan, S.; Ortak, M.A.; Tuna, C.; Akoguz, A.; Sertel, E.; Ustundag, B.B. Comparison of classification accuracy of co-located hyperspectral & multispectral images for agricultural purposes. In Proceedings of the 2016 Fifth International Conference on Agro-Geoinformatics (Agro-Geoinformatics), Tianjin, China, 18–20 July 2016; IEEE: New York, NY, USA, 2016; pp. 1–4.

52. Lodhi, V.; Chakravarty, D.; Mitra, P. Hyperspectral Imaging for Earth Observation: Platforms and Instruments. *J. Indian Inst. Sci.* **2018**, *98*, 429–443. [CrossRef]

53. Aasen, H.; Bolten, A. Multi-temporal high-resolution imaging spectroscopy with hyperspectral 2D imagers - From theory to application. *Remote Sens. Environ.* **2018**, *205*, 374–389. [CrossRef]

54. Jia, X.; Li, S.; Ke, S.; Hu, B. Overview of spaceborne hyperspectral imagers and the research progress in bathymetric maps. In Proceedings of the Second Target Recognition and Artificial Intelligence Summit Forum. International Society for Optics and Photonics, Shenyang, China, 28–30 August 2019; SPIE-INT SOC Optical Engineering: Bellingham, WA, USA, 2020.

55. Headwall Hyperspectral Sensors. Available online: https://www.headwallphotonics.com/hyperspectral-sensors (accessed on 8 May 2020).

56. Pullanagari, R.R.; Kereszturi, G.; Yule, I. Integrating Airborne Hyperspectral, Topographic, and Soil Data for Estimating Pasture Quality Using Recursive Feature Elimination with Random Forest Regression. *Remote Sens.* **2018**, *10*, 1117. [CrossRef]

57. Verger, A.; Baret, F.; Camacho, F. Optimal modalities for radiative transfer-neural network estimation of canopy biophysical characteristics: Evaluation over an agricultural area with CHRIS/PROBA observations. *Remote Sens. Environ.* **2011**, *115*, 415–426. [CrossRef]

58. Antony, R.; Ray, S.S.; Panigrahy, S. Discrimination of wheat crop stage using CHRIS/PROBA multi-angle narrowband data. *Remote Sens. Lett.* **2011**, *2*, 71–80. [CrossRef]

59. Casa, R.; Castaldi, F.; Pascucci, S.; Palombo, A.; Pignatti, S. A comparison of sensor resolution and calibration strategies for soil texture estimation from hyperspectral remote sensing. *Geoderma* **2013**, *197*, 17–26. [CrossRef]

60. Kumar, A.S.K.; Samudraiah, D.R.M. Hyperspectral Imager Onboard Indian Mini Satellite-1. In *Optical Payloads for Space Missions*; Qian, S., Ed.; John Wiley & Sons: Hoboken, NJ, USA, 2015; Volume 6, pp. 141–160.

61. IMS-1 (Indian Microsatellite-1). Available online: https://directory.eoportal.org/web/eoportal/satellite-missions/i/ims-1 (accessed on 31 March 2020).

62. Raval, M.S. Hyperspectral Imaging: A Paradigm in Remote Sensing. *CSI Commun.* **2014**, *7*, 7–9.

63. Khobragade, A.N.; Raghuwanshi, M.M. Contextual Soft Classification Approaches for Crops Identification Using Multi-sensory Remote Sensing Data: Machine Learning Perspective for Satellite Images. In *Artificial Intelligence Perspectives and Applications*; Springer: Cham, Switzerland, 2015; pp. 333–346.

64. Hyperspectral Imager for the Coastal Ocean. Available online: http://hico.coas.oregonstate.edu/ (accessed on 1 April 2020).

65. Krutz, D.; Müller, R.; Knodt, U.; Günther, B.; Walter, I.; Sebastian, I.; Säuberlich, T.; Reulke, R.; Carmona, E.; Eckardt, A.; et al. The Instrument Design of the DLR Earth SensingImaging Spectrometer (DESIS). *Sensors* **2019**, *19*, 1622. [CrossRef]

66. ISS Utilization: HISUI (Hyperspectral Imager Suite). Available online: https://eoportal.org/web/eoportal/satellite-missions/content/-/article/iss-utilization-hisui-hyperspectral-imager-suite-#launch (accessed on 1 April 2020).

67. Pignatti, S.; Palombo, A.; Pascucci, S.; Romano, F.; Santini, F.; Simoniello, T.; Umberto, A.; Vincenzo, C.; Acito, N.; Diani, M.; et al. The PRISMA hyperspectral mission: Science activities and opportunities for agriculture and land monitoring. In Proceedings of the 2013 IEEE International Geoscience and Remote Sensing Symposium-IGARSS, Melbourne, VIC, Australia, 21–26 July 2013; pp. 4558–4561.

68. EnMap Hyperspectral Imager. Available online: http://www.enmap.org/index.html (accessed on 1 December 2019).

69. Feingersh, T.; Ben-Dor, E. SHALOM—A Commercial Hyperspectral Space Mission. In *Optical Payloads for Space Missions*; Qian, S.E., Ed.; John Wiley & Sons, Ltd.: Hoboken, NJ, USA, 2015; pp. 247–263.

70. Pandey, P.C.; Manevski, K.; Srivastava, P.K.; Petropoulos, G.P. The Use of Hyperspectral Earth Observation Data for Land Use/Cover Classification: Present Status, Challenges, and Future Outlook. In *Hyperspectral Remote Sensing of Vegetation*, 2nd ed.; Thenkabail, P.S., Lyon, J.G., Huete, A., Eds.; CRC Press: Boca Raton, FL, USA, 2018; Volume 4.

71. HyspIRI Mission Study. Available online: https://hyspiri.jpl.nasa.gov/ (accessed on 1 August 2020).

72. Malec, S.; Rogge, D.; Heiden, U.; Sanchez-Azofeifa, A.; Bachmann, M.; Wegmann, M. Capability of Spaceborne Hyperspectral EnMAP Mission for Mapping Fractional Cover for Soil Erosion Modeling. *Remote Sens.* **2015**, *7*, 11776–11800. [CrossRef]

73. Siegmann, B.; Jarmer, T.; Beyer, F.; Ehlers, M. The Potential of Pan-Sharpened EnMAP Data for the Assessment of Wheat LAI. *Remote Sens.* **2015**, *7*, 12737–12762. [CrossRef]

74. Locherer, M.; Hank, T.; Danner, M.; Mauser, W. Retrieval of Seasonal Leaf Area Index from Simulated EnMAP Data through Optimized LUT-Based Inversion of the PROSAIL Model. *Remote Sens.* **2015**, *7*, 10321–10346. [CrossRef]

75. Bachmann, M.; Makarau, A.; Segl, K.; Richter, R. Estimating the Influence of Spectral and Radiometric Calibration Uncertainties on EnMAP Data Products—Examples for Ground Reflectance Retrieval and Vegetation Indices. *Remote Sens.* **2015**, *7*, 10689–10714. [CrossRef]

76. Castaldi, F.; Palombo, A.; Santini, F.; Pascucci, S.; Pignatti, S.; Casa, R. Evaluation of the potential of the current and forthcoming multispectral and hyperspectral imagers to estimate soil texture and organic carbon. *Remote Sens. Environ.* **2016**, *179*, 54–65. [CrossRef]

77. Castaldi, F.; Palombo, A.; Pascucci, S.; Pignatti, S.; Santini, F.; Casa, R. Reducing the Influence of Soil Moisture on the Estimation of Clay from Hyperspectral Data: A Case Study Using Simulated PRISMA Data. *Remote Sens.* **2015**, *7*, 15561–15582. [CrossRef]

78. Ghasrodashti, E.; Karami, A.; Heylen, R.; Scheunders, P. Spatial Resolution Enhancement of Hyperspectral Images Using Spectral Unmixing and Bayesian Sparse Representation. *Remote Sens.* **2017**, *9*, 541. [CrossRef]

79. Yang, J.; Li, Y.; Chan, J.; Shen, Q. Image Fusion for Spatial Enhancement of Hyperspectral Image via Pixel Group Based Non-Local Sparse Representation. *Remote Sens.* **2017**, *9*, 53. [CrossRef]

80. Zhao, Y.; Yang, J.; Chan, J.C. Hyperspectral Imagery Super-Resolution by Spatial-Spectral Joint Nonlocal Similarity. *IEEE J. STARS* **2014**, *7*, 2671–2679. [CrossRef]

81. Loncan, L.; Almeida, L.B.; Bioucas-Dias, J.M.; Briottet, X.; Chanussot, J.; Dobigeon, N.; Fabre, S.; Liao, W.; Licciardi, G.A.; Simões, M.; et al. Hyperspectral pansharpening: A review. *IEEE Geosci. Remote Sens. Mag.* **2015**, *3*, 27–46. [CrossRef]

82. Asner, G.P.; Heidebrecht, K.B. Imaging spectroscopy for desertification studies: Comparing aviris and eo-1 hyperion in argentina drylands. *IEEE Trans. Geosci. Remote* **2003**, *41*, 1283–1296. [CrossRef]

83. Weng, Y.; Gong, P.; Zhu, Z. A Spectral Index for Estimating Soil Salinity in the Yellow River Delta Region of China Using EO-1 Hyperion Data. *Pedosphere* **2010**, *20*, 378–388. [CrossRef]

84. Mulla, D.J. Twenty five years of remote sensing in precision agriculture: Key advances and remaining knowledge gaps. *Biosyst. Eng.* **2013**, *114*, 358–371. [CrossRef]

85. Jacquemoud, S.; Baret, F.; Andrieu, B.; Danson, F.M.; Jaggard, K. Extraction of vegetation biophysical parameters by inversion of the PROSPECT + SAIL models on sugar beet canopy reflectance data. Application to TM and AVIRIS sensors. *Remote Sens. Environ.* **1995**, *52*, 163–172. [CrossRef]

86. Gat, N.; Erives, H.; Fitzgerald, G.J.; Kaffka, S.R.; Maas, S.J. Estimating sugar beet yield using AVIRIS-derived indices. In *Summaries of the 9th JPL Airborne Earth Science Workshop. Unpaginated CD*; Jet Propulsion Laboratory: Pasadena, CA, USA, 2000.

87. Estep, L.; Terrie, G.; Davis, B. Crop stress detection using AVIRIS hyperspectral imagery and artificial neural networks. *Int. J. Remote Sens.* **2004**, *25*, 4999–5004. [CrossRef]

88. Cheng, Y.; Ustin, S.L.; Riano, D.; Vanderbilt, V.C. Water content estimation from hyperspectral images and MODIS indexes in Southeastern Arizona. *Remote Sens. Environ.* **2008**, *112*, 363–374. [CrossRef]

89. Palacios-Orueta, A.; Ustin, S.L. Remote Sensing of Soil Properties in the Santa Monica Mountains I. Spectral Analysis. *Remote Sens. Environ.* **1998**, *65*, 170–183. [CrossRef]

90. Gat, N.; Erives, H.; Maas, S.J.; Fitzgerald, G.J. Application of low altitude AVIRIS imagery of agricultural fields in the San Joaquin Valley, CA, to precision farming. In *The 8th JPL Airborne Earth Science Workshop*; Academia: Pasadena, CA, USA, 1999; pp. 145–150. Available online: https://www.researchgate.net/publication/2434575_Application_Of_Low_Altitude_Aviris_Imagery_Of_Agricultural_Fields_In_The_San_Joaquin_Valley_Ca_To_Precision_Farming (accessed on 11 July 2020.).

91. Nigam, R.; Tripathy, R.; Dutta, S.; Bhagia, N.; Nagori, R.; Chandrasekar, K.; Kot, R.; Bhattacharya, B.K.; Ustin, S. Crop type discrimination and health assessment using hyperspectral imaging. *Curr. Sci.* **2019**, *116*, 1108–1123. [CrossRef]

92. Shivers, S.W.; Roberts, D.A.; McFadden, J.P. Using paired thermal and hyperspectral aerial imagery to quantify land surface temperature variability and assess crop stress within California orchards. *Remote Sens. Environ.* **2019**, *222*, 215–231. [CrossRef]

93. Ran, Q.; Li, W.; Du, Q.; Yang, C. Hyperspectral image classification for mapping agricultural tillage practices. *J. Appl. Remote Sens.* **2015**, *9*, 97298. [CrossRef]

94. Shivers, S.W.; Roberts, D.A.; McFadden, J.P.; Tague. C. Using Imaging Spectrometry to Study Changes in Crop Area in California's Central Valley during Drought. *Remote Sens.* **2018**, *10*, 1556. [CrossRef]

95. Haboudane, D.; Miller, J.R.; Tremblay, N.; Zarco-Tejada, P.J.; Dextraze, L. Integrated narrow-band vegetation indices for prediction of crop chlorophyll content for application to precision agriculture. *Remote Sens. Environ.* **2002**, *81*, 416–426. [CrossRef]

96. Liu, J.; Miller, J.R.; Haboudane, D.; Pattey, E.; Hochheim, K. Crop fraction estimation from casi hyperspectral data using linear spectral unmixing and vegetation indices. *Can. J. Remote Sens.* **2008**, *34*, S124–S138. [CrossRef]

97. Goel, P.K.; Prasher, S.O.; Landry, J.; Patel, R.M.; Viau, A.A. Hyperspectral image classification to detect weed infestations and nitrogen status in corn. *Trans. ASAE* **2003**, *46*, 539.

98. Richter, K.; Hank, T.; Mauser, W. Preparatory analyses and development of algorithms for agricultural applications in the context of the EnMAP hyperspectral mission. In Proceedings of the Remote Sensing for Agriculture, Ecosystems, and Hydrology XII. International Society for Optics and Photonics, Toulouse, France, 22 October 2010; pp. 782407–7824011.

99. Jarmer, T. Spectroscopy and hyperspectral imagery for monitoring summer barley. *Int. J. Remote Sens.* **2013**, *34*, 6067–6078. [CrossRef]

100. Thomas, U.; Philippe, D.; Christian, B.; Franz, R.; Frédéric, M.; Martin, S.; Miriam, M.; Lucien, H. Retrieving the Bioenergy Potential from Maize Crops Using Hyperspectral Remote Sensing. *Remote Sens.* **2013**, *5*, 254–273.

101. Mewes, T.; Franke, J.; Menz, G. Spectral requirements on airborne hyperspectral remote sensing data for wheat disease detection. *Precis. Agric.* **2011**, *12*, 795–812. [CrossRef]

102. Hbirkou, C.; Pätzold, S.; Mahlein, A.; Welp, G. Airborne hyperspectral imaging of spatial soil organic carbon heterogeneity at the field-scale. *Geoderma* **2012**, *175–176*, 21–28. [CrossRef]

103. Cilia, C.; Panigada, C.; Rossini, M.; Meroni, M.; Busetto, L.; Amaducci, S.; Boschetti, M.; Picchi, V.; Colombo, R. Nitrogen Status Assessment for Variable Rate Fertilization in Maize through Hyperspectral Imagery. *Remote Sens.* **2014**, *6*, 6549–6565. [CrossRef]

104. Ambrus, A.; Burai, P.; Lénárt, C.; Enyedi, P.; Kovács, Z. Estimating biomass of winter wheat using narrowband vegetation indices for precision agriculture. *J. Cent. Eur. Green Innov.* **2015**, *3*, 13–22.

105. Oppelt, N.; Mauser, W. Hyperspectral monitoring of physiological parameters of wheat during a vegetation period using AVIS data. *Int. J. Remote Sens.* **2004**, *25*, 145–159. [CrossRef]

106. Bannari, A.; Pacheco, A.; Staenz, K.; McNairn, H.; Omari, K. Estimating and mapping crop residues cover on agricultural lands using hyperspectral and IKONOS data. *Remote Sens. Environ.* **2006**, *104*, 447–459. [CrossRef]

107. Van Wesemael, B.; Tychon, B.; Bartholomeus, H.; Kooistra, L.; van Leeuwen, M.; Stevens, A.; Ben-Dor, E. Soil Organic Carbon mapping of partially vegetated agricultural fields with imaging spectroscopy. *Int. J. Appl. Earth Obs.* **2011**, *13*, 81–88.

108. Finn, M.P.; Lewis, M.D.; Bosch, D.D.; Giraldo, M.; Yamamoto, K.; Sullivan, D.G.; Kincaid, R.; Luna, R.; Allam, G.K.; Kvien, C.; et al. Remote Sensing of Soil Moisture Using Airborne Hyperspectral Data. *Gisci. Remote Sens.* **2011**, *48*, 522–540. [CrossRef]

109. Xie, Q.; Huang, W.; Liang, D.; Chen, P.; Wu, C.; Yang, G.; Zhang, J.; Huang, L.; Zhang, D. Leaf Area Index Estimation Using Vegetation Indices Derived From Airborne Hyperspectral Images in Winter Wheat. *IEEE J. STARS* **2014**, *7*, 3586–3594. [CrossRef]

110. Castaldi, F.; Chabrillat, S.; Jones, A.; Vreys, K.; Bomans, B.; van Wesemael, B. Soil Organic Carbon Estimation in Croplands by Hyperspectral Remote APEX Data Using the LUCAS Topsoil Database. *Remote Sens.* **2018**, *10*, 153. [CrossRef]

111. Luo, S.; Wang, C.; Xi, X.; Zeng, H.; Li, D.; Xia, S.; Wang, P. Fusion of Airborne Discrete-Return LiDAR and Hyperspectral Data for Land Cover Classification. *Remote Sens.* **2016**, *8*, 3. [CrossRef]

112. Mart, L.; Tard, A.; Pal, V.; Arbiol, R. Atmospheric correction algorithm applied to CASI multi-height hyperspectral imagery. *Parameters* **2006**, *1*, 4.

113. AVIRIS Data—New Data Acquisitions. Available online: https://aviris.jpl.nasa.gov/data/newdata.html (accessed on 1 August 2020).

114. Lu, B.; He, Y. Species classification using Unmanned Aerial Vehicle (UAV)-acquired high spatial resolution imagery in a heterogeneous grassland. *ISPRS J. Photogramm.* **2017**, *128*, 73–85. [CrossRef]

115. Casa, R.; Pascucci, S.; Pignatti, S.; Palombo, A.; Nanni, U.; Harfouche, A.; Laura, L.; Di Rocco, M.; Fantozzi, P. UAV-based hyperspectral imaging for weed discrimination in maize. In *Precision Agriculture '19*; Stafford, J.V., Ed.; Wageningen Academic Publishers: Wageningen, The Netherlands, 2019; pp. 24–35.

116. Dao, P.D.; He, Y.; Lu, B. Maximizing the quantitative utility of airborne hyperspectral imagery for studying plant physiology: An optimal sensor exposure setting procedure and empirical line method for atmospheric correction. *Int. J. Appl. Earth Obs.* **2019**, *77*, 140–150. [CrossRef]

117. Capolupo, A.; Kooistra, L.; Berendonk, C.; Boccia, L.; Suomalainen, J. Estimating plant traits of grasslands from UAV-acquired hyperspectral images: A comparison of statistical approaches. *ISPRS Int. J. Geo Inf.* **2015**, *4*, 2792–2820. [CrossRef]

118. Lu, B.; He, Y. Optimal spatial resolution of Unmanned Aerial Vehicle (UAV)-acquired imagery for species classification in a heterogeneous grassland ecosystem. *Gisci. Remote Sens.* **2018**, *55*, 205–220. [CrossRef]

119. Bohnenkamp, D.; Behmann, J.; Mahlein, A. In-Field Detection of Yellow Rust in Wheat on the Ground Canopy and UAV Scale. *Remote Sens.* **2019**, *11*, 2495. [CrossRef]

120. Habib, A.; Han, Y.; Xiong, W.; He, F.; Zhang, Z.; Crawford, M. Automated Ortho-Rectification of UAV-Based Hyperspectral Data over an Agricultural Field Using Frame RGB Imagery. *Remote Sens.* **2016**, *8*, 796. [CrossRef]

121. Honkavaara, E.; Saari, H.; Kaivosoja, J.; Pölönen, I.; Hakala, T.; Litkey, P.; Mäkynen, J.; Pesonen, L. Processing and assessment of spectrometric, stereoscopic imagery collected using a lightweight UAV spectral camera for precision agriculture. *Remote Sens.* **2013**, *5*, 5006–5039. [CrossRef]

122. Saari, H.; Pellikka, I.; Pesonen, L.; Tuominen, S.; Heikkila, J.; Holmlund, C.; Makynen, J.; Ojala, K.; Antila, T. Unmanned Aerial Vehicle (UAV) operated spectral camera system for forest and agriculture applications. In Proceedings of the Remote Sensing for Agriculture, Ecosystems, and Hydrology XIII. International Society for Optics and Photonics, Prague, Czech Republic, 6 October 2011; Volume 8174.

123. Honkavaara, E.; Kaivosoja, J.; Mäkynen, J.; Pellikka, I.; Pesonen, L.; Saari, H.; Salo, H.; Hakala, T.; Marklelin, L.; Rosnell, T. Hyperspectral reflectance signatures and point clouds for precision agriculture by light weight UAV imaging system. *ISPRS Ann. Photogramm. Remote Sens. Spat. Inf. Sci.* **2012**, *7*, 353–358. [CrossRef]

124. Yue, J.; Yang, G.; Li, C.; Li, Z.; Wang, Y.; Feng, H.; Xu, B. Estimation of Winter Wheat Above-Ground Biomass Using Unmanned Aerial Vehicle-Based Snapshot Hyperspectral Sensor and Crop Height Improved Models. *Remote Sens.* **2017**, *9*, 708. [CrossRef]

125. Pölönen, I.; Saari, H.; Kaivosoja, J.; Honkavaara, E.; Pesonen, L. Hyperspectral imaging based biomass and nitrogen content estimations from light-weight UAV. In Proceedings of the Remote Sensing for Agriculture, Ecosystems, and Hydrology XV. International Society for Optics and Photonics, Dresden, Germany, 16 October 2013; p. 88870J.

126. Kaivosoja, J.; Pesonen, L.; Kleemola, J.; Pölönen, I.; Salo, H.; Honkavaara, E.; Saari, H.; Mäkynen, J.; Rajala, A. A case study of a precision fertilizer application task generation for wheat based on classified hyperspectral data from UAV combined with farm history data. In Proceedings of the SPIE Remote Sensing, Dresden, Germany, 24–26 September 2013; pp. 1–11.

127. Akhtman, Y.; Golubeva, E.; Tutubalina, O.; Zimin, M. Application of hyperspectural images and ground data for precision farming. *Geogr. Environ. Sustain.* **2017**, *10*, 117–128. [CrossRef]

128. Izzo, R.R.; Lakso, A.N.; Marcellus, E.D.; Bauch, T.D.; Raqueno, N.G.; van Aardt, J. An initial analysis of real-time sUAS-based detection of grapevine water status in the Finger Lakes Wine Country of Upstate New York. In *Proceedings of the Autonomous Air and Ground Sensing Systems for Agricultural Optimization and Phenotyping IV*; International Society for Optics and Photonics: Baltimore, MD, USA, 2019.

129. Scherrer, B.; Sheppard, J.; Jha, P.; Shaw, J.A. Hyperspectral imaging and neural networks to classify herbicide-resistant weeds. *J. Appl. Remote Sens.* **2019**, *13*, 044516. [CrossRef]

130. Yue, J.; Feng, H.; Jin, X.; Yuan, H.; Li, Z.; Zhou, C.; Yang, G.; Tian, Q. A Comparison of Crop Parameters Estimation Using Images from UAV-Mounted Snapshot Hyperspectral Sensor and High-Definition Digital Camera. *Remote Sens.* **2018**, *10*, 1138. [CrossRef]

131. Dalponte, M.; Orka, H.O.; Gobakken, T.; Gianelle, D.; Naesset, E. Tree Species Classification in Boreal Forests with Hyperspectral Data. *IEEE Trans. Geosci. Remote* **2013**, *51*, 2632–2645. [CrossRef]

132. Aasen, H.; Bendig, J.; Bolten, A.; Bennertz, S.; Willkomm, M.; Bareth, G. Introduction and preliminary results of a calibration for full-frame hyperspectral cameras to monitor agricultural crops with UAVs. *Int. Arch. Photogramm. Remote Sens. Spat. Inf. Sci.* **2014**, *XL-7*, 1–8. [CrossRef]

133. Zhu, W.; Sun, Z.; Huang, Y.; Lai, J.; Li, J.; Zhang, J.; Yang, B.; Li, B.; Li, S.; Zhu, K.; et al. Improving Field-Scale Wheat LAI Retrieval Based on UAV Remote-Sensing Observations and Optimized VI-LUTs. *Remote Sens.* **2019**, *11*, 2456. [CrossRef]

134. Zhao, J.; Zhong, Y.; Hu, X.; Wei, L.; Zhang, L. A robust spectral-spatial approach to identifying heterogeneous crops using remote sensing imagery with high spectral and spatial resolutions. *Remote Sens. Environ.* **2020**, *239*, 111605. [CrossRef]

135. Zarco-Tejada, P.J.; González-Dugo, V.; Berni, J.A.J. Fluorescence, temperature and narrow-band indices acquired from a UAV platform for water stress detection using a micro-hyperspectral imager and a thermal camera. *Remote Sens. Environ.* **2012**, *117*, 322–337. [CrossRef]

136. Lu, B.; He, Y.; Liu, H.H.T. Mapping vegetation biophysical and biochemical properties using unmanned aerial vehicles-acquired imagery. *Int. J. Remote Sens.* **2018**, *39*, 5265–5287. [CrossRef]

137. Malmir, M.; Tahmasbian, I.; Xu, Z.; Farrar, M.B.; Bai, S.H. Prediction of soil macro- and micro-elements in sieved and ground air-dried soils using laboratory-based hyperspectral imaging technique. *Geoderma* **2019**, *340*, 70–80. [CrossRef]

138. Van de Vijver, R.; Mertens, K.; Heungens, K.; Somers, B.; Nuyttens, D.; Borra-Serrano, I.; Lootens, P.; Roldan-Ruiz, I.; Vangeyte, J.; Saeys, W. In-field detection of Altemaria solani in potato crops using hyperspectral imaging. *Comput. Electron. Agric.* **2020**, *168*, 105106. [CrossRef]

139. Eddy, P.R.; Smith, A.M.; Hill, B.D.; Peddle, D.R.; Coburn, C.A.; Blackshaw, R.E. Hybrid segmentation - Artificial Neural Network classification of high resolution hyperspectral imagery for Site-Specific Herbicide Management in agriculture. *Photogramm. Eng. Remote Sens.* **2008**, *74*, 1249–1257. [CrossRef]

140. Feng, H.; Chen, G.; Xiong, L.; Liu, Q.; Yang, W. Accurate Digitization of the Chlorophyll Distribution of Individual Rice Leaves Using Hyperspectral Imaging and an Integrated Image Analysis Pipeline. *Front. Plant Sci.* **2017**, *8*, 1238. [CrossRef]

141. Asaari, M.S.M.; Mishra, P.; Mertens, S.; Dhondt, S.; Inzé, D.; Wuyts, N.; Scheunders, P. Close-range hyperspectral image analysis for the early detection of stress responses in individual plants in a high-throughput phenotyping platform. *ISPRS J. Photogramm.* **2018**, *138*, 121–138. [CrossRef]

142. Zhu, W.; Li, J.; Li, L.; Wang, A.; Wei, X.; Mao, H. Nondestructive diagnostics of soluble sugar, total nitrogen and their ratio of tomato leaves in greenhouse by polarized spectra–hyperspectra Introduction to the pls Package l data fusion. *Int. J. Agric. Biol. Eng.* **2020**, *13*, 189–197.

143. Morel, J.; Jay, S.; Féret, J.; Bakache, A.; Bendoula, R.; Carreel, F.; Gorretta, N. Exploring the potential of PROCOSINE and close-range hyperspectral imaging to study the effects of fungal diseases on leaf physiology. *Sci. Rep.* **2018**, *8*, 1–13. [CrossRef] [PubMed]

144. Nagasubramanian, K.; Jones, S.; Singh, A.K.; Sarkar, S.; Singh, A.; Ganapathysubramanian, B. Plant disease identification using explainable 3D deep learning on hyperspectral images. *Plant Methods* **2019**, *15*, 98. [CrossRef] [PubMed]

145. Lopatin, J.; Fassnacht, F.E.; Kattenborn, T.; Schmidtlein, S. Mapping plant species in mixed grassland communities using close range imaging spectroscopy. *Remote Sens. Environ.* **2017**, *201*, 12–23. [CrossRef]

146. Behmann, J.; Mahlein, A.; Paulus, S.; Dupuis, J.; Kuhlmann, H.; Oerke, E.; Plümer, L. Generation and application of hyperspectral 3D plant models: Methods and challenges. *Mach. Vis. Appl.* **2016**, *27*, 611–624. [CrossRef]

147. Antonucci, F.; Menesatti, P.; Holden, N.M.; Canali, E.; Giorgi, S.; Maienza, A.; Stazi, S.R. Hyperspectral Visible and Near-Infrared Determination of Copper Concentration in Agricultural Polluted Soils. *Commun. Soil Sci. Plan.* **2012**, *43*, 1401–1411. [CrossRef]

148. Wan, P.; Yang, G.; Xu, B.; Feng, H.; Yu, H. Geometric Correction Method of Rotary Scanning Hyperspectral Image in Agriculture Application. In Proceedings of the Conferences of the Photoelectronic Technology Committee of the Chinese Society of Astronautics, Beijing, China, 13–15 May 2014.

149. Yeh, Y.; Chung, W.; Liao, J.; Chung, C.; Kuo, Y.; Lin, T. Strawberry foliar anthracnose assessment by hyperspectral imaging. *Comput. Electron. Agric.* **2016**, *122*, 1–9. [CrossRef]

150. Liu, Y.; Wang, T.; Ma, L.; Wang, N. Spectral calibration of hyperspectral data observed from a hyperspectrometer loaded on an Unmanned Aerial Vehicle platform. *IEEE J. Sel. Top. Appl. Earth Obs. Remote Sens.* **2014**, *7*, 2630–2638.

151. Miglani, A.; Ray, S.S.; Pandey, R.; Parihar, J.S. Evaluation of EO-1 hyperion data for agricultural applications. *J. Indian Soc. Remote* **2008**, *36*, 255–266. [CrossRef]

152. Amato, U.; Antoniadis, A.; Carfora, M.F.; Colandrea, P.; Cuomo, V.; Franzese, M.; Pignatti, S.; Serio, C. Statistical Classification for Assessing PRISMA Hyperspectral Potential for Agricultural Land Use. *IEEE J. STARS* **2013**, *6*, 615–625. [CrossRef]

153. Thenkabail, P.S.; Gumma, M.K.; Teluguntla, P.; Mohammed, I.A. Hyperspectral remote sensing of vegetation and agricultural crops. *Photogramm. Eng. Remote Sens. J. Am. Soc. Photogramm.* **2014**, *80*, 697–709.

154. Wang, Y.; Yao, H.; Zhao, S. Auto-encoder based dimensionality reduction. *Neurocomputing* **2016**, *184*, 232–242. [CrossRef]

155. Hsu, P.; Tseng, Y.; Gong, P. Dimension Reduction of Hyperspectral Images for Classification Applications. *Geogr. Inf. Sci.* **2002**, *8*, 1–8. [CrossRef]

156. Abdolmaleki, M.; Fathianpour, N.; Tabaei, M. Evaluating the performance of the wavelet transform in extracting spectral alteration features from hyperspectral images. *Int. J. Remote Sens.* **2018**, *39*, 6076–6094. [CrossRef]

157. Cao, X.; Yao, J.; Fu, X.; Bi, H.; Hong, D. An Enhanced 3-D Discrete Wavelet Transform for Hyperspectral Image Classification. *IEEE Geosci. Remote Soc.* **2020**, 1–5. [CrossRef]

158. Prabhakar, T.V.N.; Geetha, P. Two-dimensional empirical wavelet transform based supervised hyperspectral image classification. *ISPRS J. Photogramm.* **2017**, *133*, 37–45. [CrossRef]

159. Geng, X.; Sun, K.; Ji, L.; Zhao, Y. A Fast Volume-Gradient-Based Band Selection Method for Hyperspectral Image. *IEEE Trans. Geosci. Remote* **2014**, *52*, 7111–7119. [CrossRef]

160. Wang, C.; Gong, M.; Zhang, M.; Chan, Y. Unsupervised Hyperspectral Image Band Selection via Column Subset Selection. *IEEE Geosci. Remote Soc.* **2015**, *12*, 1411–1415. [CrossRef]

161. Wang, Q.; Lin, J.; Yuan, Y. Salient Band Selection for Hyperspectral Image Classification via Manifold Ranking. *IEEE Trans. Neural Netw. Learn. Syst.* **2016**, *27*, 1279–1289. [CrossRef]

162. Thenkabail, P.S.; Smith, R.B.; De Pauw, E. Hyperspectral vegetation indices and their relationships with agricultural crop characteristics. *Remote Sens. Environ.* **2000**, *71*, 158–182. [CrossRef]
163. Nevalainen, O.; Hakala, T.; Suomalainen, J.; Kaasalainen, S. Nitrogen concentration estimation with hyperspectral LiDAR. *ISPRS Ann. Photogramm. Remote Sens. Spat. Inf. Sci.* **2013**, *2*, 205–210. [CrossRef]
164. Huang, W.; Lamb, D.W.; Niu, Z.; Zhang, Y.; Liu, L.; Wang, J. Identification of yellow rust in wheat using in-situ spectral reflectance measurements and airborne hyperspectral imaging. *Precis. Agric.* **2007**, *8*, 187–197. [CrossRef]
165. Tong, A.; He, Y. Estimating and mapping chlorophyll content for a heterogeneous grassland: Comparing prediction power of a suite of vegetation indices across scales between years. *ISPRS J. Photogramm.* **2017**, *126*, 146–167. [CrossRef]
166. Haboudane, D.; Tremblay, N.; Miller, J.R.; Vigneault, P. Remote estimation of crop chlorophyll content using spectral indices derived from hyperspectral data. *IEEE T. Geosci. Remote* **2008**, *46*, 423–437. [CrossRef]
167. Main, R.; Cho, M.A.; Mathieu, R.; O'Kennedy, M.M.; Ramoelo, A.; Koch, S. An investigation into robust spectral indices for leaf chlorophyll estimation. *ISPRS J. Photogramm.* **2011**, *66*, 751–761. [CrossRef]
168. Peng, Y.; Gitelson, A.A. Remote estimation of gross primary productivity in soybean and maize based on total crop chlorophyll content. *Remote Sens. Environ.* **2012**, *117*, 440–448. [CrossRef]
169. Croft, H.; Chen, J.M.; Zhang, Y. The applicability of empirical vegetation indices for determining leaf chlorophyll content over different leaf and canopy structures. *Ecol. Complex.* **2014**, *17*, 119–130. [CrossRef]
170. Zhou, X.; Huang, W.; Kong, W.; Ye, H.; Luo, J.; Chen, P. Remote estimation of canopy nitrogen content in winter wheat using airborne hyperspectral reflectance measurements. *Adv. Space Res.* **2016**, *58*, 1627–1637. [CrossRef]
171. Yue, J.; Feng, H.; Yang, G.; Li, Z. A comparison of regression techniques for estimation of above-ground winter wheat biomass using near-surface spectroscopy. *Remote Sens.* **2018**, *10*, 66. [CrossRef]
172. Hansen, P.M.; Schjoerring, J.K. Reflectance measurement of canopy biomass and nitrogen status in wheat crops using normalized difference vegetation indices and partial least squares regression. *Remote Sens. Environ.* **2003**, *86*, 542–553. [CrossRef]
173. Nguyen, H.T.; Lee, B. Assessment of rice leaf growth and nitrogen status by hyperspectral canopy reflectance and partial least square regression. *Eur. J. Agron.* **2006**, *24*, 349–356. [CrossRef]
174. Pedregosa, F.; Varoquaux, G.; Gramfort, A.; Michel, V.; Thirion, B.; Grisel, O.; Blondel, M.; Prettenhofer, P.; Weiss, R.; Dubourg, V.; et al. Scikit-learn: Machine learning in Python. *Mach. Learn.* **2011**, *12*, 2825–2830.
175. Mevik, B.; Wehrens, R. *Introduction to the PLS Package. Help Sect. "Pls" Package R Studio Softw*; R Found. Stat. Comput.: Vienna, Austria, 2015; pp. 1–23.
176. Asner, G.P.; Martin, R.E.; Anderson, C.B.; Knapp, D.E. Quantifying forest canopy traits: Imaging spectroscopy versus field survey. *Remote Sens. Environ.* **2015**, *158*, 15–27. [CrossRef]
177. Kiala, Z.; Odindi, J.; Mutanga, O. Potential of interval partial least square regression in estimating leaf area index. *S. Afr. J. Sci.* **2017**, *113*, 40–48. [CrossRef]
178. Wang, Z.; Kawamura, K.; Sakuno, Y.; Fan, X.; Gong, Z.; Lim, J. Retrieval of Chlorophyll-a and Total Suspended Solids Using Iterative Stepwise Elimination Partial Least Squares (ISE-PLS) Regression Based on Field Hyperspectral Measurements in Irrigation Ponds in Higashihiroshima, Japan. *Remote Sens.* **2017**, *9*, 264. [CrossRef]
179. Mehmood, T.; Ahmed, B. The diversity in the applications of partial least squares: An overview. *J. Chemometr.* **2016**, *30*, 4–17. [CrossRef]
180. Jacquemoud, S.; Baret, F. PROSPECT—A model of leaf optical-properties spectra. *Remote Sens. Environ.* **1990**, *34*, 75–91. [CrossRef]
181. Jacquemoud, S.; Bacour, C.; Poilve, H.; Frangi, J.P. Comparison of four radiative transfer models to simulate plant canopies reflectance: Direct and inverse mode. *Remote Sens. Environ.* **2000**, *74*, 471–481. [CrossRef]
182. Casa, R.; Jones, H.G. Retrieval of crop canopy properties: A comparison between model inversion from hyperspectral data and image classification. *Int. J. Remote Sens.* **2004**, *25*, 1119–1130. [CrossRef]
183. Richter, K.; Hank, T.; Atzberger, C.; Locherer, M.; Mauser, W. Regularization strategies for agricultural monitoring: The EnMAP vegetation analyzer (AVA). In Proceedings of the 2012 IEEE International Geoscience and Remote Sensing Symposium, Munich, Germany, 22–27 July 2012; pp. 6613–6616.
184. Wu, C.; Wang, L.; Niu, Z.; Gao, S.; Wu, M. Nondestructive estimation of canopy chlorophyll content using Hyperion and Landsat/TM images. *Int. J. Remote Sens.* **2010**, *31*, 2159–2167. [CrossRef]

185. Darvishzadeh, R.; Atzberger, C.; Skidmore, A.; Schlerf, M. Mapping grassland leaf area index with airborne hyperspectral imagery: A comparison study of statistical approaches and inversion of radiative transfer models. *ISPRS J. Photogramm.* **2011**, *66*, 894–906. [CrossRef]

186. Breiman, L. Random forests. *Mach. Learn.* **2001**, *45*, 5–32. [CrossRef]

187. Were, K.; Bui, D.T.; Dick, O.B.; Singh, B.R. A comparative assessment of support vector regression, artificial neural networks, and random forests for predicting and mapping soil organic carbon stocks across an Afromontane landscape. *Ecol. Indic.* **2015**, *52*, 394–403. [CrossRef]

188. Gao, J.; Nuyttens, D.; Lootens, P.; He, Y.; Pieters, J.G. Recognising weeds in a maize crop using a random forest machine-learning algorithm and near-infrared snapshot mosaic hyperspectral imagery. *Biosyst. Eng.* **2018**, *170*, 39–50. [CrossRef]

189. Siegmann, B.; Jarmer, T. Comparison of different regression models and validation techniques for the assessment of wheat leaf area index from hyperspectral data. *Int. J. Remote Sens.* **2015**, *36*, 4519–4534. [CrossRef]

190. Adam, E.; Deng, H.; Odindi, J.; Abdel-Rahman, E.M.; Mutanga, O. Detecting the Early Stage of Phaeosphaeria Leaf Spot Infestations in Maize Crop Using In Situ Hyperspectral Data and Guided Regularized Random Forest Algorithm. *J. Spectrosc.* **2017**, *2017*, 1–8. [CrossRef]

191. Kamilaris, A.; Prenafeta-Boldú, F.X. Deep learning in agriculture: A survey. *Comput. Electron. Agric.* **2018**, *147*, 70–90. [CrossRef]

192. Yuan, Q.; Shen, H.; Li, T.; Li, Z.; Li, S.; Jiang, Y.; Xu, H.; Tan, W.; Yang, Q.; Wang, J.; et al. Deep learning in environmental remote sensing: Achievements and challenges. *Remote Sens. Environ.* **2020**, *241*, 111716. [CrossRef]

193. Sharma, A.; Liu, X.; Yang, X. Land cover classification from multi-temporal, multi-spectral remotely sensed imagery using patch-based recurrent neural networks. *Neural Netw.* **2018**, *105*, 346–355. [CrossRef]

194. Zhang, C.; Sargent, I.; Pan, X.; Li, H.; Gardiner, A.; Hare, J.; Atkinson, P.M. Joint Deep Learning for land cover and land use classification. *Remote Sens. Environ.* **2019**, *221*, 173–187. [CrossRef]

195. Rezaee, M.; Mahdianpari, M.; Zhang, Y.; Salehi, B. Deep Convolutional Neural Network for Complex Wetland Classification Using Optical Remote Sensing Imagery. *IEEE J. STARS* **2018**, *11*, 3030–3039. [CrossRef]

196. Xu, Y.; Wu, L.; Xie, Z.; Chen, Z. Building Extraction in Very High Resolution Remote Sensing Imagery Using Deep Learning and Guided Filters. *Remote Sens.* **2018**, *10*, 144. [CrossRef]

197. Kuwata, K.; Shibasaki, R. Estimating crop yields with deep learning and remotely sensed data. In Proceedings of the 2015 IEEE International Geoscience and Remote Sensing Symposium (IGARSS), Milan, Italy, 26–31 July 2015; pp. 858–861.

198. Mohanty, S.P.; Hughes, D.P.; Salathé, M. Using Deep Learning for Image-Based Plant Disease Detection. *Front. Plant Sci.* **2016**, *7*, 1419. [CrossRef] [PubMed]

199. Ji, S.; Zhang, C.; Xu, A.; Shi, Y.; Duan, Y. 3D Convolutional Neural Networks for Crop Classification with Multi-Temporal Remote Sensing Images. *Remote Sens.* **2018**, *10*, 75. [CrossRef]

200. Ndikumana, E.; Ho Tong Minh, D.; Baghdadi, N.; Courault, D.; Hossard, L. Deep Recurrent Neural Network for Agricultural Classification using multitemporal SAR Sentinel-1 for Camargue, France. *Remote Sens.* **2018**, *10*, 1217. [CrossRef]

201. Singh, A.K.; Ganapathysubramanian, B.; Sarkar, S.; Singh, A. Deep Learning for Plant Stress Phenotyping: Trends and Future Perspectives. *Trends Plant Sci.* **2018**, *23*, 883–898. [CrossRef]

202. Chlingaryan, A.; Sukkarieh, S.; Whelan, B. Machine learning approaches for crop yield prediction and nitrogen status estimation in precision agriculture: A review. *Comput. Electron. Agric.* **2018**, *151*, 61–69. [CrossRef]

203. Song, X.; Zhang, G.; Liu, F.; Li, D.; Zhao, Y.; Yang, J. Modeling spatio-temporal distribution of soil moisture by deep learning-based cellular automata model. *J. Arid Land* **2016**, *8*, 734–748. [CrossRef]

204. Moharana, S.; Dutta, S. Estimation of water stress variability for a rice agriculture system from space-borne hyperion imagery. *Agr. Water Manag.* **2019**, *213*, 260–269. [CrossRef]

205. Yang, C. Airborne Hyperspectral Imagery for Mapping Crop Yield Variability. *Geogr. Compass* **2009**, *3*, 1717–1731. [CrossRef]

206. Zimdahl, R.L. *Six Chemicals That Changed Agriculture*; Academic Press: Cambridge, MA, USA, 2015.

207. Goel, P.K.; Prasher, S.O.; Landry, J.A.; Patel, R.M.; Bonnell, R.B.; Viau, A.A.; Miller, J.R. Potential of airborne hyperspectral remote sensing to detect nitrogen deficiency and weed infestation in corn. *Comput. Electron. Agric.* **2003**, *38*, 99–124. [CrossRef]

208. Quemada, M.; Gabriel, J.; Zarco-Tejada, P. Airborne Hyperspectral Images and Ground-Level Optical Sensors As Assessment Tools for Maize Nitrogen Fertilization. *Remote Sens.* **2014**, *6*, 2940–2962. [CrossRef]

209. Koppe, W.; Laudien, R.; Gnyp, M.L.; Jia, L.; Li, F.; Chen, X.; Bareth, G. Deriving winter wheat characteristics from combined radar and hyperspectral data analysis. In Proceedings of the Geoinformatics, Wuhan, China, 28–29 October 2006; Remotely Sensed Data and Information. SPIE-INT SOC Optical Engineering: Bellingham, WA, USA, 2006.

210. Castaldi, F.; Castrignano, A.; Casa, R. A data fusion and spatial data analysis approach for the estimation of wheat grain nitrogen uptake from satellite data. *Int. J. Remote Sens.* **2016**, *37*, 4317–4336. [CrossRef]

211. Zheng, H.; Zhou, X.; Cheng, T.; Yao, X.; Tian, Y.; Cao, W.; Zhu, Y. Evaluation of a uav-based hyperspectral frame camera for monitoring the leaf nitrogen concentration in rice. In Proceedings of the IEEE International Symposium on Geoscience and Remote Sensing IGARSS, Beijing, China, 10–15 July 2016; pp. 7350–7353.

212. Zhou, K.; Cheng, T.; Zhu, Y.; Cao, W.; Ustin, S.L.; Zheng, H.; Yao, X.; Tian, Y. Assessing the Impact of Spatial Resolution on the Estimation of Leaf Nitrogen Concentration Over the Full Season of Paddy Rice Using Near-Surface Imaging Spectroscopy Data. *Front. Plant Sci.* **2018**, *9*, 964. [CrossRef] [PubMed]

213. Nasi, R.; Viljanen, N.; Kaivosoja, J.; Alhonoja, K.; Hakala, T.; Markelin, L.; Honkavaara, E. Estimating Biomass and Nitrogen Amount of Barley and Grass Using UAV and Aircraft Based Spectral and Photogrammetric 3D Features. *Remote Sens.* **2018**, *10*, 1082. [CrossRef]

214. Nigon, T.J.; Mulla, D.J.; Rosen, C.J.; Cohen, Y.; Alchanatis, V.; Knight, J.; Rud, R. Hyperspectral aerial imagery for detecting nitrogen stress in two potato cultivars. *Comput. Electron. Agric.* **2015**, *112*, 36–46. [CrossRef]

215. Chen, S.; Chen, C.; Wang, C.; Yang, I.; Hsiao, S. Evaluation of nitrogen content in cabbage seedlings using hyper-spectral images. In Proceedings of the Optics East, Boston, MA, USA, 9–12 September 2007; p. L7610.

216. Miphokasap, P.; Wannasiri, W. Estimations of Nitrogen Concentration in Sugarcane Using Hyperspectral Imagery. *Sustainability* **2018**, *10*, 1266. [CrossRef]

217. Malmir, M.; Tahmasbian, I.; Xu, Z.; Farrar, M.B.; Bai, S.H. Prediction of macronutrients in plant leaves using chemometric analysis and wavelength selection. *J. Soil. Sediment.* **2020**, *20*, 249–259. [CrossRef]

218. Lowe, A.; Harrison, N.; French, A.P. Hyperspectral image analysis techniques for the detection and classification of the early onset of plant disease and stress. *Plant Methods* **2017**, *13*, 80. [CrossRef]

219. Kingra, P.K.; Majumder, D.; Singh, S.P. Application of Remote Sensing and Gis in Agriculture and Natural Resource Management Under Changing Climatic Conditions. *Agric. Res. J.* **2016**, *53*, 295. [CrossRef]

220. Karimi, Y.; Prasher, S.O.; McNairn, H.; Bonnell, R.B.; Dutilleul, P.; Goel, R.K. Classification accuracy of discriminant analysis, artificial neural networks, and decision trees for weed and nitrogen stress detection in corn. *Trans. ASAE* **2005**, *48*, 1261–1268. [CrossRef]

221. Zhang, Y.; Slaughter, D.C.; Staab, E.S. Robust hyperspectral vision-based classification for multi-season weed mapping. *ISPRS J. Photogramm.* **2012**, *69*, 65–73. [CrossRef]

222. Eddy, P.R.; Smith, A.M.; Hill, B.D.; Peddle, D.R.; Coburn, C.A.; Blackshaw, R.E. Weed and crop discrimination using hyperspectral image data and reduced bandsets. *Can. J. Remote Sens.* **2014**, *39*, 481–490. [CrossRef]

223. Liu, B.; Li, R.; Li, H.; You, G.; Yan, S.; Tong, Q. Crop/Weed Discrimination Using a Field Imaging Spectrometer System. *Sensors* **2019**, *19*, 5154. [CrossRef] [PubMed]

224. LÓPEZ-Granados, F. Weed detection for site-specific weed management: Mapping and real-time approaches. *Weed Res.* **2011**, *51*, 1–11. [CrossRef]

225. Thomas, S.; Kuska, M.T.; Bohnenkamp, D.; Brugger, A.; Alisaac, E.; Wahabzada, M.; Behmann, J.; Mahlein, A. Benefits of hyperspectral imaging for plant disease detection and plant protection: A technical perspective. *J. Plant Dis. Protect.* **2018**, *125*, 5–20. [CrossRef]

226. Bauriegel, E.; Giebel, A.; Geyer, M.; Schmidt, U.; Herppich, W.B. Early detection of Fusarium infection in wheat using hyper-spectral imaging. *Comput. Electron. Agric.* **2011**, *75*, 304–312. [CrossRef]

227. Zhang, N.; Pan, Y.; Feng, H.; Zhao, X.; Yang, X.; Ding, C.; Yang, G. Development of Fusarium head blight classification index using hyperspectral microscopy images of winter wheat spikelets. *Biosyst. Eng.* **2019**, *186*, 83–99. [CrossRef]

228. Mahlein, A.; Oerke, E.; Steiner, U.; Dehne, H. Recent advances in sensing plant diseases for precision crop protection. *Eur. J. Plant Pathol.* **2012**, *133*, 197–209. [CrossRef]

229. Casa, R.; Castaldi, F.; Pascucci, S.; Basso, B.; Pignatti, S. Geophysical and Hyperspectral Data Fusion Techniques for In-Field Estimation of Soil Properties. *Vadose Zone J.* **2013**, *12*, vzj2012.0201. [CrossRef]
230. Casa, R.; Castaldi, F.; Pascucci, S.; Pignatti, S. Potential of hyperspectral remote sensing for field scale soil mapping and precision agriculture applications. *Ital. J. Agron.* **2012**, *7*, 43. [CrossRef]
231. Gedminas, L.; Martin, S. Soil Organic Matter Mapping Using Hyperspectral Imagery and Elevation Data. In *IEEE Aerospace Conference Proceedings*; IEEE: Big Sky, MT, USA, 2019.
232. Song, X.; Yan, G.; Wan, J.; Liu, L.; Xue, X.; Li, C.; Huang, W. Use of airborne hyperspectral imagery to investigate the influence of soil nitrogen supplies and variable-rate fertilization to winter wheat growth. In Proceedings of the SPIE, Florence, Italy, 11 October 2007.
233. Wang, W.; Li, Z.; Wang, C.; Zheng, D.; Du, H. Prediction of Available Potassium Content in Cinnamon Soil Using Hyperspectral Imaging Technology. *Spectrosc. Spect. Anal.* **2019**, *39*, 1579–1585.
234. McCann, C.; Repasky, K.S.; Lawrence, R.; Powell, S. Multi–temporal mesoscale hyperspectral data of mixed agricultural and grassland regions for anomaly detection. *ISPRS J. Photogramm.* **2017**, *131*, 121–133. [CrossRef]

remote sensing

Article

Detection of Canopy Chlorophyll Content of Corn Based on Continuous Wavelet Transform Analysis

Junyi Zhang [1,2], Hong Sun [1], Dehua Gao [1], Lang Qiao [1], Ning Liu [1], Minzan Li [1,*] and Yao Zhang [1]

[1] Key Laboratory of Modern Precision Agriculture System Integration Research, Ministry of Education, China Agricultural University, Beijing 100083, China; junyizh@cau.edu.cn (J.Z.); sunhong@cau.edu.cn (H.S.); dehua_gao@cau.edu.cn (D.G.); b20193080667@cau.edu.cn (L.Q.); ningliu@cau.edu.cn (N.L.); zhangyao@cau.edu.cn (Y.Z.)

[2] College of Energy and Intelligence Engineering, Henan University of Animal Husbandry and Economy, Zhengzhou 450046, China

* Correspondence: limz@cau.edu.cn

Received: 23 July 2020; Accepted: 20 August 2020; Published: 24 August 2020

Abstract: The content of chlorophyll, an important substance for photosynthesis in plants, is an important index used to characterize the photosynthetic rate and nutrient grade of plants. The real-time rapid acquisition of crop chlorophyll content is of great significance for guiding fine management and differentiated fertilization in the field. This study used the method of continuous wavelet transform (CWT) to process the collected visible and near-infrared spectra of a corn canopy. This task was conducted to extract the valuable information in the spectral data and improve the sensitivity of chlorophyll content assessment. First, a Savitzky–Golay filter and standard normal variable processing were applied to the spectral data to eliminate the influence of random noise and limit drift on spectral reflectance. Second, CWT was performed on the spectral reflection curve with 10 frequency scales to obtain the wavelet energy coefficient of the spectral data. The characteristic bands related to chlorophyll content in the spectral data and the wavelet energy coefficients were screened using the maximum correlation coefficient and the local correlation coefficient extrema, respectively. A partial least-square regression model was established. Results showed that the characteristic bands selected via local correlation coefficient extrema in a wavelet energy coefficient created a detection model with optimal accuracy. The determination coefficient ($R_c{}^2$) of the calibration set was 0.7856, and the root-mean-square error (*RMSE*) of the calibration set (*RMSEC*) was 3.0408. The determination coefficient ($R_v{}^2$) of the validation set is was 0.7364, and the *RMSE* of the validation set (*RMSEV*) was 3.3032. Continuous wavelet transform is a process of data dimension enhancement which can effectively extract the sensitive variables from spectral datasets and improve the detection accuracy of models.

Keywords: canopy spectra; chlorophyll content; continuous wavelet transform (CWT); correlation coefficient; partial least square regression (PLSR)

1. Introduction

Corn, one of the major food crops in the world, provides an important guarantee of food security and economic development. Proper nitrogen application is one of the keys to a good harvest of corn [1,2]. A number of studies have shown that the leaf chlorophyll content (LCC) can be used to predict the nitrogen requirement of crops [3,4]. Chlorophyll content is an important indicator of crop photosynthesis ability and nutrition level. Variable fertilization can be achieved using nitrogen fertilizers according to different chlorophyll contents through accurate monitoring of the chlorophyll content of corn leaves [5,6]. Appropriate fertilization can ensure that crops receive adequate nitrogen and avoid soil and water pollution caused by excessive fertilization [7]. This mechanism is the key to improving the photosynthetic performance of crops, thereby regulating the growth and development of corn and increasing the input–output ratio of corn fertilizers [8]. The topic of detecting chlorophyll content of corn is one of the active areas in field management research today. Thus, this study aimed to detect the chlorophyll content in field crops to evaluate the growth status and providing guidance for fertilization.

The traditional chlorophyll detection method is an analytical chemical method, which has high precision. However, the process is complex, time-consuming, and may damage crops. This method cannot meet the requirements of rapid and nondestructive testing on site. Spectral analytical technology has been widely used in qualitative and quantitative analysis of the physicochemical parameters of farmland crops because of its fast, nondestructive, and nonpolluting characteristics. Kapp-Junior et al. [9] developed a novel regression model able to produce a prescription of the required nitrogen (N) for maize by combining spectral reflectance data and agronomic efficiency. Lu et al. [10] used hyperspectral techniques to analyze the vertical distribution of nitrogen in corn. Zhang et al. [11] used leaf characteristic spectra to forecast apple sugar content. These studies highlight the feasibility and efficiency of evaluating crop nutrients via spectroscopy. Therefore, rapid detection of the chlorophyll content was conducted in the present research by using hyperspectral technology during the growth stages.

Most current studies on the detection of chlorophyll content via spectral analysis have focused on the exploration of spectral characteristics to quantitate the intensity and position of molecular absorption or reflection [12,13]. The two types of methods to quantify the spectral absorption and reflectance of specific matters include multivariate statistical analysis and region positioning calculation. First, multivariate statistical methods are used to select and enhance the parameters of spectral reflectance, derivative spectrum, and vegetation index using maximum correlation coefficient analysis. Liang et al. [14] compared fifty hyperspectral vegetation indices, such as the photochemical reflectance index and canopy chlorophyll index, to identify the most appropriate vegetation indices for crop LCC and canopy chlorophyll content (CCC) inversion. Xu et al. [15] used simulated datasets from the PROSAIL model to establish a 2D-matrix-based relationship between leaf chlorophyll and red-edge relative indices ($RERI_{(705nm)}$ and $RERI_{(783nm)}$). The leaf chlorophyll content can be retrieved using the two vegetation indices from observations on the basis of the matrix. Neto et al. [16] created a sunflower leaf chlorophyll model with the spectral reflectance in the band of 500–1039 nm by using partial least-squares regression (PLSR). Rei et al. [17] used two methods, namely machine-learning algorithms and the inversion of a radiative transfer model, to detect the LCC of tea. Second, the characteristic spectral positions show changes with the local correlation extreme values, which generally include the red edge and green peak. Li et al. [18] used spectral reflectance to construct red-edge spectral parameters and newly developed red-edge region parameters to detect the chlorophyll content in rapeseed leaves. Sun et al. [19] indicated that the blue edge, red valley, and eight other spectral parameters could be used to reflect the chlorophyll content of potato crops. Zheng et al. [20] developed a model of chlorophyll content in potato leaves on the basis of the red-edge location. The mentioned characteristic parameters selected by maximum or local extreme correlation have been widely used as sensitive spectral variables for detecting the chlorophyll content.

However, the canopy reflectance spectrum and sensitive bands of crops are easily affected by external interferences of dynamically changing soil background, vegetation canopy geometry, and atmospheric conditions during the growth periods [21–23]. Numerous studies have attempted to improve the detection models of chlorophyll content by eliminating the irrelevant and noise information of the spectral data [24,25]. With regard to the ideas and principles of radiation transmission, the combination of the PROSPECT leaf optical property model and SAIL (Scattering by arbitrarily inclined leaves) model, also referred to as PROSAIL, has been used to develop methods for retrieval of vegetation biophysical properties. Mridha et al. [26] used the broadband canopy radiation transfer model PROSAIL to invert the leaf area index (LAI), LCC, CCC, and leaf equivalent water thickness of the biophysical variables in soybeans. Botha et al. [27] evaluated the ability of the PROSAIL canopy-level reflectance model to detect LCC of spring wheat (*Triticum aestivum* L.) during the growth stages between pretillering (Zadoks growth stage (ZGS15)) and booting (ZGS50). Lunagaria et al. discussed the spectral sensitivity of crop canopy parameters using a theoretical model. The results indicated that the reflectance in the visible range was important for chlorophyll retrieval. Reflectance in the near-infrared range has importance for retrieval of leaf inclination angle, dry matter, and LAI. Accordingly, the influencing factors are difficult to reduce, and the modeling results are challenging to improve because of the external and internal interferences affecting the field canopy spectrum [28]. The primary concern of this research area is to overcome such challenges to detect the chlorophyll content by hyperspectral technology during the growth stages.

Although scholars have tried to use preprocessing methods (such as continuum removal, first-order differentiation, and high-pass filtering) to eliminate the noise and enhance the characteristic signals caused by certain factors (such as sample background and stray light), challenges and problems in effectively removing the interference signals, especially random and low-frequency signals, exist during dynamic growth periods [29–31]. We addressed the primary concern using continuous wavelet transform (CWT) to overcome such problems. CWT, with its rich wavelet base function, multiresolution analysis, time-frequency localization, and other advantages, has received increasing attention in image and signal analysis, decomposition, compression, and denoising [32–37]. This method can effectively separate low-frequency signals from high-frequency signals and extract the weak information from the spectral signal. Chen et al. [38] studied the CWT, taking 265 leaves of 47 plants as the sample spectrum and effectively inverting the water content in the sample, with a high precision of up to 75%. Li et al. proposed a new technique (WREP) to extract red-edge positions (REPs) on the basis of the application of CWT to the reflectance spectra. The results demonstrated that WREP obtained the best detection accuracy for LCC and CCC compared with the traditional techniques. High scales of wavelet decomposition were favorable for the detection of CCC and low scales for the detection of LCC [39]. These studies highlight that the CWT can be used to improve the modeling results of LCC detections. However, great uncertainty still exists regarding the effects of this method to help improve LCC detection during growth periods in which the spectral characteristics are dynamically changed and influenced by soil background, vegetation canopy geometry, and atmospheric conditions. Similar to spectral wavelengths, whether to use the local correlation extremum method or the maximum correlation coefficient method to select sensitive wavelet features is worth discussing. Thus, this study aimed to clarify and create a model to monitor the CCC of corn on the basis of CWT during the growth stages.

This study focused on the relationship between LCC and corn canopy spectral reflectance to propose an efficient method to evaluate the chlorophyll content of corn. The main aims of this study were as follows: (1) to comparative analyze the advantages of maximum correlation value and local correlation extreme value in selecting feature variables; (2) to use CWT to decompose the original spectral data and extract the weak information in the spectrum to detect the chlorophyll content.

2. Materials and Methods

2.1. Experiments and Materials

The experiments were conducted in Hengshui City, Hebei Province, China. Seventy-two sampling areas were present in the test field, as shown in Figure 1, with six fertilization levels to ensure the gradient of chlorophyll content. The nitrogen fertilizer was pure nitrogen, and the phosphorus fertilizer was P_2O_5. The spectral data were collected in three growth periods on the basis of the growth time and conditions, namely, G1 (6 leaf stage), G2 (9 leaf stage), and G3 (12 leaf stage). During growth periods, one leaf sample was collected in each sampling area, so that two hundred and sixteen leaf samples were collected.

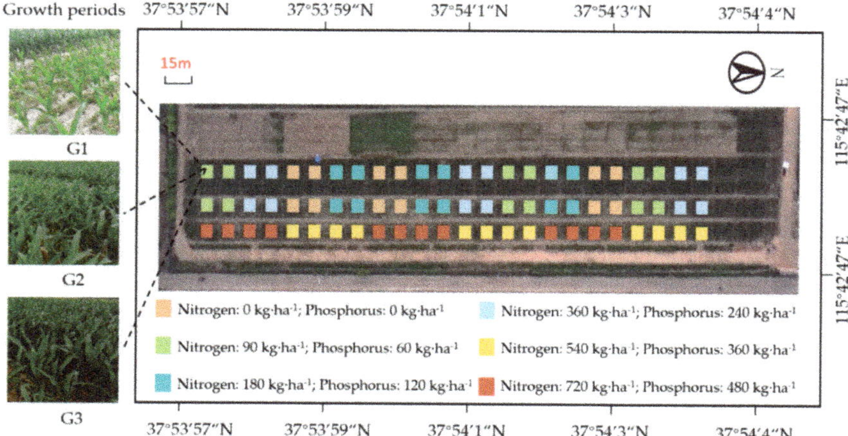

Figure 1. Locations and treatments of the experiment.

The overall process of analyzing reflection spectrum data and chlorophyll content is shown in Figure 2; this mainly included the collection of spectrum data, preprocessing of the spectral data, selection of the characteristic variables, and establishment of a detection model for the chlorophyll content. The selection of the characteristic variables was conducted on the basis of comparison of the characteristic wavelengths and wavelet features selected by maximum correlation coefficients and local extrema of the correlation coefficients.

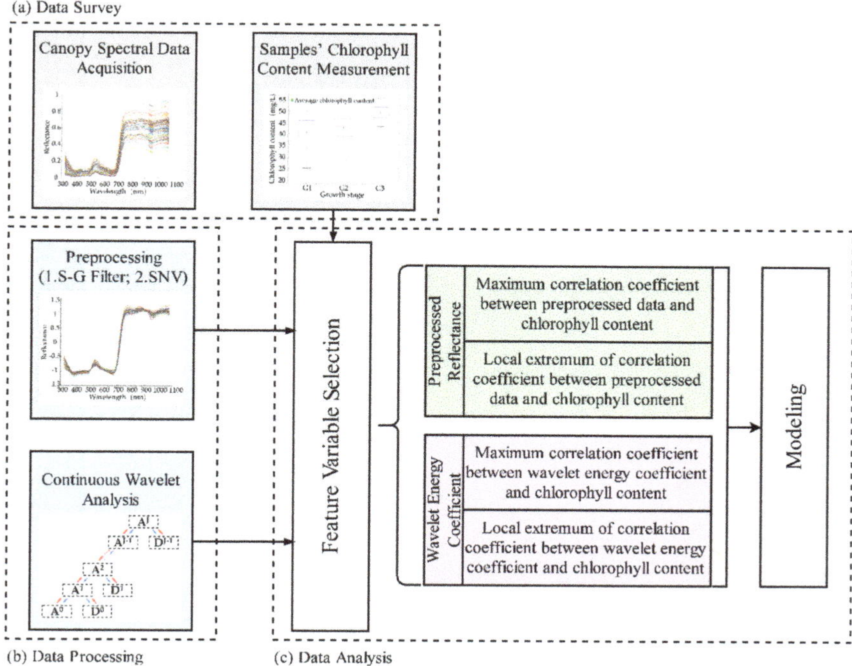

Figure 2. Data analysis flow chart.

2.2. Field Spectrum Data Collection and Chlorophyll Content Measurement

The ASD FieldSpec® HandHeld 2 was used in the field to measure the canopy reflectance of corn. This tool is a hand-held spectrometer with a wavelength range of 325–1075 nm, wavelength accuracy of 1 nm, and spectral resolution of <3.0 nm at 700 nm [40]. Sample leaves were randomly chosen in each sample area to measure the spectral reflectance. Sample leaves were then sealed for subsequent chlorophyll extraction experiment. Spectrum reflectance data were collected three times above the leaf during the spectrum measurement. The averaged reflectance was taken as an original spectral datum.

The chlorophyll content of the sample leaves was measured in the laboratory via SHIMADZU UV2450 spectrophotometry. The spectrophotometry measurement wavelength range was 190–900 nm and the band width was 0.1–5 nm. The spectral resolution was 0.1 nm and the stray light was lower than 0.015%. The main stems of corn leaves were removed, and the leaves were shredded and evenly mixed. Approximately 0.4 g crushed leaf samples were soaked in 25 mL acetone and absolute ethanol mixture for 24 hours, and the mixture ratio was 2:1. During soaking, the mixed solutions were shaken three times to accelerate the chlorophyll extraction. The absorbances of extract solution at 645 and 663 nm were measured with a UV spectrophotometer. The concentrations of chlorophyll *a* (Chl_a) and chlorophyll *b* (Chl_b) were calculated using the following equations:

$$Chl_a = 12.72 \times A_{663} - 2.59 \times A_{645}, \tag{1}$$

$$Chl_b = 22.88 \times A_{645} - 4.67 \times A_{663}, \tag{2}$$

$$Chl_T = Chl_a\left(\text{mg L}^{-1}\right) + Chl_b\left(\text{mg L}^{-1}\right), \tag{3}$$

where A_{645} and A_{633} are the absorbances of the extract solution at 645 and 663 nm, respectively, and Chl_T is the total chlorophyll [41].

2.3. Spectrum Data Preprocessing

The corn canopy spectrum collected in the field environment contained noise information due to the uneven surface of the sample, random noise, different optical paths, and light scattering. First, a Savitzky–Golay (S-G) filter was used to smoothen the reflection spectrum, and the smoothing window was set to 13 [42]. The S-G filter is based on the principle of least squares. Multiple fitting was performed to the original signal in the correction window and the final conversion result was calculated by the multiplicity of the fitting. Using m (m is odd) continuous wavelength points as the window, the data points inside the smooth window were fitted by p-order polynomial function, and the polynomial equation combination was obtained. The smoothing coefficient was obtained using least-square fitting, and the corrected spectral value of the center point of the window was calculated. By successively moving the position of the smoothing window and repeating the above polynomial fitting steps, the spectra after S-G filtering were obtained.

Second, the standard normal variable (SNV) method was used to process the smoothed spectral curve to reduce the influence of the scattering effect [43]. Standard normal variable correction is often used to eliminate the effects of different particle sizes, surface scattering, and optical path differences in NIR diffuse reflectance spectra. The SNV correction of sample spectra were independent of each other and did not involve the spectral information related to the sample set. First, the sample spectrum was centralized, which means that the mean value of spectral reflectance of each spectral data was subtracted from the sample. The standard deviation of the sample reflectance was then used to scale up. After SNV correction, the spectral mean of each sample became 0 and the variance became 1. The SNV correction spectrum of sample j is as follows:

$$A_{j,SNV} = \left(A_j - \overline{A_j}\right)/\sigma_j, \tag{4}$$

where $\overline{A_j}$ is the mean value of the spectrum of sample j and σ_j is the standard deviation of the spectrum of sample j.

Wavelet analysis is one of the potential technologies used in the extraction of weak hyperspectral information. Wavelet transform is a function combination that decomposes a complex signal into simple subsignal components. The spectral signal can be decomposed into subsignals of different frequencies when applied to the analysis of crop spectral data. We effectively used the overall structural characteristics of spectral information and extracted the weak information hidden in the spectral signal. Moreover, we searched for the optimal combination of the subsignal components to detect the chlorophyll content of the crop canopy.

2.4. Sample-Set Division Algorithm According to Sample-Set Partitioning Based on the Joint X–Y Distance (SPXY)

In this study, the SPXY algorithm proposed by Galvão et al. was used to divide the modeling and verification sets [44]. This approach is a method for dividing the sample set on the basis of the statistical perspective, and it comprehensively considers the difference between the spectrum and the property parameters to select the modeling set. The SPXY algorithm first calculates the Euclidean distance between the spectrum data of all samples using Equation (5). The algorithm then selects the two with the largest distance as the first two samples in the modeling set.

$$d_x(p,q) = \sqrt{\sum_i^I \left[x_p(i) - x_q(i) \right]^2}; \; p,q \in [1,N],$$

(5)

where $x_p(i)$ and $x_q(i)$ are the spectral parameters of samples p and q at i wavelength, respectively; I is the number of wavelengths in the spectrum; and N is the number of samples.

The Euclidean distances between the remaining and selected samples were calculated. The sample with the next longest Euclidean distance was selected as the third sample in the modeling set. We repeated the above-mentioned steps until the number of selected samples was equal to the predetermined number.

The nature property factor $d_y(p,q)$ was considered as Equation (6) on the basis of the above-mentioned formula.

$$d_y(p,q) = \sqrt{\left(y_p - y_q \right)^2}; \; p,q \in [1,N],$$

(6)

where y_p and y_q are the property parameters of samples p and q, respectively.

Variables $d_x(p,q)$ and $d_y(p,q)$ were divided by their maximum values in the dataset to ensure that the sample had the same weight in the spectral and property spaces. The standardized xy distance formula was as follows:

$$d_{xy}(p,q) = \frac{d_x(p,q)}{max_{p,q\in[1,N]}d_x(p,q)} + \frac{d_y(p,q)}{max_{p,q\in[1,N]}d_y(p,q)}.$$

(7)

2.5. Spectrum Characteristic Variable Selection Method

2.5.1. Characteristic Variable Selection

(1) Maximum Correlation Coefficient Method

The characteristic variables needed to be filtered to simplify the model and improve its accuracy. The correlation analytical method is widely used to select variables highly correlated with chlorophyll content on the basis of the maximum correlation coefficient method. However, the selection result showed multicollinearity between the adjacent wavelength variables [45].

(2)   Local Extremum of Correlation Coefficient Method

The local extremum of the correlation coefficient method was proposed to improve the variable selection strategy of correlation analysis to solve multicollinearity between the adjacent wavelength variables: the correlation coefficient of spectral reflectance and chlorophyll content were calculated, and the correlation coefficient curve was drawn. Thereafter, the local extreme points of the correlation curve (the zero-crossing positions of the correlation first-derivative curve) were calculated as the chlorophyll characteristic wavelengths [46]. The maximum correlation wavelengths were also selected on the basis of the maximum correlation coefficient method for comparative analysis.

2.5.2. Continuous Wavelet Analysis

From the perspective of signal processing, wavelet analysis can be used to perform data analysis in the frequency and time domains and extract available information from the signal [47]. The reflection spectrum analysis is highly similar to electronic signal analysis. Accordingly, CWT can be used to decompose the reflection spectrum curve at different frequency scales to generate a series of wavelet energy coefficients. The CWT process is shown in Equation (8).

$$W_f(a,b) = \frac{1}{\sqrt{a}}f(\lambda)\psi(\frac{\lambda-b}{a})d\lambda, \tag{8}$$

where a is the frequency scale factor which is set to 2^n ($n = 1, 2, \ldots , 10$) gradients, and translation factor b is the center wavelength of the mother wavelet function. The mother wavelet function $\psi(\lambda)$ uses the second-order Gaussian function. $f(x)$ is a 1D reflection spectrum, and the wavelet coefficient $W_f(a,b)$ (denoted as $WF_{a,b}$) is 2D data, including frequency scale (1, 2, \ldots , 10) and wavelength (325–1075 nm).

Correlation analysis of the wavelet energy coefficient and chlorophyll content was performed. The local extreme value of the correlation was calculated as the sensitive wavelet feature. The wavelet coefficient with the highest correlation was selected for comparative analysis.

The extreme point was calculated using the peak function in MATLAB software. The parameters of the peaks' function were set as follows: the correlation extreme value peak distance of the spectral reflectance was set to 10, and the minimum peak value was set to 0.1; the correlation extreme value peak distance of the wavelet coefficient was set to 30, and the minimum peak value was set to 0.3.

2.6. Establishing Chlorophyll Content Detection Model Based on PLSR

PLSR gradually became a widely used modeling method in spectral analysis after it was proposed by Geladi in 1986 [48–50]. PLSR can solve the problems of autocorrelation and multicollinearity between variables on the basis of the method of principal component extraction. PLSR was used to perform principal component decomposition simultaneously on the spectral reflectance matrix and LCC matrix. During decomposition, PLSR correlated the spectral and chlorophyll content matrixes and established a linear regression model between the two to detect the chlorophyll content of corn leaves. The leave-one-out cross-validation (LOOCV) method was used for internal interactive verification, and the optimal number of characteristic variables was determined by root-mean-square error of cross-validation (RMSECV). The model evaluation indicators were the validation coefficient of validation set model (R_v^2) and the root-mean-square error of validation set ($RMSEV$).

3. Results

3.1. Analysis of Canopy Spectral Response of Corn during Growth Periods

The original reflectance spectrum of the corn canopy is shown in Figure 3a. The figure demonstrates serious noise-point information in the spectral curve. The noise point of the spectral curve was significantly reduced after the S-G filtering (Figure 3b). The scattering effect of the sample reflection spectrum was significantly improved after the SNV correction (Figure 3c). The average spectral curves of three growth stages are shown in Figure 3d.

In general, 400–500 and 611–710 nm were two low-reflectance regions in the visible light band due to the strong absorption of blue and red light by leaf pigments. The absorption valleys appeared near 400 and 680 nm. Approximately 520–610 nm was a high-reflection area due to the strong reflection by the leaf pigments of green light. The reflection peak appeared near 550 nm. In the near-infrared region, the reflectance sharply increased from 711 nm to 760 nm due to the large cavity of the reflective surface in the spongy tissue structure of the mesophyll, thereby showing a "rapid climb" trend. The 761–1000 nm region was a strong reflection area, and the curve was close to horizontal, thereby showing a "high reflection platform". A weak absorption valley appeared around 970 nm due to the absorption of water.

Figure 3d demonstrates that the different growth periods varied in the four spectral ranges of 325–400, 401–700, 761–970, and 971–1075 nm. The spectral reflectance increased with the growth period in the ranges of 325–400 and 761–970 nm. The reflectance decreased with the growth period in the ranges of 401–700 and 971–1075 nm.

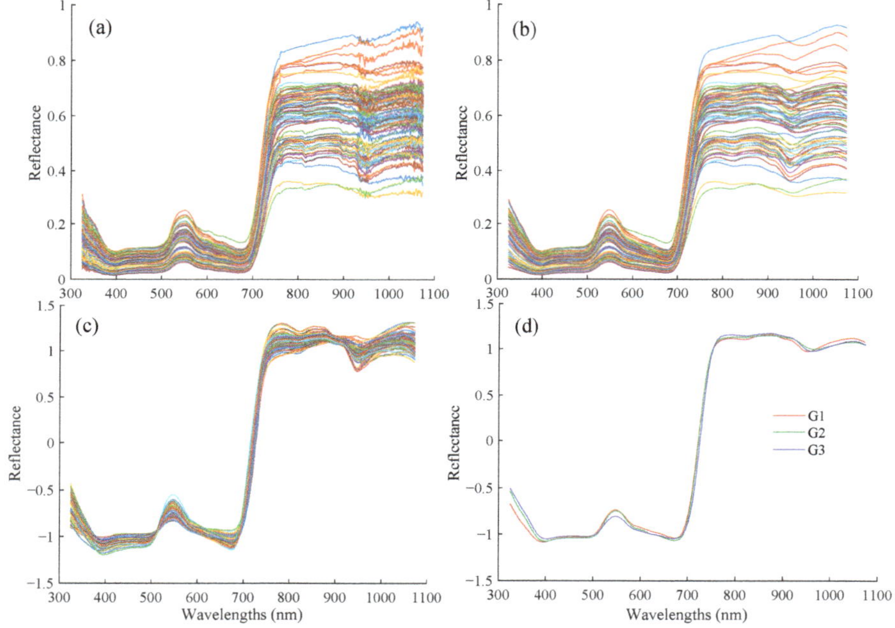

Figure 3. Corn canopy reflectance spectral curve. (**a**) Original canopy reflectance spectra; (**b**) canopy reflectance spectral after the S-G filtering; (**c**) canopy reflectance spectra after the S-G filtering and SNV; (**d**) average canopy spectra of three growth stages.

3.2. Statistical Analysis and Sample-Set Division

The trend of the average chlorophyll content with the growth period is shown in Figure 4, which demonstrated an increase from G1 to G3. From G1 to G3, the variation ranges of the chlorophyll content between samples gradually concentrated. The SPXY algorithm was used to divide the sample set according to the ratio of 2:1. The division result is shown in Table 1. One hundred and forty-four samples were included in the modeling set to establish a chlorophyll content detection model, and 72 samples in the verification set to test the performance of the detection model. The range of chlorophyll content of the samples in the modeling set was larger than that in the verification set. Thus, the sample set obtained by the SPXY algorithm was reasonable and was used for subsequent modeling.

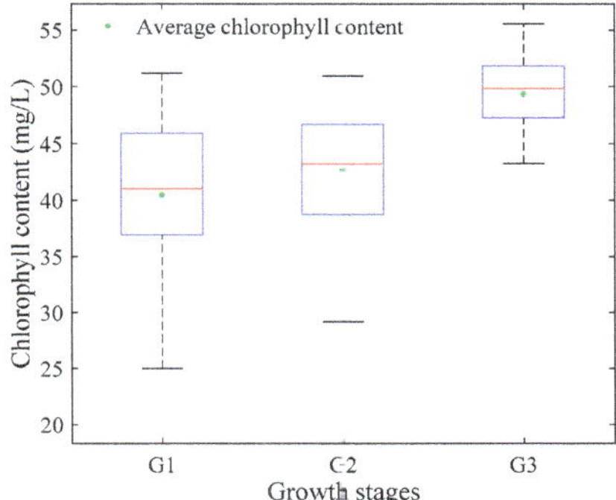

Figure 4. Statistical box line graph of chlorophyll content of the corn growth stages.

Table 1. Statistical results of the calibration set and validation set (%).

Sample Set	Sample Size	Maximum	Minimum	Average	Standard Deviation
Total sample	216	55.58	20.11	44.19	6.51
Modeling set	144	55.58	20.11	44.11	6.59
Verification set	72	55.22	25.31	44.35	6.38

3.3. Correlation Analysis of the Chlorophyll Content and Spect/ral Reflectance

3.3.1. Characteristic Wavelength Selection Based on the Maximum Correlation Coefficient Method

The correlation curve of the chlorophyll content and spectral reflectance is shown in Figure 5. The chlorophyll content was positively correlated with spectral reflectance in the blue (450–500 nm) and red (620–780 nm) regions, and negatively correlated in the green region. This result is consistent with the absorption characteristics of chlorophyll in visible light. The absolute value of the correlation coefficient between the chlorophyll content and the spectral reflectance was higher than 0.5 in the five bands, namely, 376–504, 518–596, 671–681, 698–746, and 880–913 nm. The correlation coefficient gradually decreased.

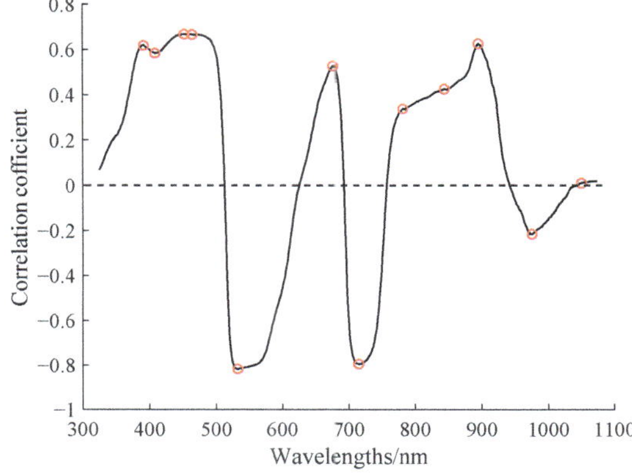

Figure 5. Correlation curve between the spectral reflectance and the chlorophyll content.

The top 15 wavelengths with the maximum correlations were selected as the chlorophyll-sensitive wavelengths (denoted as CA bands). The results were concentrated in the green light region of 529–543 nm. The CA bands were only divided into narrow bands. Accordingly, information redundancy may occur due to autocorrelation of information caused by adjacent narrow-band wavelengths.

3.3.2. Characteristic Wavelength Selection Based on the Local Extrema of the Correlation Coefficient

Twelve extreme points (denoted as CA peak bands) of correlation existed between the spectral reflectance and the chlorophyll content in five highly correlated bands. Approximately 392 and 409 nm were present in the violet region, 453 and 465 nm in the blue region, 533 nm in the green region, 677 nm in the red region, 715 and 782 nm in the red edge, and 845, 896, 976, and 1051 nm in the near-infrared region. The selected wavelengths are shown in Figure 4 with red circles.

3.4. Correlation Analysis of Chlorophyll Content and Wavelet Energy Coefficient

CWT was performed with 10 frequency scales on the spectral reflection curve. We calculated the correlation coefficients of the wavelet energy coefficient and chlorophyll content at each frequency scale. Thereafter, we took the absolute value of the correlation coefficient results and drew the distribution map of the correlation coefficients at different scales. The result is shown in Figure 6.

The wavelet energy coefficient bands with high correlation with chlorophyll content ($|r| > 0.5$) are shown in Table 2. Most of high-correlation wavelengths were concentrated in the visible light area of 325–700 nm. Selected numbers of wavelet energy bands were reduced with the scale increase. In the low-frequency scales from 1 to 4, the high-correlation bands had narrow wavelength ranges, more numbers, and clear division of intervals. In the mid- and high-frequency scales from 5 to 8, selected bands had wider wavelength ranges, less numbers, and ambiguous division of intervals. At the low-frequency scale, the high correlation bands of scale 1 and 2 were consistent, and the high correlation bands of scale 3 and 4 were consistent. With the increase of frequency scale, the correlation between the wavelet energy coefficient and chlorophyll decreased gradually from scale 5, and the boundary between the high-correlation band and the low-correlation band became ambiguous. It can also be seen from Figure 6 that as the frequency scale increased, the correlation between the wavelet energy coefficient and the chlorophyll content gradually decreased. No high-correlation wavelet energy coefficient exists on the high-frequency scale of 9–10. Therefore, the optimal features of wavelet energy should be selected on a scale of 1 to 5.

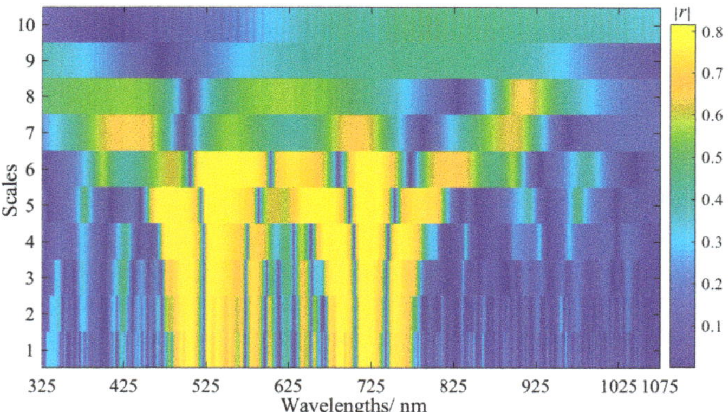

Figure 6. Distribution of the absolute values of the correlation coefficients between the wavelet energy coefficient and the chlorophyll content.

Table 2. Wavelet energy coefficient bands with high correlation with chlorophyll content ($|r| > 0.5$).

Scale	Wavelengths
1	480–516 nm, 525–565 nm, 575–590 nm, 599–604 nm,635–645 nm, 671–700 nm, 710–740 nm, 745–775 nm.
2	480–516 nm, 525–565 nm, 575–590 nm, 599–604 nm,635–645 nm, 671–700 nm, 710–740 nm, 745–775 nm.
3	480–516 nm, 525–565 nm, 575–590 nm, 635–645 nm,671–700 nm, 710–740 nm, 745–775 nm.
4	480–516 nm, 525–565 nm, 575–590 nm, 635–645 nm,671–700 nm, 710–740 nm, 745–775 nm.
5	455–500 nm, 520–585 nm, 605–675 nm, 700–750 nm,758–811 nm.
6	455–500 nm, 520–585 nm, 605–675 nm, 700–750 nm,788–858 nm, 886–901 nm.
7	376–470 nm, 532–582 nm, 665–740 nm, 854–918 nm.
8	326–466 nm, 555–695 nm, 883–944 nm.

3.4.1. Sensitive Wavelet Feature Selection Based on the Maximum Correlation Coefficient

Fifty wavelet energy coefficients with high correlations were selected as the chlorophyll-sensitive wavelet features (denoted as CA-WFs). These result are shown in Table 3. The absolute values of the correlation coefficients of CA-WFs and chlorophyll content were all higher than 0.8.

Table 3. Location and frequency scale parameters of the chlorophyll-sensitive wavelet features (CA-WFs) with $|r|$ higher than 0.8.

Wavelet Feature	Scale	Wavelengths
CA-WF1	1	504–505 (2)
CA-WF2	2	527
CA-WF3	2	687
CA-WF4	3	527–529 (3)
CA-WF5	3	685–688 (4)
CA-WF6	3	716–718 (3)
CA-WF7	4	528–534 (7)
CA-WF8	4	679–685 (7)
CA-WF9	4	715–720 (6)
CA-WF10	5	472–483 (12)
CA-WF11	5	717–720 (4)

3.4.2. Sensitive Wavelet Feature Selection Based on the Local Extrema of the Correlation Coefficient

Fifty-five chlorophyll-sensitive wavelet energy coefficients were selected as sensitive wavelet features (denoted as CA peak WFs) on the basis of the extreme values of the correlation coefficient. The absolute values of the correlation coefficients of CA peak WFs and chlorophyll content were all higher than 0.5. The distribution of CA peak WFs at different frequency scales is shown in Figure 7.

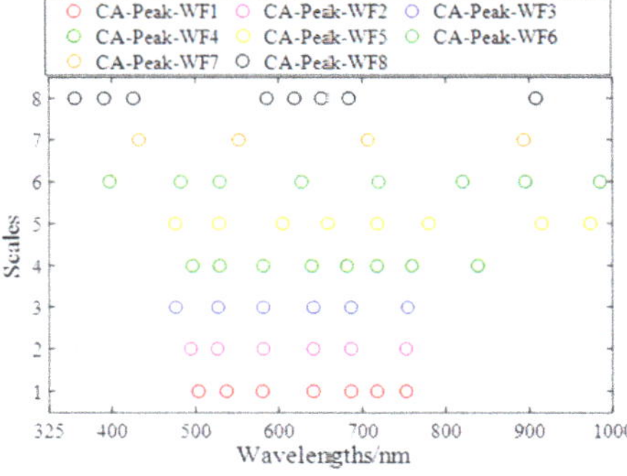

Figure 7. Result of the sensitive wavelet feature selection based on the local correlation coefficients.

The wavelength position analysis indicated that all sensitive wavelet variables of the CA Peak WF1, CA Peak WF2, and CA Peak WF3 were distributed in the visible light region, reflecting the leaf pigment information. CA Peak WF4 contained eight sensitive variables, seven of which were located in the visible light region, and another variable was located at 839 nm. CA Peak WF5 contained eight sensitive variables, six of which were located in the visible light region, and the other two variables were located at 915 and 973 nm. The 915 nm wavelength can reflect CH_2 methylene groups, and 973 nm can present the leaf moisture information. CA Peak WF6 contained eight sensitive variables, five of which were located in the visible light region, and the other three variable positions were 821, 896, and 985 nm. CA Peak WF7 and CA Peak WF8 contained four and eight variables, respectively; each of them had one sensitive variable located in the near-infrared region at 894 and 909 nm. The remaining sensitive variables were located in the visible region.

3.5. Establishment of Chlorophyll Content Detection Model with PLSR

The PLSR algorithm was used to establish a chlorophyll content detection model on the basis of the spectral characteristic variables of the CA bands, CA peak bands, CA-WFs, and CA peak WFs. Both models used LOOCV for internal cross-validation to eliminate the influence of spectral information redundancy and multicollinearity on the model accuracy. The modeling results are shown in Table 4 and the verification results are shown in Figure 8. The comparison of the four detection models showed that the PLSR chlorophyll content detection model based on CA peak WFs had the optimal performance. The decision coefficient ($R_c{}^2$) of the modeling set was 0.7856, the *RMSEC* of the modeling set was 3.0408, the decision coefficient ($R_v{}^2$) of the verification set was 0.7364, and the *RMSEV* of the verification set was 3.3032.

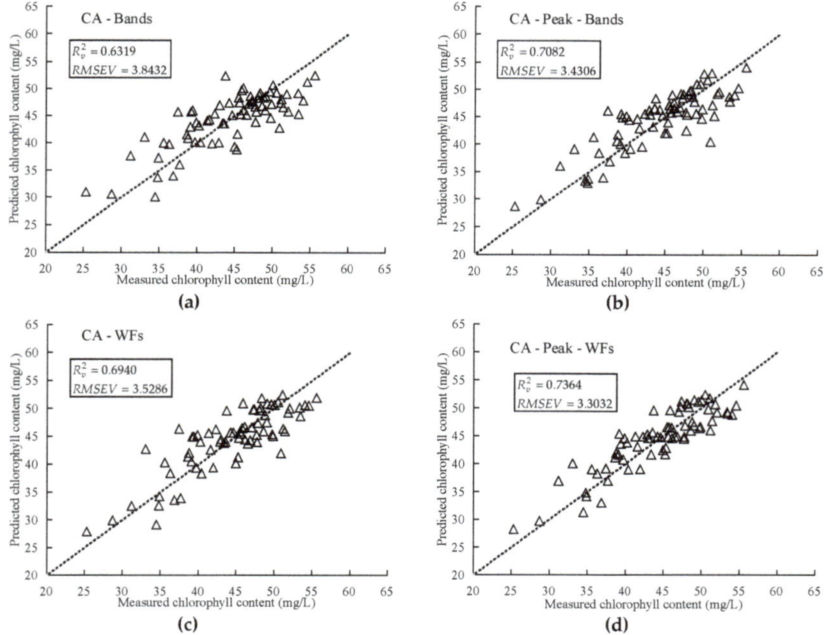

Figure 8. Results of the PLSR detection model for the chlorophyll content. (**a**) Results of the maximum correlation coefficient method; (**b**) Results of the local extremum of correlation coefficient method; (**c**) Results of the maximum correlation coefficient method with wavelet energy coefficient; (**d**) Results of the local extremum of correlation coefficient method with wavelet energy coefficient.

Table 4. Result statistics of PLSR detection model for the chlorophyll content.

Characteristic Variable	Number of Variables	Number of Principal Components	Modeling Set		Verification Set	
			$R_c{}^2$	*RMSEC*	$R_v{}^2$	*RMSEV*
CA bands	15	4	0.6959	3.6214	0.6319	3.8432
CA peak bands	12	3	0.7622	3.2015	0.7082	3.4306
CA-WFs	50	17	0.7820	3.0661	0.6940	3.5286
CA peak WFs	55	13	0.7856	3.0408	0.7364	3.3032

3.6. Chlorophyll Distribution

The spectral reflectance of the corn canopy in three growing stages was introduced into the chlorophyll content detection model. The distribution of chlorophyll in the field during the three growing periods was obtained (Figure 9).

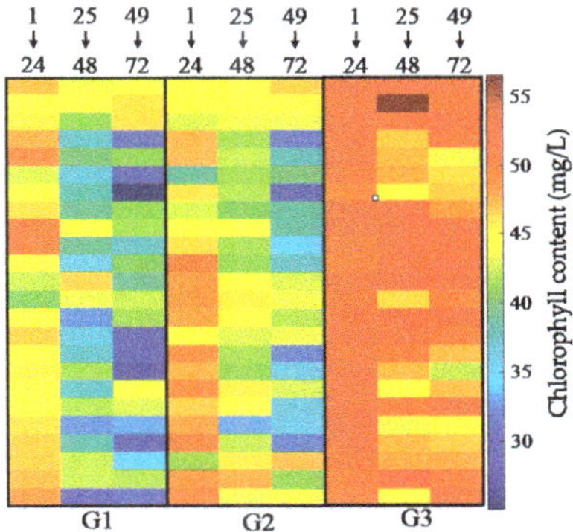

Figure 9. Field chlorophyll distribution in three growth stages.

4. Discussion

4.1. Sensitive Spectral Wavelengths

This study demonstrated that spectral measurements can be used for corn CCC detection. In the range of visible light, the chlorophyll content in the blue and red light regions was positively correlated with spectral reflectance, and reflection peaks were present; in the green light region, the chlorophyll content was negatively correlated with spectral reflectance, and an absorption valley was present [16,51,52]. When the maximum correlation coefficient method was used to filter the sensitive wavelengths, the 15 selected wavelengths were all in the absorption valley of green light, and a serious multicollinearity existed between them. The sensitive wavelengths screened by the local extremum of correlation coefficient method were relatively dispersed and distributed in the visible and near-infrared regions. In the visible light region, the chlorophyll was is characterized by the reflection peak in the red and blue light regions and the absorption valley in the green light region. The extremum characteristic wavelengths in the near-infrared region reflected the composition of other substances [53–55]. The reflectance of the 896 nm band indicates CH_3 methyl groups, that in the 976 nm band reflects the moisture content, and that in the 1051 nm band reflects CH_2 methylene groups [56–58]. These substances are relative to the canopy structure of the corn crop. Therefore, the characteristic wavelengths in the NIR region will improve the robustness of the chlorophyll detection model. The CA peak bands were more evenly distributed and had less redundant information compared with CA bands.

4.2. Continuous Wavelet Analysis

Each wavelet feature contains the information of scale and wavelength position, which corresponds to the state of the generating wavelet function in the process of CWT, namely the scaling factor and the position of shift. The physical meaning of the wavelet feature can be explained by plotting the generating wavelet function corresponding to the wavelet feature that is sensitive to the biochemical parameters. In this study, the Gaussian second derivative was chosen as the generating function of CWT. Each wavelet feature reflects the similarity between the generating wavelet function and the reflectivity spectrum at a specific wavelength position and scale. The absorption characteristics of biochemical parameters at different positions and intensities in the reflectivity spectrum were detected. This situation can be seen as the result of smoothing the spectrum at a particular wavelength and finding the second derivative. The spectral bending degree caused by the different absorption intensities of the biochemical parameters in various bands can be characterized. These two understandings of CWT can be combined to explain the physical meaning of wavelet transform.

The absorption of chlorophyll in the green light band was weaker than that in red and blue light regions. In the reflectivity spectrum, a reflection peak was formed in the green light band. The wavelet features sensitive to chlorophyll were located near the reflection peak of the green light band. Different chlorophyll contents can affect the shape and size of the reflection peak. These changes are easily captured by wavelet features in low scales. A small wavelet feature of medium and high scale near the green light band covered the whole visible band, thus providing the amplitude information of reflectance. The wavelet features at red edge and near-infrared band were stable at the leaf level, thereby indicating that this region is important for chlorophyll monitoring.

Wang et al. used a Mexican hat as the generating wavelet function to obtain the correlation between the wavelet coefficient and the SPAD value of wheat leaves, which can reflect the amount of chlorophyll content. The results showed that the wavelet features sensitive to chlorophyll were located in the red-edge band of 720–740 nm, which was consistent with the research results in References [59–62]. The multiple scattering of light inside the leaves leads to high reflectivity in the near-infrared region, resulting in the rapid increase of spectral reflectance of green vegetation at 680–750 nm, the "red edge" of the area. In the vegetation reflectance spectrum curve, the red edge is one of the most obvious spectral features, and it is an important indicator band used to describe the chlorophyll state of vegetation. The red-edge position (REP) is the wavelength position at which the reflectance of vegetation increases fastest in this interval, and it is also the inflection point of the first derivative of spectral reflectance in this interval. The position of the red edge is an important spectral parameter for detecting the chlorophyll content of vegetation. Liao et al. extracted the wavelet characteristics sensitive to the chlorophyll content of maize leaves in different layers on the basis of the canopy spectrum. The wavelet features sensitive to the chlorophyll content in the upper leaves were distributed in green light and red-edge bands. Meanwhile, the wavelet features sensitive to chlorophyll content in the middle and lower leaves were all located in the red-edge bands, and those in the green light bands disappeared [63]. This phenomenon could have been due to the strong absorption of chlorophyll in the visible range that made it difficult for the green light to penetrate to the middle and lower layers; the visible band of the canopy spectrum mainly contained information regarding chlorophyll in the upper leaves [64]. The comparison of existing literature indicated that the wavelet features at the green light band at the blade level perform better among different datasets. The wavelet features at the red edge perform better among the different datasets at the canopy level.

4.3. Sensitive Wavelet Features

The frequency-scale analysis indicated that CA-WFs were mainly distributed on the low frequency scale (1–4), and the other two types were distributed on the mid-frequency scale (5). The wavelength position analysis indicated that CA-WF1 and CA-WF10 were distributed in the blue light region, and CA-WF 3, CA-WF 5, CA-WF 6, CA-WF 8, CA-WF9, and CA-WF11 were distributed in the red light area. These regions are the strong absorption bands of leaf chlorophyll. CA-WF2, CA-WF4, and CA-WF7 were distributed in the green light regions. These regions are strong reflection bands of leaf chlorophyll. All wavelet features were distributed in the visible light region, which represented the pigment information of the leaves. CA-WFs selected using the maximum correlation coefficient method also had redundant information. For example, the 12 characteristic wavelet coefficients of CA-WF10 were distributed in 472–483 nm, and had serious variable autocorrelation. These autocorrelation problems also existed in CA-WF7, CA-WF8, and CA-WF9.

The frequency-scale analysis indicated that CA peaks WF1–WF4 were distributed in the low-frequency scale (1–4). Meanwhile, CA peaks WF5–WF7 were distributed in the medium-frequency scale (5–7), and CA peak WF8 was located in the high frequency scale (8). Most wavelet features were in the middle- and low-frequency scales, which was consistent with the result of Yao et al [47]. The wavelet features in middle and low frequencies could effectively detect the water content of wheat leaves.

The analysis from the distribution of characteristic wavelet features indicated that all the variables in CA-WFs were distributed in the visible light area and could only reflect the pigment information in the leaves. CA peak WFs were more evenly distributed compared with CA-WFs. The sensitive variable WF4—839 nm in the near-infrared region—reflected the molecular structure of RNHR. WF5—915 nm reflected the CH_2 methylene group, and WF5—973 nm showed the moisture information. WF6—985 nm reflected the starch material, WF7—894 nm showed the CH_3 methyl group, and WF8—909 nm provided protein information. CA peak WFs comprehensively reflected the material structural information of corn leaves, thereby improving the stability of the chlorophyll detection model.

4.4. Chlorophyll Content Detection Model

From the perspective of the method of selecting feature variables, the R_v^2 values of the detection models established by using as feature variables the CA bands and CA-WFs were 0.6319 and 0.6940, respectively. The R_v^2 values of the detection models established by using as feature variables the CA peak bands and CA peak WFs were 0.7082 and 0.7364, respectively. The distributions of CA bands and CA-WFs were relatively concentrated. A high degree of autocorrelation existed between variables. Meanwhile, CA peak bands and CA peak WFs were evenly distributed and comprehensively reflected information.

Comparing the CA peak bands and the CA peak WFs, under the same variable selection method, the R_v^2 of the detection model established using the CA peak WFs (0.7364) was larger than that using CA peak bands (0.7082). The CWT, a process of dimensionality-increasing operation, could dig out the spectral variable information of chlorophyll. The CA peak WFs provided more variable information related to chlorophyll content. Finally, the PLSR model established using CA peak WFs was preferred to detect the chlorophyll content of corn crops.

Comparing the detection models established using spectral reflectance (R_c^2 = 0.77) [14] and spectral index (R_c^2 = 0.70) [15], the wavelet features (R_c^2 = 0.7856) showed better detection capability, which further illustrated that the CWT can deeply mine the information in spectral data.

The model comparison demonstrated that the data after continuous wavelet decomposition can be used to effectively extract valuable information in the spectral reflectance through the dimensionality-increasing operation. In terms of the physical and chemical parameter inversion, the middle- and low-frequency wavelet features highlighted the characteristics of crop pigment and water absorption. In combination with the local extremum of correlation coefficient method, the interference of multicollinearity was eliminated, and the degree of information redundancy was reduced. The detection model established by combining these two methods showed advantages in accuracy and error elimination.

4.5. Chlorophyll Distribution in the Field

The distribution of chlorophyll in Figure 9 demonstrates that the chlorophyll concentration in the canopy of plants gradually increased with the advancement of growth period. The field observation during the field experiment also conformed to this conclusion. The green leaves slowly become dark with the gradual growth of the corn, as shown in Figure 1. The chlorophyll content gradient in the test field remained unchanged over the three growth periods. The figure shows that the chlorophyll contents in the four regions of No. 28–31, 39–48, 52–55, and 63–72 were lower than those in other regions. Some studies have shown a significant positive correlation between the chlorophyll and the nitrogen contents of plants, and chlorophyll can reflect the nitrogen demand of plants to a certain extent [65–68]. Therefore, the chlorophyll detection model based on spectral reflectance can play a guiding role in smart field management and differentiated fertilization.

4.6. Future Work

In this study, the advantages of local correlation coefficient extrema in screening feature variables were compared, the ability of CWT to extract weak spectral information was discussed, and a chlorophyll content detection model for corn canopy based on wavelet energy coefficient was constructed. Advanced methods for crop characteristics' detection like computer vision techniques should be involved [69]. In this experiment, hyperspectral images of sample leaves were collected. It is necessary to establish a more accurate and reliable chlorophyll detection model through hyperspectral images. Furthermore, a larger dataset should be used to verify the adaptability of the algorithm.

5. Conclusions

In this study, corn canopy spectrum data were collected for three growing stages. First, an S-G filter and SNV correction were applied to the reflectance spectra. Subsequently, the dynamic migration of the canopy spectral characteristics and the chlorophyll content dynamic changes in the three growing periods of G1 to G3 were analyzed. Extraction of the chlorophyll characteristic variables was carried out. Finally, the PLSR detection model of the maize chlorophyll content was established. The conclusions were as follows.

The noise point of the spectral curve was significantly reduced. The scattering effect of the reflection spectrum was significantly reduced after the preprocessing steps of S-G filtering and SNV correction. The reflectance spectrum increased in the 325–400 and 761–970 nm regions as the growth stage advanced and the growth period shifted. The reflectance decreased in the 401–700 and 971–1075 nm regions as the growth stage advanced.

The characteristic variables selected on the basis of the local extrema of correlation coefficients were more evenly distributed compared with the maximum correlation coefficient method. This method weakened the autocorrelation and information redundancy of variables and reflected comprehensive information. The wavelet coefficient obtained by performing CWT on the reflectance spectra was used to efficiently analyze the information of chlorophyll and leaf structure substances in a deep and comprehensive way. The spectral feature extraction method based on CWT highlighted the spectral reflectance features of a specific scale, while suppressing the noncorrelated spectral features and noise of other spectral bands with high flexibility. The proposed method effectively improved the matching accuracy of spectral features.

The highest R_v^2 was for the detection model established using the CA peak WFs. The results showed that the CA peak WFs had excellent detection capability for chlorophyll content. CWT combined with the local extremum of the correlation coefficient method is a potentially accurate and efficient strategy for detecting the chlorophyll content of corn crops.

Author Contributions: Conceptualization, J.Z. and M.L.; methodology, J.Z.; software, D.G.; validation, L.Q., N.L.; formal analysis, J.Z.; investigation, H.S.; resources, H.S. and Y.Z.; data curation, J.Z.; writing—original draft preparation, J.Z.; writing—review and editing, H.S.; supervision, M.L.; project administration, H.S.; funding acquisition, M.L. All authors have read and agreed to the published version of the manuscript.

Funding: The project was supported by the National Key Research and Development Program of China (Grant No. 2016YFD0200600-2016YFD0200602), the National Natural Science Foundation of China (Grant No. 31971785 and 31501219), the National Key Research and Development Program of China (Grant No. 2018YFD0300505-1), the Fundamental Research Funds for the Central Universities (Grant No. 2020TC036) and the Graduate Training Project of China Agricultural University (JG2019004 and YW2020007).

Acknowledgments: The authors wish to thank students in Key Laboratory of Modern Precision Agriculture System Integration Research for the help of Sample processing and chemical analysis. Also, many thanks to Dry Farming Institute of Hebei Academy of Agricultural and Forestry Sciences for providing experimental site. We would like to thank Zizheng Xing, Zhiyong Zhang, Ning Liu, Longsheng Cheng, and Song Li for their help with field data collection.

Conflicts of Interest: The authors declare no conflict of interest.

References

1. Johansson, E.; Haby, L.; Prieto-Linde, M.L.; Svensson, S.E. Influence of fertilizer placement on yield and protein composition in spring malting barley. *J. Soil Sci. Plant Nutr.* **2013**, *13*, 895–904. [CrossRef]
2. Bragagnolo, J.; Amado, T.J.C.; Bortolotto, R.P. Use efficiency of variable rate of nitrogen prescribed by optical sensor in corn. *Rev. Ceres* **2016**, *1*, 103–111. [CrossRef]
3. Tolera, A.; Tolessa, D.; Dagne, W. Effects of Varieties and Nitrogen Fertilizer on Yield and Yield Components of Maize on Farmers Field in Mid Altitude Areas of Western Ethiopia. *Int. J. Agron.* **2017**, *2017*, 1–13.
4. Cerrato, M.E.; Blackmer, A.M. Comparison of Models for Describing; Corn Yield Response to Nitrogen Fertilizer. *Agron. J.* **1990**, *82*, 138–143. [CrossRef]
5. Scharf, P.C.; Wiebold, W.J.; Lory, J.A. Corn yield response to nitrogen fertilizer timing and deficiency level. *Agron. J.* **2002**, *94*, 435–441. [CrossRef]
6. Gholamhoseini, M.; Aghaalikhani, M.; Sanavy, S.M.; Mirlatifi, S.M. Interactions of irrigation, weed and nitrogen on corn yield, nitrogen use efficiency and nitrate leaching. *Agric. Water Manag.* **2013**, *126*, 9–18. [CrossRef]
7. Roebeling, P.C. Using the soil and water assessment tool to estimate dissolved inorganic nitrogen water pollution abatement cost functions in central Portugal. *J. Environ. Qual.* **2014**, *43*, 168–176. [CrossRef]
8. Averill, C.; Dietze, M.C.; Bhatnagar, J.M. Continental-scale nitrogen pollution is shifting forest mycorrhizal associations and soil carbon stocks. *Glob. Chang. Biol.* **2018**, *24*, 4544–4553. [CrossRef]
9. Kapp, C.; Caires, E.F.; Guimarães, A.M.; Auler, A.C. Regression modeling nitrogen fertilization requirement for maize crop by combining spectral reflectance and agronomic efficiency. *J. Plant Nutr.* **2020**, *43*, 2152–2163. [CrossRef]
10. Lu, Y.L.; Bai, Y.L.; Ma, D.L.; Wang, L.; Yang, L.P. Nitrogen Vertical Distribution and Status Estimation Using Spectral Data in Maize. *Commun. Soil Sci. Plant Anal.* **2018**, *49*, 526–536.
11. Zhang, Y.; Zheng, L.H.; Li, M.Z.; Xiao, C.Y. Forecasting Apple Sugar Content Based on Leaf Characteristic Spectra in Different Phenological Phases. *Chin. J. Anal. Chem.* **2015**, *43*, 862–870.
12. Igamberdiev, R.M.; Bill, R.; Schubert, H.; Lennartz, B. Analysis of Cross-Seasonal Spectral Response from Kettle Holes: Application of Remote Sensing Techniques for Chlorophyll Estimation. *Remote Sens.* **2012**, *4*, 3481–3500. [CrossRef]
13. Wen, P.; He, J.; Ning, F.; Wang, R.; Zhang, Y.H.; Li, J. Estimating leaf nitrogen concentration considering unsynchronized maize growth stages with canopy hyperspectral technique. *Ecol. Indic.* **2019**, *107*, 1–16. [CrossRef]
14. Liang, L.; Qin, Z.H.; Zhao, S.H.; Di, L.P.; Zhang, C.; Deng, M.X.; Lin, H.; Zhang, L.P.; Wang, L.J.; Liu, Z.X. Estimating crop chlorophyll content with hyperspectral vegetation indices and the hybrid inversion method. *Int. J. Remote Sens.* **2016**, *37*, 2923–2949. [CrossRef]
15. Xu, M.Z.; Liu, R.G.; Chen, J.M.; Liu, Y.; Shang, R.; Ju, W.M.; Wu, C.Y.; Huang, W.J. Retrieving leaf chlorophyll content using a matrix-based vegetation index combination approach. *Remote Sens. Environ.* **2019**, *224*, 60–73. [CrossRef]
16. Neto, A.J.S.; Lopes, D.C.; Pinto, F.A.C.; Zolnier, S. Vis/NIR spectroscopy and chemometrics for non-destructive estimation of water and chlorophyll status in sunflower leaves. *Biosyst. Eng.* **2017**, *155*, 124–133. [CrossRef]
17. Sonobe, R.; Sano, T.; Horie, H. Using spectral reflectance to estimate leaf chlorophyll content of tea with shading treatments. *Biosyst. Eng.* **2018**, *175*, 168–182. [CrossRef]
18. Li, L.T.; Ren, T.; Ma, Y.; Wei, Q.Q.; Wang, S.Q.; Li, X.K.; Cong, R.H.; Liu, S.S.; Lu, J.W. Evaluating chlorophyll density in winter oilseed rape (*Brassica napus* L.) using canopy hyperspectral red-edge parameters. *Comput. Electron. Agric.* **2016**, *126*, 21–31. [CrossRef]

19. Sun, H.; Liu, N.; Xing, Z.; Zhang, Z.; Li, M.; Wu, J. Parameter Optimization of Potato Spectral Response Characteristics and Growth Stage identification. *Spectrosc. Spectr. Anal.* **2019**, *39*, 1870–1877.

20. Zheng, T.; Liu, N.; Wu, L.; Li, M.Z.; Sun, H.; Zhang, Q.; Wu, J.Z. Estimation of Chlorophyll Content in Potato Leaves Based on Spectral Red Edge Position. *IFAC Pap. OnLine* **2018**, *51*, 602–606. [CrossRef]

21. Liu, N.; Zhao, R.; Qiao, L.; Zhang, Y.; Li, M.; Sun, H.; Xing, Z.; Wang, X. Growth Stages Classification of Potato Crop Based on Analysis of Spectral Response and Variables Optimization. *Sensors* **2020**, *20*, 3995. [CrossRef] [PubMed]

22. Zhang, Y.; Guanter, L.; Berry, J.A.; Christiaan, T.; Yang, X.; Tang, J.; Zhang, F. Model-based analysis of the relationship between sun-induced chlorophyll fluorescence and gross primary production for remote sensing applications. *Remote Sens. Environ.* **2016**, *187*, 145–155. [CrossRef]

23. Thomas, C.A.; Binzel, R.P. Identifying meteorite source regions through near-Earth object spectroscopy. *Icarus* **2010**, *205*, 419–429. [CrossRef]

24. Liu, N.; Wu, L.; Chen, L.; Sun, H.; Dong, Q.; Wu, J. Spectral Characteristics Analysis and Water Content Detection of Potato Plants Leaves. *IFAC Pap. OnLine* **2018**, *51*, 541–546. [CrossRef]

25. Liu, N.; Liu, G.; Sun, H. Real-Time Detection on SPAD Value of Potato Plant Using an In-Field Spectral Imaging Sensor System. *Sensors* **2020**, *20*, 3430. [CrossRef] [PubMed]

26. Mridha, N.; Sahoo, R.N.; Sehgal, V.K.; Krishna, G.; Pargal, S.; Pradhan, S.; Gupta, V.K.; Kumar, D.N. Comparative Evaluation of Inversion Approaches of the Radiative Transfer Model for Estimation of Crop Biophysical Parameters. *Int. Agrophys.* **2015**, *29*, 201–212. [CrossRef]

27. Botha, E.J.; Leblon, B.; Zebarth, B.J.; Watmough, J. Non-destructive estimation of wheat leaf chlorophyll content from hyperspectral measurements through analytical model inversion. *Int. J. Remote Sens.* **2010**, *31*, 1679–1697. [CrossRef]

28. Lunagaria, M.M.; Patel, H.R. Evaluation of PROSAIL inversion for retrieval of chlorophyll, leaf dry matter, leaf angle, and leaf area index of wheat using spectrodirectional measurements. *Int. J. Remote Sens.* **2019**, *40*, 8125–8145. [CrossRef]

29. Zhang, T.; Yu, L.; Yi, J.; Nie, Y.; Zhou, Y. Determination of Soil Organic Matter Content Based on Hyperspectral Wavelet Energy Features. *Spectrosc. Spectr. Anal.* **2019**, *39*, 3217–3222.

30. Zhang, R.; Li, Z.F.; Pan, J.J. Coupling discrete wavelet packet transformation and local correlation maximization improving prediction accuracy of soil organic carbon based on hyperspectral reflectance. *Trans. Chin. Soc. Agric. Eng.* **2017**, *33*, 175–181.

31. Chen, H.Y.; Zhao, G.X.; Li, X.C.; Lu, W.L.; Sui, L. Application of Wavelet Analysis for Estimation of Soil Available Potassium Content with Hyperspectral Reflectance. *Sci. Agric. Sin.* **2012**, *45*, 1425–1431.

32. Pinto, L.A.; Galvão, R.K.H.; Araújo, M.C.U. Influence of wavelet transform settings on NIR and MIR spectrometric analyses of diesel, gasoline, corn and wheat. *J. Braz. Chem. Soc.* **2011**, *22*, 179–186. [CrossRef]

33. Yu, Q.H.; Yang, G.J.; Wang, C.C. Chlorophyll inversion of winter wheat based on ground hyperspectral data and PROSAIL model. *Sci. Surv. Mapp.* **2019**, *44*, 96–102, 136.

34. Yao, X.; Si, H.Y.; Cheng, T.; Jia, M.; Chen, Q.; Tian, Y.C.; Zhu, Y.; Cao, W.X.; Chen, C.Y.; Cai, J.Y.; et al. Hyperspectral Estimation of Canopy Leaf Biomass Phenotype per Ground Area Using a Continuous Wavelet Analysis in Wheat. *Front. Plant Sci.* **2018**, *9*, 1360. [CrossRef] [PubMed]

35. He, R.Y.; Li, H.; Qiao, X.J.; Jiang, J.B. Using wavelet analysis of hyperspectral remote-sensing data to estimate canopy chlorophyll content of winter wheat under stripe rust stress. *Int. J. Remote Sens.* **2018**, *39*, 4059–4076. [CrossRef]

36. Huang, Y.; Tian, Q.J.; Wang, L.; Geng, J.; Lyu, C.G. Estimating canopy leaf area index in the late stages of wheat growth using continuous wavelet transform. *J. Appl. Remote Sens.* **2014**, *8*, 83517. [CrossRef]

37. Wang, Z.L.; Chen, J.X.; Fan, Y.F.; Cheng, Y.J.; Wu, X.L.; Zhang, J.W.; Wang, B.B.; Wang, X.C.; Yong, T.W.; Liu, W.G.; et al. Evaluating photosynthetic pigment contents of maize using UVE-PLS based on continuous wavelet transform. *Comput. Electron. Agric.* **2020**, *169*, 105160. [CrossRef]

38. Chen, P.F.; Haboudane, D.; Tremblay, N.; Wang, J.H.; Vigneault, P.; Li, B.G. New spectral indicator assessing the efficiency of crop nitrogen treatment in corn and wheat. *Remote Sens. Environ.* **2010**, *114*, 1987–1997. [CrossRef]

39. Li, D.; Cheng, T.; Zhou, K.; Zheng, H.B.; Yao, X.; Tian, Y.C.; Zhu, Y.; Cao, W.X. WREP: A wavelet-based technique for extracting the red edge position from reflectance spectra for estimating leaf and canopy chlorophyll contents of cereal crops. *ISPRS J. Photogramm. Remote Sens.* **2017**, *129*, 103–117. [CrossRef]

40. Malvern Panalytical. Available online: https://www.malvernpanalytical.com (accessed on 29 June 2020).

41. Ding, Y.J.; Zhang, J.J.; Sun, H.; Li, X.H. Sensitive Bands Extraction and Prediction Model of Tomato Chlorophyll in Glass Greenhouse. *Spectrosc. Spectr. Anal.* **2017**, *37*, 194–199.

42. Savitzky, A.; Golay, M.J.E. Smoothing and Differentiation of Data by Simplified Least Squares Procedures. *Anal. Chem.* **1964**, *36*, 1627–1639. [CrossRef]

43. Fearn, T.; Riccioli, C.; Garrido-Varo, A.; Guerrero-Ginel, J.E. On the geometry of SNV and MSC. *Chemom. Intell. Lab. Syst.* **2009**, *96*, 22–26. [CrossRef]

44. Galvão, R.K.H.; Araujo, M.C.U.; José, G.E.; Pontes, M.J.C.; Silva, E.C.; Saldanha, T.C.B. A method for calibration and validation subset partitioning. *Talanta* **2005**, *67*, 736–740. [CrossRef] [PubMed]

45. Toebe, M.; Filho, A.C. Multicollinearity in path analysis of maize (*Zea mays* L.). *J. Cereal Sci.* **2013**, *57*, 453–462. [CrossRef]

46. Guo, J.Q.; Li, Y.; Wang, H.S.; Zhou, H.F. Correlation coefficient extreme method for analyzing the 2-D data of water quality experiment of river stream. *J. Hydroelectr. Eng.* **2010**, *29*, 102–106.

47. Cheng, T.; Rivard, B.; Sánchez-Azofeifa, A. Spectroscopic determination of leaf water content using continuous wavelet analysis. *Remote Sens. Environ.* **2011**, *115*, 659–670. [CrossRef]

48. Geladi, P.; Kowalski, B.R. Partial least-squares regression: A tutorial. *Anal. Chim. Acta* **1986**, *185*, 1–17. [CrossRef]

49. Fan, X.W.; Liu, Y.B.; Tao, J.M.; Weng, Y.L. Soil Salinity Retrieval from Advanced Multi-Spectral Sensor with Partial Least Square Regression. *Remote Sens.* **2015**, *24*, 488–511. [CrossRef]

50. Yue, J.B.; Feng, H.K.; Yang, G.J.; Li, Z.H. A Comparison of Regression Techniques for Estimation of Above-Ground Winter Wheat Biomass Using Near-Surface Spectroscopy. *Remote Sens.* **2018**, *10*, 66. [CrossRef]

51. Singh, S.K.; Houx, J.H.; Maw, M.J.W.; Fritschi, F.B. Assessment of growth, leaf N concentration and chlorophyll content of sweet sorghum using canopy reflectance *Field Crops Res.* **2017**, *209*, 47–57. [CrossRef]

52. Cherepanov, D.A.; Gostev, F.E.; Shelaev, I.V.; Aibush, A.V.; Semenov, A.Y.; Mamedov, M.D.; Shuvalov, V.A.; Nadtochenko, V.A. Visible and Near Infrared Absorption Spectrum of the Excited Singlet State of Chlorophyll a. *High Energy Chem.* **2020**, *54*, 145–147. [CrossRef]

53. Zeng, L.Z.; Wang, Y.Q.; Zhou, J. Spectral analysis on origination of the bands at 437 nm and 475.5 nm of chlorophyll fluorescence excitation spectrum in Arabidopsis chloroplasts. *Luminescence* **2016**, *31*, 769–774. [CrossRef] [PubMed]

54. Yendrek, C.R.; Tomaz, T.; Montes, C.M.; Cao, Y.Y.; Morse, A.M.; Brown, P.J.; McIntyre, L.M.; Leakey, A.D.B.; Ainsworth, E.A. High-Throughput Phenotyping of Maize Leaf Physiological and Biochemical Traits Using Hyperspectral Reflectance. *Plant Physiol.* **2016**, *173*, 614–626. [CrossRef] [PubMed]

55. Gholizadeh, H.; Robeson, S.M.; Rahman, A.F. Comparing the performance of multispectral vegetation indices and machine-learning algorithms for remote estimation of chlorophyll content: A case study in the Sundarbans mangrove forest. *Int. J. Remote Sens.* **2015**, *36*, 3114–3133. [CrossRef]

56. Hu, Q.S. Molecular Tracers Indicate Organic Aerosols in the Marine Boundary Layer and the History of Penguin Colonies. Ph.D. Thesis, University of Science and Technology of China, Hefei, China, May 2014.

57. Hebbar, K.B.; Subramanian, P.; Sheena, T.L.; Shwetha, K.; Sugatha, P.; Arivalagan, M.; Varaprasad, P.V. Chlorophyll and nitrogen determination in coconut using a non-destructive method. *J. Plant Nutr.* **2016**, *39*, 1610–1619. [CrossRef]

58. Shao, Q.; Yu, Z.Y.; Li, X.G.; Li, W. Agronomic traits investigation and IR spectrum analysis of a novel yellow leaf mutant in muskmelon (*Cucumis melo* L.). *J. Northeast Agric. Univ.* **2013**, *44*, 106–111.

59. Wang, H.F.; Huo, Z.G.; Zhou, G.S.; Liao, Q.H.; Feng, H.K.; Wu, L. Estimating leaf SPAD values of freeze-damaged winter wheat using continuous wavelet analysis. *Plant Physiol. Biochem.* **2016**, *98*, 39–45. [CrossRef]

60. Devadas, R.; Lamb, D.W.; Simpfendorfer, S.; Backhouse, D. Evaluating ten spectral vegetation indices for identifying rust infection in individual wheat leaves. *Precis. Agric.* **2009**, *10*, 459–470. [CrossRef]

61. Gitelson, A.A.; Gritz, Y.; Merzlyak, M.N. Relationships between leaf chlorophyll content and spectral reflectance and algorithms for non-destructive chlorophyll assessment in higher plant leaves. *J. Plant Physiol.* **2003**, *160*, 271–282. [CrossRef]

62. Li, X.Y.; Liu, G.S.; Yang, Y.F.; Zhao, C.H.; Yu, Q.W.; Song, S.X. Relationship between hyperspectral parameters and physiological and biochemical indexes of flue-cured tobacco leaves. *Agric. Sci. China* **2007**, *6*, 665–672. [CrossRef]
63. Liao, Q.H.; Wang, J.H.; Yang, G.J.; Zhang, D.Y.; Li, H.L.; Fu, Y.; Li, Z. Comparison of spectral indices and wavelet transform for estimating chlorophyll content of maize from hyperspectral reflectance. *J. Appl. Remote Sens.* **2013**, *7*, 073575. [CrossRef]
64. Sims, D.A.; Gamon, J.A. Estimation of vegetation water content and photosynthetic tissue area from spectral reflectance: A comparison of indices based on liquid water and chlorophyll absorption features. *Remote Sens. Environ.* **2003**, *84*, 526–537. [CrossRef]
65. Fitzgerald, G.; Rodriguez, D.; O'Leary, G. Measuring and predicting canopy nitrogen nutrition in wheat using a spectral index—The canopy chlorophyll content index (CCCI). *Field Crops Res.* **2010**, *116*, 318–324. [CrossRef]
66. Shi, J.Y.; Zou, X.B.; Zhao, J.W.; Wang, K.L.; Chen, Z.W.; Huang, X.W.; Zhang, D.T.; Holmes, M. Nondestructive diagnostics of nitrogen deficiency by cucumber leaf chlorophyll distribution map based on near infrared hyperspectral imaging. *Sci. Hortic.* **2012**, *138*, 190–197.
67. Szeles, A.V.; Megyes, A.; Nagy, J. Irrigation and nitrogen effects on the leaf chlorophyll content and grain yield of maize in different crop years. *Agric. Water Manag.* **2012**, *107*, 133–144. [CrossRef]
68. Padilla, F.M.; Pena-Fleitas, M.T.; Gallardo, M.; Thompson, R.B. Threshold values of canopy reflectance indices and chlorophyll meter readings for optimal nitrogen nutrition of tomato. *Ann. Appl. Biol.* **2015**, *166*, 271–285. [CrossRef]
69. Mavridou, E.; Vrochidou, E.; Papakostas, G.A.; Pachidis, T.; Kaburlasos, V.G. Machine Vision Systems in Precision Agriculture for Crop Farming. *J. Imaging* **2019**, *5*, 89. [CrossRef]

 *remote sensing*

Article

Nutrient Prediction for Tef (*Eragrostis tef*) Plant and Grain with Hyperspectral Data and Partial Least Squares Regression: Replicating Methods and Results across Environments

K. Colton Flynn [1,*,†], Amy E. Frazier [2] and Sintayehu Admas [3]

[1]   USDA-ARS, PA, Grassland Soil and Water Research Laboratory, 808 East Blackland Road, Temple, TX 76502, USA

[2]   Spatial Analysis Research Center (SPARC), School of Geographical Sciences and Urban Planning, Arizona State University, Tempe, AZ 85281, USA; amy.frazier@asu.edu

[3]   Ethiopian Biodiversity Institute, PO Box 30726 Addis Ababa, Ethiopia; sintayehu.admas@ebi.gov.et

*   Correspondence: colton.flynn@usda.gov

†   The U.S. Department of Agriculture (USDA) prohibits discrimination in all its programs and activities on the basis of race, color, national origin, age, disability, and where applicable, sex, marital status, familial status, parental status, religion, sexual orientation, genetic information, political beliefs, reprisal, or because all or part of an individual's income is derived from any public assistance program. (Not all prohibited bases apply to all programs.) Persons with disabilities who require alternative means for communication of program information (Braille, large print, audiotape, etc.) should contact USDA's TARGET Center at (202) 720-2600 (voice and TDD). To file a complaint of discrimination, write to USDA, Director, Office of Civil Rights, 1400 Independence Avenue, S.W., Washington, D.C. 20250-9410, or call (800) 795-3272 (voice) or (202) 720-6382 (TDD). USDA is an equal opportunity provider and employer.

Received: 29 July 2020; Accepted: 30 August 2020; Published: 4 September 2020

Abstract: Achieving reproducibility and replication (R&R) of scientific results is tantamount for science to progress, and it is also necessary for ensuring the self-correcting mechanism of the scientific method. Topics of R&R have sailed to the forefront of research agenda in many fields recently but have received less attention in remote sensing in general and specifically for studies utilizing hyperspectral data. Given the extremely local environments in which many hyperspectral studies are conducted (e.g., agricultural field plots), purposeful attention to the repeatability of findings across study locales can help ensure methods are generalizable. This study undertakes an investigation of the nutrient content of tef (*Eragrostis tef*), an understudied plant that is growing in importance due to both food and forage benefits, but does so within the context of the replicability of methods and findings across two study sites situated in different international and environmental contexts. The aims are to (1) determine whether calcium, magnesium, and protein of both the plant and grain can be predicted using hyperspectral data with partial least squares (PLS) regression with waveband selection, and (2) compare the replicability of models across differing environments. Results suggest the method can produce high nutrient prediction accuracy for both the plant and grain in individual environments, but selection of wavebands for nutrient prediction was not comparable across study areas. The findings suggest that the method must be calibrated in each location, thereby reducing the potential to extrapolate methods to different areas. Our findings highlight the need for greater attention to methods and results replication in remote sensing, specifically hyperspectral analyses, in order for scientific findings to be repeatable beyond the plot level.

Keywords: reproducibility; replicability; hyperspectral; waveband selection; partial least squares; Ethiopia; *Eragrostis tef*

---

1. Introduction

The reproducibility and replication (R&R) of scientific findings has recently moved to the forefront of research agenda in many fields [1–5] since it has been discovered that findings often cannot be reproduced or replicated [5,6]. While the two "R's"—reproducibility and replicability—are intertwined, there are key differences between their goals. Adopting the definitions from the National Science Foundation [7] and the National Academy of Science, Engineering, and Medicine [8], we define reproducibility as the ability of a researcher to duplicate the results of a prior study using the same data and methods as the original investigator. In short, if a researcher makes the data and methods/code available, another researcher should be able to produce the exact same results. In comparison, replicability is the ability of a researcher to duplicate results using similar methods but with new data.

Achieving R&R is critical for advancing scientific discoveries, yet neither topic has received much attention in geography and the spatial sciences, where investigations tend to be observational instead of experimental or theoretical [9]. R&R has received even less attention in remote sensing (but see [10] for an early take), even though the field is uniquely positioned to contribute to R&R on several fronts. First, there is a rich archive of publicly available remote sensing datasets (e.g., Landsat), supporting opportunities for reproducibility [9,11]. Second, remote sensing studies, and in particular hyperspectral studies, are often situated in extremely local contexts (e.g., agricultural plots) due to the need for ground reference data and the high labor and time costs of operating equipment. Yet, an implicit goal of science is to develop widely applicable methodologies and generalizable findings that can be applied in different contexts. Thus, working toward the replicability of methods and findings across different study areas is important for advancing remote sensing science.

Despite the myriad opportunities for remote sensing scientists to explore R&R issues, very few formal efforts have been documented. One reason is likely because remote sensing scientists often work with large datasets and perform complex spectral and spatial manipulations [12–16], which makes R&R difficult if processing code is not made available. Until recently, many scientific publications did not require code to be submitted as part of the manuscript review process, although this is changing. Replication in remote sensing is also hindered by attributes of local environments, which makes the transfer of results from one landscape to another difficult. However, if we are to develop methodologies that are transferrable across space, it is necessary to begin developing and implementing protocols for testing the R&R of remote sensing studies. One way to do this is to incorporate multi-field, multi-environment analyses into studies to self-test the replicability of methods and results.

Precision agriculture is one field where immediate gains can be made toward testing the replicability of methods while also contributing a larger understanding of the extent of R&R issues in remote sensing. Since the overall goal of precision agriculture is to decrease the ambiguity of decisions required on agricultural lands that are often highly variable [17], the ability to transfer methods and findings from one environment or location to another requires them to be replicable [18]. However, most studies capture data in a single region or location (often in a single crop field) under uniform conditions [12,15], thus limiting their generalizability across environmental or geographical contexts. Furthermore, the implicit assumption is that methods and findings are extendable beyond the single field in which they were tested, but in most cases, no such evidence is provided. Many studies lack basic explanation for environmental variances such as soil, hydrology, and topography that can cause reflectance variations, thereby altering results across space [16]. Ultimately, remote sensing methodologies are of little practical value for precision agriculture if they are developed, tested, and applicable in a single location where these multiple and often confounding factors are held constant.

Partial least squares (PLS) regression has become an accepted technique in vegetation studies using hyperspectral data for estimating a range of biophysical and biochemical properties [19–23]. In situations where the number of independent variables is large and the variables are collinear, which is common with hyperspectral data, multiple linear regression will often overfit the model [24,25]. PLS regression standardizes model construction from the preprocessed hyperspectral data via latent variables, from which the predictive capabilities of the model can be tested. Recently, variations of PLS regression using a waveband selection procedure [13] have been proposed and adopted, but there has been little effort to test the replicability of these methods across environments to determine whether results might be transferable.

The objective of this study is to investigate the replicability of PLS regression methods, including PLS with waveband selection, for predicting nutrient content in plant and grain material across multiple environments. This study addresses gaps in the remote sensing R&R literature by replicating a methodological workflow using hyperspectral data and PLS regression for predicting nutrients in a single crop but in two varying environments in different international contexts to determine the degree to which the methods are replicable. The focus is on *Eragrostis tef* (tef), a cereal crop primarily grown in Ethiopia, although production has been expanding to other parts of the world due to its versatility and resistance to drought. Tef is ideal for studying replicability because it is grown in different international contexts and is known for being successfully cultivated across differing environments. Additionally, very few hyperspectral analyses have been performed on non-milled grains [26], so this study contributes knowledge in that realm as well.

Tef is a grass (Family: Poaceae) that has received little attention from the remote sensing and precision agricultural communities despite its versatile cultivation characteristics. Tef is thought to be one of the earliest domesticated plants [27], with the center of origin and diversity in Ethiopia/Horn of Africa [28]. It is drought and heat resistant, has a high nutrient content, and is grown for animal feed as well as a staple food crop [29]. While tef can be cultivated across many environments, it is primarily grown in Ethiopia, where it is the most commonly harvested crop, popular for its highly nutritious, gluten-free grain [30–35]. Recently, cultivation has been spreading outside the region; in the United States, tef is planted as a sequential forage crop for livestock feed but is currently only grown in a handful of locations [29,36].

2. Data Collection and Processing

2.1. Study Sites

This study focuses on four sites in two countries. The two US sites (US1: 21.45 ha; US2: 24.19 ha) are located in Hydro, Oklahoma, which is part of the Central Great Plains ecoregion. The region experiences cold winters (average temperature minimums from 4 to −12 °C) and hot summers (temperatures greater than 38 °C). Precipitation is variable, and temperature changes can be considerable across all seasons. The US sites are located within three kilometers of each other, so the environmental characteristics are similar. Both sites have similar soils (vertisols) and are located at the same elevation (474 m). The two Ethiopia sites (ET1: 0.77 ha; ET2: 1.23 ha) are also in close geographic proximity (Figure 1). The first site is located in the warm, sub-moist lowlands, while the second site is in the warm, humid lowlands. Soil composition at both sites is similar (vertisols), but the sites are at different elevations (1919 m and 2201 m, respectively).

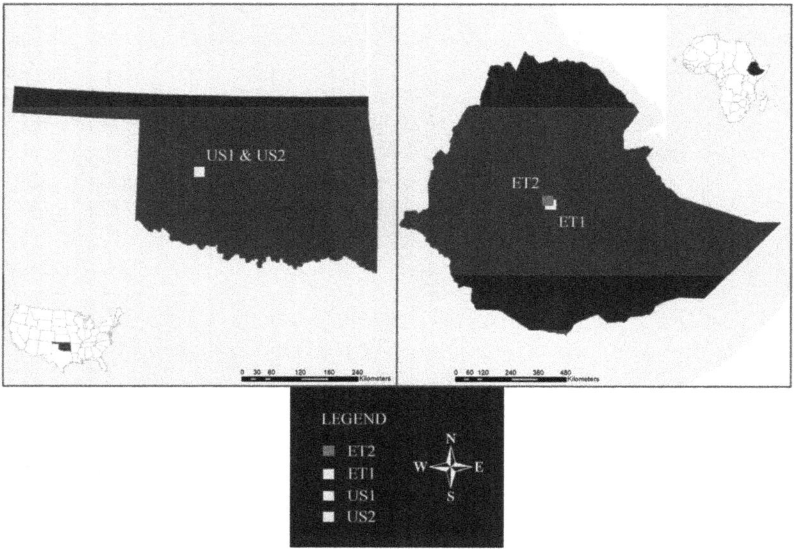

Figure 1. Locations of the four study sites.

2.2. Spectral Data Collection and Nutrient Processing

Data collection and processing methods included in situ spectral data collection and ex situ laboratory nutrient analyses. The methodology proceeded in four general phases (Figure 2), which are briefly described here and detailed below. First, canopy spectral measurements and plant samples were collected at the four sites immediately prior to peak crop maturity (seed head stage). Next, plant samples were dried and separated in the laboratory where the grain was imaged and both the plant material and grain were subjected to nutrient analyses. Third, the relationship between the nutrients and spectral components were modeled using PLS regression with both the full spectrum and a waveband selection method. Lastly, the replicability of the PLS regression methods were tested through statistical comparisons of model fits and similarity of results.

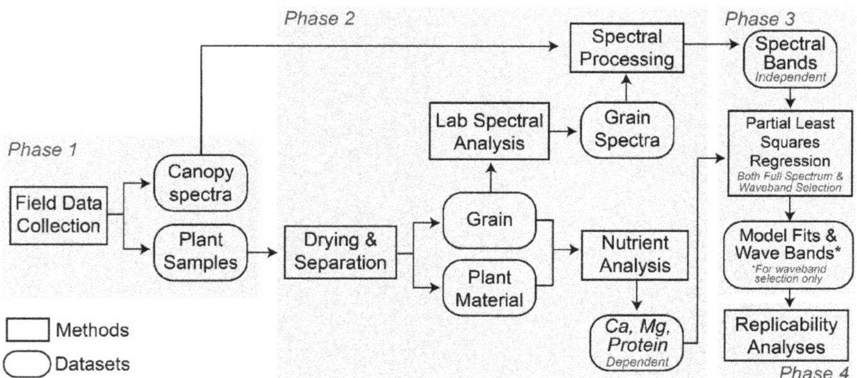

Figure 2. Methods flowchart detailing data collection and processing methods.

In Phase 1, canopy-level hyperspectral measurements and plant material were collected at each site immediately prior to harvest during peak maturity, which was late June/early July 2016 in the

US and October 2017 in Ethiopia. For each of the four sites, 40 random points were generated in ArcGIS, providing 80 possible sampling location in each country. These points were then located in the field using a hand-held GPS unit (Trimble Juno 3B, Corvallis, OR, USA). In some cases, locations could not be accessed due to sampling on the days of harvest and attempting to not interfere with the harvesting practices of the respective farmers, so the actual number of samples is below 40 for each field and below 80 for the region. At all sampling sites, canopy spectral data were captured using the same spectroradiometer (FieldSpec Pro FR: Analytical Spectral Devices [ASD], Boulder, CO, USA), measuring reflectance from 350–2500 nm with a spectral sampling width of 1.4 nm from 350–1000 nm and 2.0 nm from 1000–2500 nm. The spectroradiometer was calibrated using a Spectralon diffuse reference panel (Analytical Spectral Devices [ASD], Boulder, CO, USA) approximately every 15 min. The spectralon was held at a distance from the spectroradiometer fiber to ensure no shadows were measured. The spectroradiometer fiber was held 1.2 m above the ground; since the top of canopy was about 0.3 m above ground, the cone of acceptance was 25°, resulting in a footprint with a diameter of 0.40 m for each sample. This diameter ensured a sufficient mass of plant/grain matter was collected for nutrient testing (10 g of grain; SSSA, 1990; [37]). Reflectance data were captured between 11:00 a.m. and 2:00 p.m. local time under a cloud-free sky. It is important to mention that the same spectroradiometer was operated by the same individual in all locations to ensure the exact same data collection process was reproduced. Each location was imaged five times in succession, and spectra were averaged. Plant material in the footprint of the imaging fiber was then clipped at the base, stored in plastic bags, and placed on ice in a cooler for transport back to the laboratory.

In the laboratory, samples were dried to remove excess moisture, and the grains were separated from the plant using a traditional method of hand threshing with the assistance of a basket weaved surface. The grains from each sample, which measure approximately 1 mm in length, were aggregated in a petri dish to generate a sufficient amount to fully cover the lens of the imaging equipment. The grains were spectrally imaged in a dark room using a contact probe (Contact Probe: Analytical Spectral Devices [ASD], Boulder, CO, USA) with a halogen light source, while not equivalent to the sun, still emitted spectral wavelengths (350–2500 nm) capable of being identified using the same spectroradiometer used in the field (Figure 3). As with the in situ samples, five spectral readings were collected for each sample, and the values averaged.

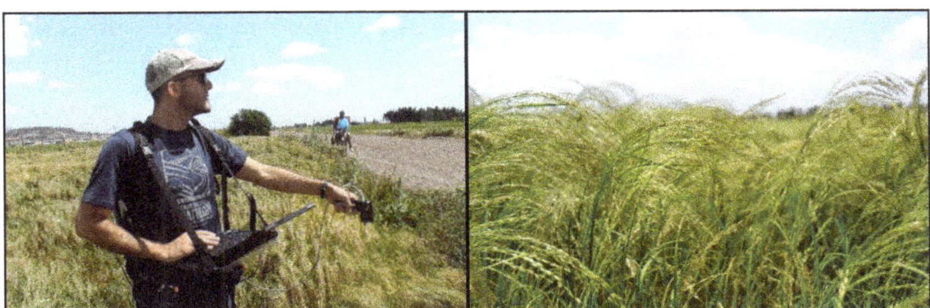

Figure 3. Use of spectroradiometer in the field (**left**) and a close-up of *Eragrostis tef* (tef; **right**).

2.3. Spectral Processing

The raw spectral curves from both the plant (in situ) and grain (ex situ) were smoothed using a Savitsky–Golay filter [38] to reduce noise (SG; Figure 4a). A third-order, 10-band moving polynomial was fitted upon the original spectral signatures [39]. Data within each 5 nm window were averaged (e.g., the value at 600 nm is average of 598–602 nm). First derivatives (hereafter, FD) were computed from the smoothed spectra (Figure 4b). Computing derivatives allows minor differences in reflectance values to be exploited and permits discrimination of key points along the spectral curves (i.e., inflections and maxima) corresponding to biophysical and biochemical components that would otherwise be difficult to detect [40,41]. Lastly, wavebands associated with atmospheric noise (1290–1495, 1705–2045, and 2355–2500 nm) and splicing points within the spectroradiometer (350–395 and 1005–1015 nm) were removed, resulting in 277 wavebands between 400–2350 nm, at a spectral resolution of 5 nm. The reflectance/derivative values at these wavebands ultimately serve as the independent variables for the statistical analyses discussed below.

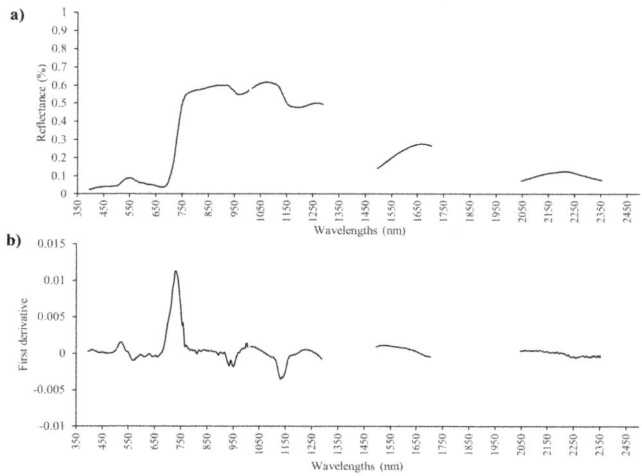

Figure 4. Typical canopy spectra (noisy atmospheric windows removed) for tef (*Eragrostis tef*) showing (**a**) Savitsky–Golay filtered reflectance curve and (**b**) first derivative (FD) transformation.

2.4. Nutrient Analysis

Since tef serves as both a food (grain) and forage (plant) crop, nutrient analyses were performed on both the plant and grain. All samples were analyzed for calcium (Ca), magnesium (Mg), and protein content (Table 1). For details on Ca and Mg laboratory procedures, readers are directed to the Soil Science Society of America [42] plant analysis guidelines. For details on protein analysis, readers are directed to the National Forage Testing Association's (NFTA) Forage Analysis Procedures [37]. For the US samples, nutrient analyses were performed at the Oklahoma State University Soil, Water, and Forage Analytical Laboratory. For the ET samples, analyses were performed by the Ethiopian Public Health Institute of Addis Ababa. These labs were chosen for their rigor and location as samples could not be transported across country boundaries. Moreover, the procedures used by both institutions followed these established standards [37,42], with aim to minimize impact on results. Ca and Mg values are expressed in ppm mg/kg, while protein values are expressed in percent (%) of total sample weight. All are expressed as dry matter weight. These nutrient data from the plant material and grains ultimately serve as the dependent variable in the PLS regression analyses (discussed below).

Table 1. Number of samples of each component collected in each study location.

Component	Nutrient	Number of Samples (*n*)	
		United States	**Ethiopia**
	Ca	67	78
Plant	Mg	67	79
	Protein	67	79
	Ca	66	78
Grain	Mg	66	79
	Protein	65	79

3. Analytical Methods

Partial least squares (PLS) regression was implemented to assess the relationship between reflectance (independent variable) and nutrient content (dependent variable) of both the plant and grain. PLS, which can also stand for projection to latent structures [43], was selected over other forms of regression because it accounts for overfitting errors common when analyzing hyperspectral data [13,44]. Briefly, PLS regression finds a set of components (called latent factors) from X, a matrix of predictors collected on the observations, that best predict Y, a matrix of dependent observations [43]. These latent factors, or latent vectors, are orthogonal, and thus explain as much of the covariance between X and Y as possible, often resulting in a smaller number of variables than principal component regression. PLS regression extracts X-scores from the latent variables to construct a model to predict the Y-scores. In PLS, the X- and Y-scores are subject to redundancy analysis that seeks directionality in factor space until the most accurate prediction is found [25,45]. When implementing PLS regression with hyperspectral data, it is important to ensure the number of latent variables does not far exceed the number of independent variables being used, as overfitting can occur [13].

3.1. Specification of the PLS Regression Model Using the Full Spectrum

The PLS equation follows a standard regression (Equation (1)):

$$\hat{y} = \beta_1 x_1 + \beta_2 x_2 + \cdots + \beta_k x_k + \varepsilon \tag{1}$$

where the response variable $\hat{y}$ is the nutrient value, and the predictor variables x_1 to x_k are the reflectance (SG) or derivative (FD) values for bands 1 to k (here, 277). β_1 to β_k are the estimated weighted regression coefficients computed directly from the PLS loadings corresponding to the model with the optimal number of latent variables, and ε is the error vector. The optimal number of latent variables (NLV) is determined through a leave-one-out (LOO) cross-validation and assessed through the minimum root mean square error:

$$RMSE_{CV} = \sqrt{\frac{\sum_{i=1}^{n}\left(\hat{y}_i^c - y_i^c\right)^2}{n}} \tag{2}$$

where $\hat{y}_i^c$ represents the predicted response when the model is built without sample i, y_i^c represents the measured nutrient value for sample i, and n represents the number of samples used in the calibration [46]. Twelve, standard PLS regressions were performed, each corresponding to a dataset in Table 1, and each including the full set of 277 bands (PLS-Full).

3.2. PLS Regression with Waveband Selection

A modified form of PLS regression, known as the waveband selection (or iterative stepwise elimination) method, was developed to eliminate noisy and unhelpful predictors in hyperspectral studies [13,44]. Instead of including all 277 wavebands, the number is reduced iteratively [44] by dropping the least important wavebands (similar to stepwise linear regression). Waveband importance (v_k) is determined as:

$$v_k = \frac{|\beta_k| s_k}{\sum_{k=1}^{K} |\beta_k| s_k} \tag{3}$$

where β_k and s_k are the regression coefficient and standard deviation corresponding to waveband k. The selection begins with all 277 wavebands, and the waveband contributing least to the model (lowest v_k) is removed. The model (Equation (1)) is then re-run with 276 variables, and so on, until the maximum predictive capability is achieved [47]. A representation of the iterative processes to determine the maximum predictive capability is shown in Figure 5. This version of the model is hereafter referred to as PLS-Wave.

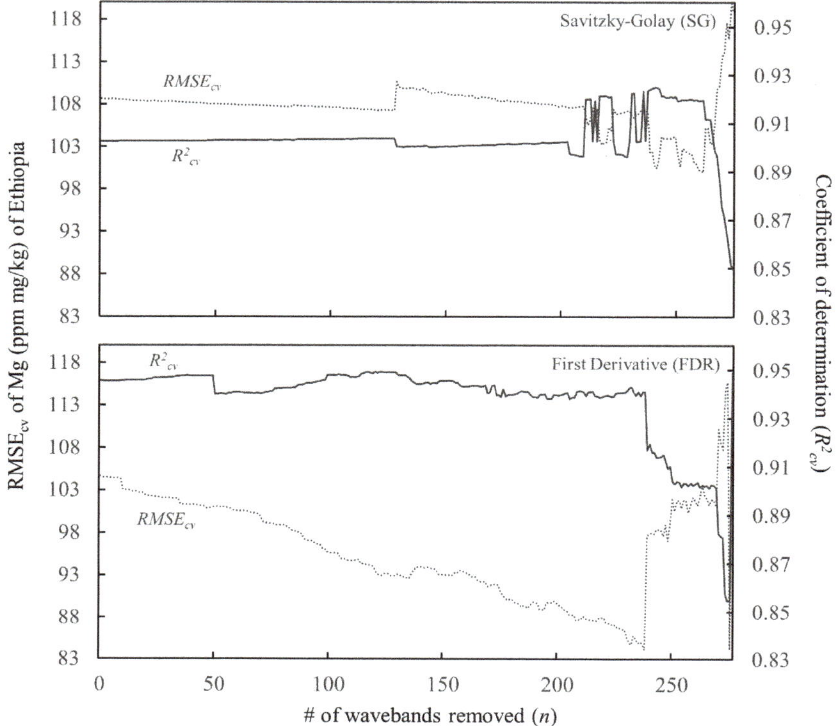

Figure 5. Representation of the iterative process to determine the number of wavebands via the waveband selection process. The number of wavebands is determined by the lowest RMSEcv value (dotted line), which in turn determines the coefficient of determination (solid line) using the given number of wavebands in the partial least square regression (PLS). Top figure is for the Savitsky–Golay filtered data. The bottom figure is for the first derivative.

3.3. Predictive Ability of PLS Regression

To test the predictive capabilities of the PLS-Full and PLS-Wave models, we implemented a bootstrapping procedure by dividing the data into calibration (65–75%) and validation (25–35%) sets replacing the data of these sets $n = 1000$ times (following [48,49]). After each separation, the models were calibrated (assessed through $RMSE_{CV}$) and then validated using root mean square error of prediction ($RMSE_P$):

$$RMSE_p = \sqrt{\frac{\sum_{i=1}^{n}\left(\hat{y}_i^v - y_i^v\right)^2}{n}} \qquad (4)$$

where $\hat{y}_i^v$ represents the predicted nutrient values, y_i^v represents the measured nutrient values, and n represents the number of samples in the validation subset [46]. Mean coefficient of determination (R^2), R^2 standard deviation (R^2 std.), and $RMSE_P$ standard deviation ($RMSE_P$ std) are also reported for the validation. PLS regressions were performed in Matlab v2016a (MathWorks, Sherborn, MA, USA).

3.4. Replication across Environments

To assess the replicability of the PLS-Full and PLS-Wave regression methodologies for predicting nutrient content across environments, we compared model performance between the US and ET using a difference of means (*t*-test) for each component–nutrient combination (e.g., grain–calcium, plant–calcium, etc.) between the two sites from the bootstrapped results. To compare the similarity of wavebands selected by the PLS-Wave model, the Jaccard index [50] was used to measure the overlap in selected wavebands compared to the total number of wavebands selected for each site (Equation (5)):

$$J(A, B) = |A \cap B|/|A \cup B| \tag{5}$$

where *A* and *B* are the set of selected wavebands in the two locations, respectively. A Jaccard value of 1.0 indicates that the models for the two locations overlap completely in terms of the wavebands selected as important for prediction; 0 indicates the two locations share none of the same wavebands.

4. Results

4.1. Plant and Grain Nutrient Analyses

Plant nutrient values for Ca, Mg, and protein were considerably higher for the US compared to ET (Table 2). Mean Ca was five to six times higher; mean Mg almost four times higher; and mean protein nearly 2.5 times higher in the US. Results were similar for grain nutrients (Table 2), with values in the US typically two to four times higher than ET. Additionally, the standard deviation and ranges for Ca and protein at both the plant and grain level were higher in the US, except for Mg, which showed similar standard deviation and ranges across both locations for plant and grain.

Table 2. Descriptive statistics for the calcium (Ca), magnesium (Mg), and protein measured for the plant and grain samples from each location.

Nutrient		Descriptive Statistics					
United States (US)		*n*	**Min.**	**Mean**	**Max.**	**Range**	**SD**
	Ca (ppm mg/kg)	67	3760	6651	9360	5600	1371
Plant	Mg (ppm mg/kg)	67	1860	2753	3620	1760	327
	Protein (%)	67	5.74	15.68	23.52	17.78	5.42
	Ca (ppm mg/kg)	66	1620	2267	3240	1620	465
Grain	Mg (ppm mg/kg)	66	1750	2015	2510	760	186
	Protein (%)	65	12.13	17.59	34.42	22.29	4.71
Ethiopia (ET)							
	Ca (ppm mg/kg)	78	437	1223	1772	1335	195
Plant	Mg (ppm mg/kg)	79	115	740	1181	1066	297
	Protein (%)	79	3.02	5.77	9.93	6.91	1.69
	Ca (ppm mg/kg)	78	716	1283	2128	1411	460
Grain	Mg (ppm mg/kg)	79	270	553	841	571	155
	Protein (%)	79	8.16	10.84	14.57	6.41	1.43

4.2. Model Performance

4.2.1. Results for the PLS-Full Model

At the plant level (Table 3, top), the PLS-Full models varied considerably in performance between the two sites with cross-validated calibration coefficients of determination (R^2_{CV}) ranging from 0.03 (Mg US) to 0.88 (Mg ET) at the plant level. The validation coefficients of determinations (Mean R^2) closely matched the calibration values, except in the case of ET (Ca and Protein (SG)) where the validation R^2 was approximately double that of the calibration. Interestingly, when comparing the US and ET, one generally outperformed the other, but the relationship changed depending on the nutrient. Differences in filtering (SG vs. FD) were largely negligible, with the exception of protein in ET, where the model fit for the FD was about twice that for the SG data (0.41 vs. 0.18). The bootstrapped R^2 values were normally or semi-normally distributed in all cases except for Mg in the US and Ca in ET (Figure S1, Supplementary Material). Importantly, *t*-test comparisons for replicability indicate the PLS-Full regression model did not replicate well across the two environments, with significant differences in fit for every nutrient at the plant level (Table 3, top).

Table 3. Partial least squares (PLS) regression and t-test results for the full spectrum model (PLS-Full) for plant and grain. SG: Savitsky–Golay; FD: First derivative; NLV: Number of latent variables; std: Standard deviation.

	Loc.	Calibration			Validation				Comparison
		NLV	$R^2{}_{CV}$	$RMSE_{CV}$	Mean R^2	R^2 Std	Mean $RMSE_P$	$RMSE_P$ Std	t-test
Plant									
Calcium									
SG	US	2	0.53	935.79	0.52	0.12	976.86	168.52	−58.75 ***
	ET	3	0.10	229.07	0.18	0.14	221.14	42.02	
FD	US	2	0.51	954.17	0.50	0.12	1000.47	156.54	−56.12 ***
	ET	1	0.09	231.48	0.20	0.17	233.25	82.71	
Magnesium									
SG	US	3	0.03	327.93	0.06	0.08	345.49	65.33	276.75 ***
	ET	7	0.87	108.63	0.86	0.04	117.70	16.45	
FD	US	3	0.03	334.21	0.05	0.07	359.43	61.58	326.52 ***
	ET	7	0.88	104.66	0.86	0.04	117.47	16.25	
Protein									
SG	US	9	0.86	2.04	0.82	0.06	2.40	0.37	54.64 ***
	ET	5	0.18	1.34	0.38	0.12	1.42	0.16	
FD	US	4	0.84	2.13	0.81	0.04	2.38	0.29	481.18 ***
	ET	2	0.41	1.30	0.44	0.12	1.35	0.15	
Grain									
Calcium									
SG	US	7	0.88	159.89	0.87	0.04	176.55	28.41	−50.57 ***
	ET	14	0.78	216.06	0.74	0.07	253.12	31.16	
FD	US	2	0.88	163.19	0.86	0.04	183.84	28.04	−44.96 ***
	ET	8	0.80	205.09	0.78	0.05	234.28	25.71	
Magnesium									
SG	US	6	0.71	99.49	0.70	0.10	108.77	18.81	219.53 ***
	ET	15	0.71	84.09	0.65	0.10	99.31	15.57	
FD	US	3	0.73	96.49	0.72	0.10	104.89	17.96	144.8 ***
	ET	9	0.74	79.67	0.71	0.08	90.55	12.24	
Protein									
SG	US	15	0.93	1.23	0.90	0.05	1.59	0.51	−115.03 ***
	ET	5	0.47	1.03	0.49	0.14	1.09	0.18	
FD	US	5	0.91	1.43	0.89	0.04	1.65	0.47	−106.51 ***
	ET	7	0.64	0.86	0.62	0.11	0.95	0.15	

*** t-test significant at 0.001, ET served as first group.

At the grain level (Table 3, bottom), results were more consistent across the study sites and nutrients. This is believed to be the case in-part to grain spectral values measured in a controlled setting, while plants were measured in a field setting. Calibration R^2 ranged [0.47, 0.93], and validation R^2 ranged [0.49, 0.90] (Table 3). The results for ET protein using the SG data were a slight outlier though, and when considering only the FD values, the minimums increase to 0.64 and 0.62, respectively. The differences between the SG and FD filters were again negligible, except in the Mg ET case. The t-test comparisons for grain indicate the PLS-Full regression method did not replicate well across the environments.

4.2.2. Results for the PLS-Wave Model

At the plant level, the PLS-Wave models varied in performance in both environments with calibration R^2 ranging [0.11, 0.94] (Table 4, top). Differences between the SG and FD were negligible in most cases. Fits for Mg in ET and Protein in the US were high (0.88–0.94), but there was little consistency across locations. *T*-test comparisons show significant differences in model performance for all nutrients/processing methods between the two locations indicating the PLS-Wave regression model did not replicate well in terms of predicting plant nutrients across these two environments.

Table 4. Partial least squares (PLS) regression and *t*-test results for the waveband selection model (PLS-Wave) for plant and grain. SG: Savitsky–Golay; FD: First derivative; NLV: Number of latent variables; std: Standard deviation.

	Loc.	Calibration				Validation				Compare
		NLV	R^2_{CV}	$RMSE_{CV}$	# of Waves	Mean R^2	R^2 Std	Mean $RMSE_P$	$RMSE_P$ Std	*t*-test
					Plant					
					Calcium					
SG	US	3	0.56	899.24	9	0.55	0.10	954.24	113.58	−52.36 ***
	ET	2	0.12	226.93	49	0.23	0.17	221.46	74.21	
FD	US	1	0.53	929.65	9	0.55	0.10	937.88	125.67	−39.102 ***
	ET	1	0.11	228.37	2	0.25	0.19	219.72	75.70	
					Magnesium					
SG	US	2	0.08	312.41	121	0.10	0.11	323.94	56.66	−11.34 ***
	ET	10	0.92	100.04	15	0.88	0.03	109.74	14.80	
FD	US	4	0.30	277.97	50	0.28	0.14	296.41	48.51	−2.33 *
	ET	5	0.94	84.14	38	0.92	0.02	88.49	12.98	
					Protein					
SG	US	6	0.88	1.84	46	0.88	0.04	1.93	0.29	−85.95 ***
	ET	3	0.45	1.25	31	0.46	0.11	1.32	0.12	
FD	US	5	0.92	1.53	46	0.90	0.04	1.78	0.28	−72.48 ***
	ET	1	0.42	1.28	166	0.47	0.12	1.31	0.16	
					Grain					
					Calcium					
SG	US	8	0.90	149.80	14	0.88	0.03	169.35	25.42	−40.05 ***
	ET	7	0.81	200.20	43	0.79	0.06	223.68	28.61	
FD	US	2	0.91	142.23	47	0.89	0.04	156.88	25.36	17.45 ***
	ET	12	0.93	122.84	44	0.92	0.02	142.35	20.65	
					Magnesium					
SG	US	4	0.73	96.74	77	0.73	0.10	102.28	22.78	−0.7781
	ET	12	0.73	80.54	15	0.73	0.07	87.75	11.56	
FD	US	4	0.78	86.09	7	0.78	0.06	90.85	11.74	31.02 ***
	ET	8	0.87	56.29	33	0.85	0.04	63.38	9.02	
					Protein					
SG	US	12	0.95	1.00	74	0.94	0.03	1.17	0.29	−94.58 ***
	ET	6	0.52	0.99	23	0.53	0.14	1.05	0.17	
FD	US	4	0.93	1.24	33	0.92	0.03	1.43	0.33	−76.55 ***
	ET	7	0.67	0.82	159	0.65	0.11	0.91	0.14	

* *t*-test significant at 0.05, ** at 0.01, *** at 0.001.

At the grain level, the PLS-Wave models performed generally well for all nutrients across the two locations, with calibration R^2 ranging [0.52, 0.95] (Table 4, bottom). Fits for protein were lower for ET than US, but fits for Ca and Mg were similar. Bootstrapping resulted in normal and semi-normal distributions for all nutrients and processing (Figure S2, Supplementary Material). For Ca ET, a bimodal distribution was observed. Upon further investigation of the original Ca ET data, the distribution deviated from a normal distribution and instead was bimodal and skewed to the right, explaining the bimodal and skewed distribution of the R^2 bootstrapping values (Figure S2, Supplementary Material). T-test comparisons again indicate significant differences in all cases, despite similar R^2 values for some of the nutrient-component combinations, indicating the PLS-Wave regression model did not replicate well across these two environments.

In addition to significant differences in the model fits, the PLS waveband selection procedure also indicated considerable differences in the number and position of important wavebands for nutrient prediction (Figure 6). At both the plant and grain level, Jaccard indices were low (<0.2), with three instances having no overlapping bands (Table 5). For instance, nine wavebands were selected as important for predicting plant-level Ca in the US (SG): One at 760 nm, and eight from 1020–1060 nm (Figure 6). In contrast, the important bands for predicting Ca from the ET samples (SG) included fifteen bands positioned from 1045–1115 nm. The Jaccard index for this set was 0.069 (Table 5) indicating minimal overlap in the band positioning.

Table 5. Jaccard index for similarity between bands selected by partial least square regression (PLS) for the US and Ethiopia (ET) datasets for the various models.

Plant	SG	FD
Ca	0.069	0.00
Mg	0.081	0.148
Protein	0.052	0.185
Grain	**SG**	**FD**
Ca	0.00	0.110
Mg	0.00	0.046
Protein	0.025	0.104

In summary, the results from this study indicate that the PLS regression, using both the full model and the waveband selection process, did not generate replicable findings across the two study areas. We found statistically significant differences in model fits for 11 of the 12 comparisons; only the model for grain Mg using the SG filtering (Table 4) was not statistically different. Even though model fits were not comparable, it is still possible for the wavebands identified as important for prediction to be similar. However, the Jaccard index (Table 5) and plots of the important wavebands for each model (Figure 6) indicate very little overlap in the important wavebands between the US and ET samples, further diminishing the case for replicability across these two study sites.

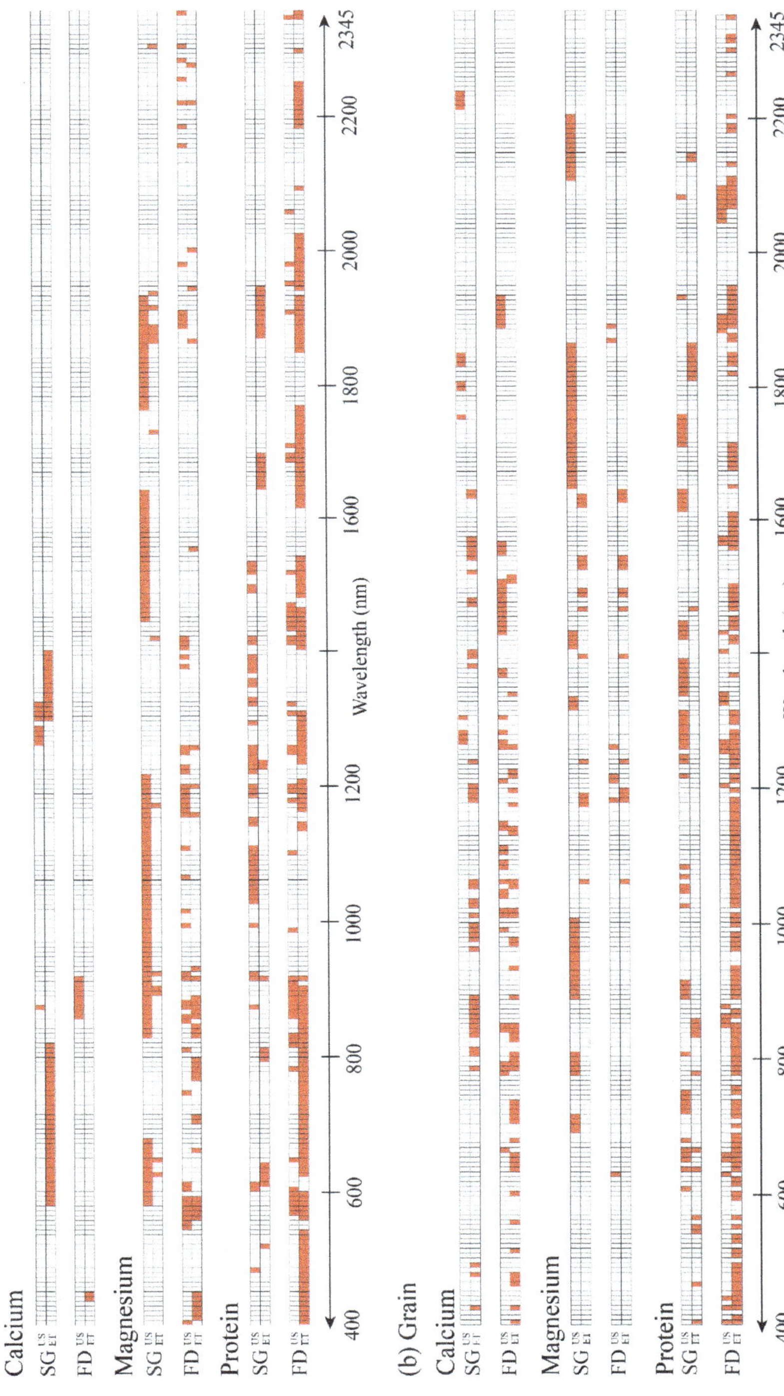

Figure 6. Selected wavebands (red) from the partial least square regression (PLS) using waveband selection for (**a**) plant, and (**b**) grain. Abbreviations: SG: Savitsky–Golay; FD: First derivative; US: United States; ET: Ethiopia.

4.3. Combined Samples for Multiple Environments

Following the single environment analytics, we tested the performance of a PLSR for the two environments (US and ET) combined. We re-ran the analytics with the same methodologies for the PLS-Full and PLS-Wave models using the combined samples (named USET), and found this generally resulted in better model fits compared to the single sites for both calibration and validation for both the PLS-Full (Table 6) and PLS-wave models (Table 7). Calibration and validation coefficients of determination (R^2) were all greater than 0.9 for all plant and grain nutrients using both SG and FD data and for both the PLS-Full and PLS-Wave models. RMSE values from the validation were slightly higher than those from the calibration, which is expected, and $RMSE_p$ values were all within 11 percent, so the models do not appear to be overfit.

Table 6. Partial least squares regression (PLS) results for the full spectrum model (PLS-Full) for plant and grain using the combined United States and Ethiopia (USET) dataset. SG: Savitsky–Golay; FD: First derivative; NLV: Number of latent variables; std: Standard deviation.

		Calibration			Validation		
	NLV	R^2_{CV}	$RMSE_{CV}$	**Mean R^2**	R^2 **Std**	**Mean $RMSE_P$**	$RMSE_P$ **Std**
Plant							
Calcium							
SG	7	0.94	674.18	0.94	0.01	693.01	70.18
FD	5	0.94	684.48	0.94	0.01	694.93	87.95
Magnesium							
SG	12	0.93	274.06	0.93	0.02	287.42	32.92
FD	6	0.93	269.72	0.93	0.02	289.64	37.94
Protein							
SG	8	0.91	1.86	0.91	0.01	1.96	0.19
FD	4	0.90	1.94	0.90	0.02	2.03	0.22
Grain							
Calcium							
SG	14	0.90	216.52	0.80	0.03	240.30	22.71
FD	11	0.91	203.73	0.90	0.02	223.02	19.95
Magnesium							
SG	14	0.97	118.87	0.97	0.01	128.61	13.50
FD	9	0.98	117.99	0.97	0.01	130.00	14.64
Protein							
SG	15	0.93	1.25	0.92	0.02	1.37	0.25
FD	9	0.93	1.23	0.93	0.02	1.34	0.24

Table 7. Partial least squares regression (PLS) results for the waveband selection model (PLS-Wave) for plant and grain using the combined Untied States and Ethiopia (USET) dataset. SG: Savitsky–Golay; FD: First derivative; NLV: Number of latent variables; std: Standard deviation.

		Calibration				Validation		
	NLV	R^2_{CV}	$RMSE_{CV}$	**# of Waves**	**Mean R^2**	R^2 **Std**	**Mean $RMSE_P$**	$RMSE_P$ **Std**
Plant								
Calcium								
SG	5	0.95	658.69	13	0.95	0.01	664.46	72.39
FD	4	0.95	664.55	38	0.94	0.01	690.87	79.27

Table 7. *Cont.*

	Calibration				Validation			
	NLV	$R^2{}_{CV}$	$RMSE_{CV}$	**# of Waves**	**Mean** R^2	R^2 **Std**	**Mean** $RMSE_P$	$RMSE_P$ **Std**
Plant								
Magnesium								
SG	9	0.94	265.07	44	0.93	0.01	276.48	31.00
FD	5	0.94	261.27	112	0.94	0.01	271.65	33.62
Protein								
SG	5	0.92	1.78	13	0.92	0.02	1.81	0.18
FD	8	0.93	1.68	52	0.92	0.02	1.79	0.16
Grain								
Calcium								
SG	11	0.91	206.43	59	0.90	0.02	221.41	21.83
FD	10	0.93	173.75	59	0.93	0.02	186.35	17.83
Magnesium								
SG	12	0.98	106.71	39	0.98	0.00	114.98	11.90
FD	7	0.98	101.05	51	0.98	0.00	103.95	9.54
Protein								
SG	10	0.94	1.12	19	0.94	0.02	1.16	0.19
FD	9	0.95	1.04	31	0.95	0.01	1.10	0.17

5. Discussion

5.1. Replicability of Scientific Methods and Findings

Demonstrating that methods and findings can be replicated across studies is critical for the self-correcting mechanism of the scientific method to function properly and is imperative for generating solutions that can be applied widely. Hyperspectral data is commonly used in precision agriculture for predicting biochemical properties of vegetation and nutrients [18,26,51–53], yet the replicability of findings is rarely tested across different study sites. Scalable science is needed to develop solutions that can be applied globally. The adoption of PLS regression has provided a solution to some of the computational challenges when working with high dimensional, hyperspectral data in remote sensing in general and precision agriculture more specifically, but whether these solutions are transferable has not been widely explored. This study used hyperspectral data and in situ samples to build PLS models to predict plant and grain nutrients for tef and test the replicability of those models for predicting across different environments.

We found significant differences in model fits for both the PLS-Full and PLS-Wave models along with differences in the number and locations of wavebands deemed important for prediction with the PLS-Wave models. Differences in the optimal number and location of wavebands for predicting nutrients via plant canopy measurements may be influenced by varying management practices, such as differing irrigation practices, which may lead to variable water content in the crops [41]. The ET fields were rain fed while the US fields were irrigated through to harvest; therefore variable plant water content may have influenced which wavebands were selected, as water can cause access noise in spectral signatures. This noise is particularly prevalent in wavebands that are sensitive to O-H bonds, including the spectral range 971–1400 nm [54–56]. While we removed a portion of this range from our data (see Figure 4), it is possible remaining wavelengths were affected by noise making replication difficult.

The number of wavebands selected in the PLS models for ET was often less than the number selected for the US (Figure 5). This difference may reflect how the PLS models incorporate water-induced noise that may have been present in the US samples. These findings suggest that understanding how hyperspectral remote sensing methods, such as PLS regression, replicate across agricultural environments may require greater controls on the conditions under which crops are being cultivated. In addition to irrigation and other management differences, changes in latitude and sun angle could have led to differences in scattering and light absorption during field data collection [57], which would impact replication comparisons. However, hyperspectral readings for the grain samples were collected in a controlled laboratory setting with the same halogen lamp, so we can assume that any differences in wavebands were not the result of external factors (i.e., sun angles, latitudes, etc.).

When comparing the nutrient content of the grains, there were clear differences between the study areas (Table 2). These large differences likely result in varying chemical property relationships within the grain, which in turn can result in differential absorption and scattering of electromagnetic energy. The large variance amongst biochemicals within the grain may result in noise for some nutrients as nutrient reflectance properties are often associated with near or similar portions of the electromagnetic spectrum [41].

In short, had we completed this study only in the US, the PLS regression method (both with and without waveband selection) would have produced favorable findings for all three nutrients at the grain level, and for protein at the plant level. Similarly, had we completed this study only in ET, PLS regression would have also produced favorable findings for all three nutrients at the grain level, and for Mg at the plant level. Yet, even where favorable results were found in both environments (all three nutrients at the grain level), the model fits were statistically different, and the wavebands selected were not similar. This finding serves as a cautionary tale that urges researchers to refrain from placing too much confidence on R^2 values, which may not translate to different areas. Even when the R^2 values were superficially high, there were statistical differences in their values. This is not to suggest that hyperspectral investigations with PLS regression are not useful or valuable but rather that caution should be used when translating findings from one site to another.

5.2. Combining Samples from Multiple Regions

A valid question arises as to whether combining samples from multiple regions can make models more robust for prediction across environments. Higher performance was obtained when the US and ET samples were combined. However, these higher R^2 values are likely artifacts of a larger spread among the dataset (larger range of levels of nutrients across locations).

It is worth noting that in three instances, the $RMSE_P$ from the combined USET dataset was greater than the minimum value of the nutrient in ET (see Table 2). These included Ca and Mg at the plant level for the PLS-Full model (Table 6) and Ca at the plant level for the PLS-Wave model (Table 7). Upon further investigation, it was determined that two of these samples were outliers (PLS-Full Mg and Ca), where the value was more than two standard deviations from the mean. Thus, these issues do not appear to be widespread or have a large impact on predictive capabilities of the models. However, it is important to keep this point in mind when combining data from differing regions with varying environmental factors and agricultural practices, where there may be large disparities in nutrient content. These disparities may result in the identification of between-site variation, requiring a greater variation of nutrient values to fully establish a generalized model.

Nevertheless, the results from the combined dataset are promising because they suggest that increasing within-sample variation can improve PLS model predictability across study areas. It is also worth mentioning that this practice could result in a reduction of accuracy of measurement, especially for field measurements, as the data may take on more noise. In this case, the use of single region analytics may be more beneficial. Nonetheless, improvement of predictability has been a similar finding in past studies that have referred to a need for global (i.e., USET) versus local (i.e., US and ET) modeling within similar measurement collections such as those in soil geochemistry studies [58,59]. Since it is unlikely though that this range of variation would be captured within a single site in many cases, it becomes necessary to combine samples from a variety of locations, and likely from a variety of study sites and research groups. To this point, we recommend building a more open and transparent culture of data sharing within the remote sensing community that can permit data sharing to advance our modeling capabilities and promote development of more generalizable predictive models. We recognize that a shift is already underway in many disciplines to promote data and code sharing, with some journals mandating these components accompany manuscript submission to allow reproducibility checks. Our findings here suggest that sharing data may have broader impacts beyond simply reproducing tests, but rather could result in more robust predictive models.

5.3. Non-Milled Grains

Since very few hyperspectral analyses have been performed on non-milled grains [26], we comment briefly on those aspects here. We generally found positive results using hyperspectral data with PLS regression to predict nutrient content of non-milled grains. While the reflectance measurements for these grain samples were collected in a controlled environment (i.e., a dark room using a contact probe), the coefficients of determination were noticeably higher for the grains compared to the plant canopy measurements collected in situ.

An interesting finding emerging from this research is that in the US, several blocks of wavebands corresponding to the VIS-NIR regions were identified as important for protein prediction when using FD values (Table S1, supplementary material). We compared these findings to other studies predicting protein from hyperspectral data [26,56,60]. We do see a considerable amount of overlap in the bands identified in this study for tef, and the bands identified in those studies for forage quality, cooked hams, and cereal grains (rice, oats, and maize), particularly for the FD transformation (Table 8). However, as noted by Talens et al. [56], several of the band ranges found to be important across multiple studies for protein prediction do fall in the range impacted by O-H bonds. So, it is difficult to determine whether these consistencies are true positives or are being affected by other factors.

Table 8. Identification of matching bands associated with tef protein content that has also been identified as important protein indicators in past hyperspectral studies. Abbreviations: S: Savitsky–Golay; F: First derivative; B: Both; US: United States; ET: Ethiopia.

Object Measured	Band(s)	US Plant	ET Plant	US Grain	ET Grain	Paper
Forage	583		F		F	
Forage	679		B	S	F	[60]
Forage	700–800 (Red Edge)	B	F	B	B	
Forage	710			F	B	
Cooked Hams	930				F	
Cooked Hams	971	S	F	S	F	
Cooked Hams	1051	F	F	S	F	[56]
Cooked Hams	1137		F	B		
Cooked Hams	1165	F			F	
Cooked Hams	1212	S	F			
Cereal Grains	1216		F			[26]
Cereal Grains	1100–1400	B	F	B	B	
Cooked Hams	1645		F		S	[56]
Cooked Hams	1682		S	F	F	
Forage	2091		F			
Forage	2128					[60]
Forage	2146			F	F	

5.4. Potential Limitations

Environmental factors such as sun angle, soil types/moisture, agronomic practices, etc., can contribute to the reflectance and nutrient differences at the plant the level. Given the number of samples collected at each location, this study is not meant to be a representative sampling for all tef varieties at varying growth stages around the world. To establish more generalized models that can be applied to a broader range of locations/tef varieties, future studies should focus on obtaining a greater number of samples across a greater number of location for multiple tef varieties.

6. Conclusions

This study investigated the replicability of PLS regression methods, including PLS with waveband selection, for predicting nutrient content in plant and grain material across multiple environments. Using *Eragrostis tef* (tef) as a target crop, this study compared PLS model fits and selected wavebands across two environments in the U.S. and Ethiopia to determine the extent to which the methods and finding are replicable. Three main findings emerge from this study:

First, the model fits and wavebands selected as important for nutrient prediction were not replicable across the two study sites in the US and Ethiopia. Eleven of the 12 comparisons had statistically different model fits across 1000 bootstrapped iterations, and the Jaccard index for similarity indicated very low similarities in the wavebands selected.

Combining samples from both environments improved model fits, suggesting that increasing within-sample variation may improve the predictability of PLS models across study areas, though caution is reserved if great disparities in sample values are great. Our recommendation is to build a more open and transparent culture of data sharing within the remote sensing community that will permit data sharing in order to advance modeling capabilities and promote development of more generalizable predictive models.

Results using PLS regression with hyperspectral data from non-milled grains were generally positive, and wavebands for protein prediction generally agreed with other studies. While more research is needed to determine whether these consistencies are true positives or are affected by other factors, this study contributes to the gap in the literature related to non-milled grains.

Supplementary Materials: The following are available online at http://www.mdpi.com/2072-4292/12/18/2867/s1, Figure S1: Bootstrapping results for plant and grain analyses with the PLS-Full model, Figure S2: Bootstrapping results for plant and grain analyses with the PLS-Wave model, Table S1: Selected wavebands from the partial least square regression (PLS) using waveband selection (PLS-Wave) for protein in plant and grain for the combined United State and Ethiopia samples (USET).

Author Contributions: Conceptualization: K.C.F., A.E.F.; Methods: K.C.F., A.E.F., S.A.; Fieldwork/Data Collection: K.C.F., A.E.F., S.A.; Data Analysis: K.C.F., A.E.F.; Writing/Editing: K.C.F., A.E.F., S.A. All authors have read and agreed to the published version of the manuscript.

Funding: This research received no external funding.

Acknowledgments: The authors would like to thank the many that supported this research including the Ethiopian Biodiversity Institute for hosting and supporting the research/field work, Oklahoma State University's Soil, Water, and Forage Analytical Laboratory and the Ethiopian Public Health Institute of Addis Ababa for providing the required nutrient analyses, Kensuke Kawamura for sharing his partial least square regression with waveband selection code, and the Fulbright U.S. Student Scholars Program for providing fiscal support. Special thanks also to Dejene Dida, Tariku Geda, Thiago Souza, Nathalia Gratchet, Andy Han, and Emily Ellis for their support and aid in field and laboratory experiments. Moreover, we would like to thank the editors for their time and consideration of the manuscript. USDA is an Equal Opportunity Provider and Employer.

Conflicts of Interest: The authors declare no conflict of interest.

References

1. Asendorph, J.B.; Conner, M.; De Fruyt, F.; De Houwer, J.; Denissen, J.J.; Fiedler, K.; Fiedler, S.; Funder, D.C.; Kliegl, R.; Nosek, B.A.; et al. Recommendations for increasing replicability in psychology. *Eur. J. Pers.* **2013**, *27*, 108–119. [CrossRef]

2. Camerer, C.F.; Dreber, A.; Forsell, E.; Ho, T.H.; Huber, J.; Johannesson, M.; Kirchler, M.; Almenberg, J.; Altmejd, A.; Chan, T.; et al. Evaluating replicability of laboratory experiments in economics. *Science* **2016**, *351*, 1433–1436. [CrossRef] [PubMed]

3. Begley, C.G.; Ioannidis, J.P. Reproducibility in science: Improving the standard for basic and preclinical research. *Circ. Res.* **2015**, *116*, 116–126. [CrossRef]

4. Ioannidis, J.P.; Stanley, T.D.; Doucouliagos, H. The power of bias in economics research. *Econ. J.* **2017**, *127*, F236–F265. [CrossRef]

5. Baker, M. Over half or psychology studies fail reproducibility test. *Nat. News* **2015**. [CrossRef]

6. Baker, M. 1500 scientists lift the lid on reproducibility. *Nature* **2016**, *533*, 452–454. [CrossRef] [PubMed]

7. Bollem, K.; Cacioppo, J.T.; Kaplan, R.; Krosnick, J.; Olds, J.L. *Social, Behavioral, and Economic Sciences Perspectives on Robust and Reliable Science*; National Science Foundation: Arlington, VA, USA, 2015. Available online: https://www.nsf.gov/sbe/SBE_Spring_2015_AC_Meeting_Presentations/Bollen_Report_on_Replicability_SubcommitteeMay_2015.pdf (accessed on 6 April 2020).

8. National Academies of Sciences, Engineering, and Medicine (NAS). *Reproducibility and Replicability in Science*; National Academies Press: Washington, DC, USA, 2013.

9. Kedron, P.J.; Frazier, A.E.; Trgovac, A.B.; Nelson, T.; Fotheringham, A.S. Reproducibility and replicability in geographical analysis. *Geogr. Anal.* **2019**. [CrossRef]

10. Anderson, K.; Milton, E.J. On the temporal stability of ground calibration targets: Implications for the reproducibility of remote sensing methodologies. *Int. J. Remote Sens.* **2006**, *27*, 3365–3374. [CrossRef]

11. Xu, C.; Holmgren, M.; Van Nes, E.H.; Hirota, M.; Chapin, F.S., III; Scheffer, M. A changing number of alternative states in the boreal biome: Reproducibility risks of replacing remote sensing products. *PLoS ONE* **2015**, *10*, e0143014. [CrossRef]

12. Ferreira, M.P.; Feret, J.; Grau, E.; Gastellu-Etchergorry, J.; Hummel do Amaral, C.; Shimabukuro, Y.; Filho, C. Retrieving structural and chemical properties of individual tree crowns in a highly diverse tropical forest with 3D radiative transfer modeling and imaging spectroscopy. *Remote Sens. Environ.* **2018**, *211*, 276–291. [CrossRef]

13. Kawamura, K.; Watanabe, N.; Sakanoue, S.; Inoue, Y. Estimating forage biomass and quality in a mixed sown pasture based on partial least squares regression with waveband selection. *Jpn. Soc. Grassl. Sci.* **2008**, *54*, 131–145. [CrossRef]

14. Nakaji, T.; Oguma, H.; Nakamura, M.; Kachina, P.; Asanok, L.; Marod, D.; Aiba, M.; Kurokawa, H.; Kosugi, Y.; Kassim, A.R.; et al. Estimation of six leaf traits of East Asian forest tree species by leaf spectroscopy and partial least square regression. *Remote Sens. Environ.* **2019**, *233*, 111381. [CrossRef]

15. Serbin, S.P.; Singh, A.; Desai, A.R.; Dubois, S.G.; Jablonski, A.D.; Kingdon, C.C.; Kruger, E.L.; Townsend, P.A. Remotely estimating photosynthetic capacity, and its response to temperature, in vegetation canopies using imaging spectroscopy. *Remote Sens. Environ.* **2015**, *167*, 78–87. [CrossRef]

16. Weiss, M.; Jacob, F.; Duveiller, G. Remote Sensing of agricultural applications: A meta-review. *Remote Sens. Environ.* **2020**, *236*, 111402. [CrossRef]

17. Schellberg, J.; Hill, M.J.; Gerhards, R.; Rothmund, M.; Braun, M. Precision agriculture on grassland: Applications, perspectives and constraints. *Eur. J. Agron.* **2008**, *29*, 59–71. [CrossRef]

18. Obermeier, W.A.; Lehnert, L.W.; Pohl, M.J.; Giaronni, S.M.; Silva, B.; Seibert, R.; Laser, H.; Moser, G.; Miller, C.; Luterbacher, J.; et al. Grassland ecosystem services in a changing environment: The potential of hyperspectral monitoring. *Remote Sens. Environ.* **2019**, *232*, 111273. [CrossRef]

19. Cho, M.A.; Skidmore, A.; Corsi, F.; Van Wieren, S.E.; Sobhan, I. Estimation of green grass/herb biomass for airborne hyperspectral imagery using spectral indices and partial least squares regression. *Int. J. Appl. Earth Obs.* **2007**, *9*, 414–424. [CrossRef]

20. Dechant, B.; Ryu, Y.; Kang, M. Making full use of hyperspectral data for gross primary productivity estimation with multivariate regression: Mechanistic insights from observations and process-based simulations. *Remote Sens. Environ.* **2019**, *234*, 111435. [CrossRef]

21. Fu, Y.; Yang, G.; Wang, J.; Song, X.; Feng, H. Winter wheat biomass estimation based on spectral indices, band depth analysis and partial least squares regression using hyperspectral measurements. *Comput. Electron. Agric.* **2014**, *100*, 51–59. [CrossRef]

22. Li, F.; Mistele, B.; Hu, Y.; Chen, X.; Schmidhalter, U. Reflectance estimation of canopy nitrogen content in winter wheat using optimized hyperspectral spectral indices and partial least squares regression. *Eur. J. Agron.* **2014**, *52*, 198–209. [CrossRef]

23. Meach-Hensold, K.; Montes, C.M.; Wu, J.; Guan, K.; Fu, P.; Ainsworth, E.A.; Pederson, T.; Moore, C.E.; Brown, K.L.; Raines, C.; et al. High-throughput field phenotyping using hyperspectral reflectance and partial least squares regression (PLSR) reveals genetic modifications to photosynthetic capacity. *Remote Sens. Environ.* **2019**, *231*, 111176. [CrossRef] [PubMed]

24. De Jong, S. SIMPLS: An alternative approach to partial least squares regression. *Chemom. Intell. Lab.* **1993**, *18*, 251–263. [CrossRef]

25. Geladi, P.; Kowalski, B. Partial least squares regression: A tutorial. *Anal. Chim. Acta* **1986**, *185*, 1–17. [CrossRef]

26. Caporaso, N.; Whitworth, M.B.; Fisk, I.D. Protein content prediction in single wheat kernels using hyperspectral imaging. *Food Chem.* **2018**, *240*, 32–42. [CrossRef]

27. Murphy, D.J. *People, Plants, and Genes: The Story of Crops and Humanity*; Oxford University Press: Oxford, UK, 2007; ISBN 9780199207145.

28. Vavilov, N. *The Origin, Variation, Immunity and Breeding of Cultivated Plants*; Chronica Botanica: Waltham, MA, USA; Stechert-Hafner: New York, NY, USA, 1951.

29. Flynn, K.C. Site suitability analysis for tef (*Eragrostis tef*) within the contiguous United States. *Comput. Electron. Agric.* **2019**, *159*, 119–128. [CrossRef]

30. Boe, A.; Somerfieldt, J.; Wynia, R.; Thiex, N. A preliminary evaluation of the forage potential of teff. *Proc. S. Dak. Acad. Sci.* **1986**, *65*, 75–82.

31. Bultosa, G.; Taylor, J.N.R. *Encyclopedia of Grain Science*; Wringley, C., Corke, H., Walker, C., Eds.; Academic Press: Oxford, UK, 2004; pp. 281–289.

32. Dekking, L.S.; Winkelaar, Y.K.; Koning, F. The Ethiopian cereal tef in celiac disease. *N. Engl. J. Med.* **2005**, *353*, 1748–1749. [CrossRef] [PubMed]

33. Gerbremariam, M.M.; Zarnkow, M.; Becker, T. Teff (*Eragrostis tef*) as a raw material for malting, brewing and manufacturing of gluten-free foods and beverages: A review. *J. Food Sci. Tech.* **2014**, *51*, 2881–2895. [CrossRef]

34. Hopman, G.D.; Dekking, E.H.A.; Blokland, M.L.J.; Wuisman, M.C.; Zuijderduin, W.M.K.F.; Schweizer, J.J. Tef in the diet of celiac patients in the Netherlands. *Scand. J. Gastroenterol.* **2008**, *43*, 277–282. [CrossRef]

35. Twidwell, E.K.; Boe, A.; Casper, D.P. Tef: A New Annual Forage Grass for South Dakota? South Dakota State University Extra Extension: Brookings, SD, USA, 2002; p. 8071.

36. Flynn, K.C.; Frazier, A.E.; Admas, S. Performance of chlorophyll prediction indices for *Eragrostis tef* at Sentinel-2 MSI and Landsat-8 OLI spectral resolutions. *Precis. Agric.* **2020**, 1–15. [CrossRef]

37. National Forage Testing Association. *Forage Analyses Procedures*; South Dakota State University Extra Extension: Brookings, SD, USA, 1993; pp. 79–81.

38. Savitzky, A.; Golay, M.J.E. Smoothing and differentiation of data by simplified least squares procedures. *Anal. Chem.* **1964**, *36*, 1627–1639. [CrossRef]

39. Dawson, C.T.; Curran, P.J. A new technique for interpolating the reflectance red edger position. *Int. J. Remote Sens.* **1998**, *19*, 2133–2139. [CrossRef]

40. Frazier, A.E.; Wang, L.; Chen, J. Two new hyperspectral indices for comparing vegetation chlorophyll content. *Geo Spat. Inf. Sci.* **2014**, *17*, 17–25. [CrossRef]

41. Kokaly, R.F.; Asner, G.P.; Ollinger, S.V.; Martin, M.E.; Wessman, C.A. Characterizing canopy biochemistry from imaging spectroscopy and its application to ecosystem studies. *Remote Sens. Environ.* **2009**, *113*, 578–591. [CrossRef]

42. Soil Science Society of America (SSSA). *Soil Testing and Plant Analysis*, 3rd ed.; Soil Science Society of America: Madison, WI, USA, 1990; pp. 404–411.

43. Western States Laboratory Proficiency Testing Program. *Soil and Plant Analytic Methods*; Version 4.00; Western Regional Extension Publication: Pullman, WA, USA, 1997; pp. 117–119.

44. Abdi, H. Partial least squares regression and projection on latent structure regression (PLS regression). *WIREs Comp. Stat.* **2010**, *2*, 97–106. [CrossRef]

45. Kawamura, K.; Ikeura, H.; Phongchanmaixay, S.; Khanthavong, P. Canopy hyperspectral sensing of paddy fields at the booting stage and PLSL regression can assess grain yield. *Remote Sens.* **2018**, *10*, 1249. [CrossRef]

46. Wold, S.; Sjostrom, M.; Eriksson, L. PLS-regression: A basic tool of chemometrics. *Chemom. Intell. Lab.* **2001**, *58*, 109–130. [CrossRef]

47. Forina, M.; Lanteri, S.; Oliveros, M.C.C. Selection of useful predictors in multivariate calibration. *Anal. Bioanal. Chem.* **2004**, *380*, 397–418. [CrossRef]

48. Boggia, R.; Forina, M.; Fossa, P.; Mosti, L. Chemometric study and validation strategies in the structure-activity relationship of new cardiotonic agents. *QSAR Comb. Sci.* **1997**, *16*, 201–213. [CrossRef]

49. Efron, B. Bootstrap methods: Another look at the jackknife. *Ann. Statist.* **1979**, *7*, 1–26. [CrossRef]

50. Mutanga, O.; Skidmore, A.K.; Prins, H.H.T. Predicting in situ pasture quality in the Kruger National Park, South Africa, using continuum-removed absorption features. *Remote Sens. Environ.* **2004**, *89*, 393–408. [CrossRef]

51. Jaccard, P. Étude comparative de la distribution florale dans une portion des Alpes et des Jura. *Bull. Soc. Vaud. Sci. Nat.* **1901**, *37*, 547–579. [CrossRef]

52. Agelet, L.E.; Armstrong, P.R.; Clariana, I.R.; Hurburgh, C.R. Measurement of single soybean seed attributes by near-infrared technologies: A comparative study. *J. Agric. Food Chem.* **2012**, *60*, 8314–8322. [CrossRef] [PubMed]

53. Martinez-Valdivieso, D.; Font, R.; Gomez, P.; Blanco-Diaz, T.; Rio-Celestino, M.D. Determining the mineral composition in Cucurbita pepo fruit using near infrared reflectance spectroscopy. *J. Sci. Food Agric.* **2014**, *94*, 3171–3180. [CrossRef]

54. Phan-Thien, K.-Y.; Golic, M.; Wright, G.C.; Lee, N.A. Feasibility of estimating peanut essential mineral by near infrared reflectance spectroscopy. *Sens. Instrum. Food Qual. Saf.* **2011**, *5*, 43–49. [CrossRef]

55. Barlocco, N.; Vadell, A.; Ballesteros, F.; Galietta, G.; Cozzolino, D. Predicting intramuscular fat, moisture and Warner-Bratzler shear force in pork muscle using near infrared reflectance spectroscopy. *Anim. Sci.* **2006**, *82*, 111–116. [CrossRef]

56. Cozzolino, D.; Murray, I. Identification of animal meat muscles by visible and near infrared reflectance spectroscopy. *LWT-Food Sci. Technol.* **2004**, *37*, 447–452. [CrossRef]

57. Talens, P.; Mora, L.; Morsy, N.; Barbin, D.F.; ElMasry, G.; Sun, D. Prediction of water and protein contents and quality classification of Spanish cooked ham using NIR hyperspectral imaging. *J. Food Eng.* **2013**, *117*, 272–280. [CrossRef]

58. Jensen, J.R. *Introductory Digital Image Processing: A Remote Sensing Perspective*; Prentice Hall: Upper Saddle River, NJ, USA, 2016.

59. Wijewardane, N.K.; Ge, Y.; Wills, S.; Libohova, Z. Predicting physical and chemical properties of US soils with a mid-infrared reflectance spectral library. *Soil Sci. Soc. Am. J.* **2018**, *82*, 722–731. [CrossRef]

60. Guerrero, C.; Stenberg, B.; Wetterlind, J.; Viscarra Rossel, R.A.; Maestre, F.T.; Mouazen, A.M.; Zornoza, R.; Ruiz-Sinoga, J.D.; Kuang, B. Assessment of soil organic carbon at local scale with spiked NIR calibrations: Effects of selection nd extra-weighting on the spiking subset. *Eur. J. Soil Sci.* **2014**, *65*, 248–263. [CrossRef]

MDPI

St. Alban-Anlage 66

4052 Basel

Switzerland

Tel. +41 61 683 77 34

Fax +41 61 302 89 18

www.mdpi.com

Remote Sensing Editorial Office

E-mail: remotesensing@mdpi.com

www.mdpi.com/journal/remotesensing